Intermediate Algebra

Ronald J. Harshbarger
The Pennsylvania State University

Harper & Row, Publishers
New York, Evanston, San Francisco, London

Sponsoring Editor: George J. Telecki
Project Editor: Lois Lombardo
Designer: Andrea Clark
Production Supervisor: Will C. Jomarrón
Compositor: Kingsport Press
Printer and Binder: Halliday Lithograph Corporation
Art Studio: J & R Technical Services Inc.

Intermediate Algebra

Copyright © 1976 by Ronald J. Harshbarger

All rights reserved. Printed in the United States of America. No part of this book may be used or reproduced in any manner whatsoever without written permission except in the case of brief quotations embodied in critical articles and reviews. For information address Harper & Row, Publishers, Inc., 10 East 53rd Street, New York, N.Y. 10022.

Library of Congress Cataloging in Publication Data
Harshbarger, Ronald J   1938–
   Intermediate algebra.

   Includes index.
   1. Algebra.   I. Title.
QA152.2.H378   512.9′042   75-19491
ISBN 0-06-042683-7

Contents

Preface vii

1 Sets and the Real-Number System 1

   Introduction 1
   1-1 Sets 1
   1-2 Natural Numbers and Integers 9
   1-3 Rational Numbers 13
   1-4 Irrational Numbers 17
   1-5 Properties of Real Numbers 20
   1-6 Order of Operations and Symbols of Grouping 24
   1-7 Addition and Subtraction of Signed Numbers 29
   1-8 Products and Quotients of Signed Numbers 34

   Chapter Test 40

2 Operations on Algebraic Expressions 42

   2-1 Combination of Like Terms 42
   2-2 Sums and Differences of Polynomials 46
   2-3 Multiplication and Division of Monomials 48
   2-4 Symbols of Grouping 54
   2-5 Products and Quotients of Polynomials 57

   Chapter Test 63

3 Factoring 65

   3-1 Factoring—Type I: Factoring Out Common Monomials 65
   3-2 Products of Two Binomials 67
   3-3 Factoring—Type II: Factoring Trinomials 69

- 3-4 Factoring—Type III: Perfect Trinomial Squares 76
- 3-5 Factoring—Type IV: Difference of Two Squares 79
- 3-6 Products and Factors Involving Cubes 81
- 3-7 Factoring Completely 83
- 3-8 Trinomials in Quadratic Form; Factoring by Grouping 85

  Chapter Test 88

4 Algebraic Fractions 89

- 4-1 Simplifying Algebraic Fractions 89
- 4-2 Products and Quotients of Fractions 93
- 4-3 Addition and Subtraction of Algebraic Fractions 98
- 4-4 Complex Fractions 106

  Chapter Test 109

5 Linear Equations and Inequalities in One Variable 110

- 5-1 Equations 110
- 5-2 Solution of Linear Equations in One Variable 115
- 5-3 Literal Equations 121
- 5-4 Word Problems 124
- 5-5 Mixture Problems 128
- 5-6 Linear Inequalities in One Variable 131
- 5-7 Equations and Inequalities Involving Absolute Values 137

  Chapter Test 141

6 Exponents and Radicals 143

- 6-1 Exponents 143
- 6-2 Zero and Negative Exponents 146
- 6-3 Roots 151
- 6-4 Fractional Exponents and Radicals 154
- 6-5 Simplifying Radicals; Multiplication 157
- 6-6 Simplifying Radicals; Division 161
- 6-7 Simplifying Radicals; Addition and Subtraction 165
- 6-8 Operations with Expressions Containing Radicals 168

  Chapter Test 172

7 Quadratic Equations in One Variable 174

- 7-1 Solution of Quadratic Equations by Factoring 174
- 7-2 Solution of Quadratic Equations by Completing the Square 177
- 7-3 Solution of Quadratic Equations by Formula 181
- 7-4 Imaginary and Complex Numbers 185
- 7-5 Quadratic Equations with Complex Solutions 188
- 7-6 Word Problems 190
- 7-7 Fractional Equations 194

Chapter Test 199

8 Linear Equations in Two Unknowns 200

- 8-1 Solution of Linear Equations in Two Variables 200
- 8-2 Slope of a Line 207
- 8-3 Using Slopes to Graph Lines 212
- 8-4 Solution of Linear Equations in Two Unknowns 217
- 8-5 Word Problems 226
- 8-6 Motion Problems 230

Chapter Test 233

9 Relations and Functions 235

- 9-1 Relations 235
- 9-2 Functions 241
- 9-3 Functional Notation 246
- 9-4 Graphing Relations and Functions 249
- 9-5 Special Functions 225
- 9-6 Composite and Inverse Functions 263

Chapter Test 269

10 Exponential and Logarithmic Functions 271

- 10-1 Exponential Functions 271
- 10-2 Logarithmic Functions 276
- 10-3 Properties of Logarithms 280
- 10-4 Computations with Logarithms 282

Chapter Test 289

11 Higher-Degree Polynomial Functions 290

- 11-1 Zeros of Polynomials 290
- 11-2 Synthetic Division 294
- 11-3 Solutions of Polynomial Equations 298
- 11-4 Rational Solutions of Polynomial Equations 301
- **Chapter Test** 307

12 Conic Sections 308

- 12-1 Parabolas 308
- 12-2 Circles 314
- 12-3 Ellipses 318
- 12-4 Hyperbolas 326
- **Chapter Test** 335

13 Sequences and Series 336

- 13-1 Sequences 336
- 13-2 Summation Notation 342
- 13-3 Series 347
- 13-4 The Binomial Formula 353
- **Chapter Test** 357

Appendix 359

$\sqrt{2}$ Is Not Rational (Proof) 359
Properties of Logarithms (Proofs) 360
Table I   Squares, Cubes, and Square Roots 361
Table II   Four-Place Logarithms 362
Table III   Exponential Functions 364

Answers to Odd-Numbered Exercises and Chapter Tests 365

Index 403

Preface

This text is designed for students who plan to continue the study of mathematics and for those who need only to improve their skills in algebra. It is intended primarily for a one-semester course for students who have had some prior experience with algebra (a one-year high school course or its equivalent). However, it is written so that a student with little or no algebra background can understand it as well.

The text begins with the study of real numbers and their properties. These properties are then used in the development of algebraic concepts. A functional approach is used in the discussion of quadratic, polynomial, exponential, and logarithmic equations. Algebraic procedures are explained with step-by-step instructions; numerous examples are given; and a large number of exercises provide practice using both concepts and skills. The learning objectives for each topic are listed at the beginning of the section; the exercises evaluate the students' attainment of the objectives. In the same way, a test included at the end of each chapter measures the attainment of the objectives for that chapter.

The text may be used in a traditional classroom setting or as part of an individualized system. The individualized system includes the text, an individualized student guide, and a data bank of computer-generated mastery tests. Answers to the odd-numbered exercises and to all the chapter test questions are included in the text.

The student guide uses a semiprogrammed format to drill and further elaborate topics treated in the text. It also contains the answers to the even-numbered exercises in the text. The data bank of mastery tests enables students to take tests on chapters when they feel they are ready. Thus, the students can progress at their own rate and receive individualized instruction as necessary.

The tests at the end of each chapter may be used as *pretests* to determine which chapters, if any, need not be discussed. Some topics may be omitted without losing continuity in the text. Specifically,

Section	3-6	Products and Factors Involving Cubes
Section	5-5	Mixture Problems
Section	5-7	Equations and Inequalities Involving Absolute Values
Section	6-8	Operations with Expressions Containing Radicals
Section	8-6	Motion Problems
Section	13-4	The Binomial Formula

are topics which if omitted will not disrupt the flow of the course.

I would like to express my gratitude to Professors Frank Kocher, Samuel Laposata, and Andris Niedra for their assistance in the preparation of this text. I would also like to thank Mr. George Telecki, mathematics editor, and Ms. Lois Lombardo, project editor, in the College Department at Harper & Row, for their valuable assistance.

RONALD J. HARSHBARGER

1
sets and the real-number system

Introduction

The study of algebra is based on number systems and their properties. We shall begin by discussing sets and the real number system. In algebra we frequently represent numbers by letters, and these symbols are called **literal numbers.** Thus we can use letters, such as a, b, c, x, y, and so on, to **represent** numbers of arithmetic. Because the letters represent numbers, we use them frequently to describe the properties of numbers. We shall review the properties of the real numbers in this chapter, using algebraic symbols to make the statements concise, yet general. For example, the statement that two real numbers added in either order result in the same sum may be stated as: For real numbers a and b, $a + b = b + a$.

1-1 Sets

OBJECTIVES: After completing this section, you should be able to:

a. Specify sets in two ways.
b. Indicate membership in a given set.
c. Determine if a set is empty.
d. Determine when one set is a subset of another.
e. Determine when two sets are equal.
f. Determine when two sets are disjoint.
g. Find the intersection of two sets.
h. Find the union of two sets.

This course is concerned with **sets** of numbers such as the set of integers and the set of real numbers. A **set** is a collection of objects. We may talk about a set of books, a set of dishes, a set of students, a set of numbers, and so on. There are two ways to tell what a given set contains. The first way is by **listing** the **elements** (or members) of the set (usually between braces). We may say that a set A contains 1, 2, 3, and 4 by writing $A = \{1, 2, 3, 4\}$. $B = \{x, y, z\}$ states that set B contains the elements x, y, and z. We can state that x is a member of B by writing $x \in B$.

If all the members of a set can be listed, it is said to be a **finite** set. $A = \{1, 2, 3, 4\}$ and $B = \{x, y, z\}$ are examples of finite sets. We cannot list all the members of an **infinite set**, but we can use dots to indicate the members of such a set. For example, $N = \{1, 2, 3, 4, \ldots\}$ is an infinite set containing the **natural numbers**. Although they are not listed specifically, we know $10 \in N$, $1121 \in N$, and $15{,}201 \in N$. But $\frac{1}{2}$ is not a member of N because $\frac{1}{2}$ is not a natural number. We indicate this by writing $\frac{1}{2} \notin N$.

Another way to specify what a given set contains is by **description**. For example, we may use $D = \{x: x \text{ is a Dodge automobile}\}$ to describe the set of all Dodge automobiles. $F = \{y: y \text{ is an odd number}\}$ is read "F is the set of all y such that y is an odd number." Thus $3 \in F$, $5 \in F$, $7 \in F$, and $9 \in F$ because they are odd numbers, and $6 \notin F$ because 6 is not an odd number.

EXAMPLE 1

Use two methods to write the following sets.

a. The set A of natural numbers less than 6.
b. The set B of natural numbers greater than 10.
c. The set C containing only 0.

Solution

a. $A = \{1, 2, 3, 4, 5\}$ or $A = \{x: x \text{ is a natural number less than 6}\}$
b. $B = \{11, 12, 13, 14, \ldots\}$ or $B = \{x: x \text{ is a natural number greater than 10}\}$
c. $C = \{0\}$ or $C = \{x: x = 0\}$

Note that set C of Example 1 contained one member, 0; set A contained five members; and set B contained an infinite number of numbers. It is possible for a set to contain no numbers. Such a set is called the **empty set** or the **null set,** and it is denoted by ∅ or by { }. The set of living veterans of the War of 1812 is empty because there are no living veterans of that war. Thus

{x: x is a living veteran of the War of 1812} = ∅

EXAMPLE 2

Which of the following sets are empty?

$A = \{0\}$

$B = \{x: x$ is a woman president of the United States$\}$

$C = \{x: x$ is a telephone$\}$

$D = \{\emptyset\}$

Solution

A contains 0; so it is not empty.
There is no woman president of the United States; so $B = \emptyset$.
C contains telephones; so C is not empty.
D contains ∅; so it is not empty.
Thus only B is empty.

If set A is the set of all girl students and set B is the set of all students, then every member of set A is also a member of set B. In such a case, we say that A is a subset of B and write $A \subseteq B$ (read "A is contained in B").

*If every member of set **A** is a member of set **B**, then **A** is said to be a subset of **B**.*

EXAMPLE 3

a. If $A = \{a, b, c, f\}$ and $B = \{a, b, c, d, f, e\}$, then A is a **subset** of B.

b. If $M = \{1, 4, \pi, 6, \sqrt{3}\}$ and $P = \{1, \pi, 6\}$, then $P \subseteq M$.

The empty set ∅ is considered to be a subset of every set.

∅ ⊆ A for all sets A

If two sets have the same members, the two sets are said to be *equal*, written $A = B$.

Set A and set B are equal if they contain the same members.

EXAMPLE 4

$X = \{1, 2, 4, 3\}$ and $Y = \{4, 3, 2, 1\}$ are equal sets because they contain the same members. (The order in which the members are written does not matter.)

Note: X is a subset of Y, and Y is a subset of X in Example 4. Thus we may say that $X = Y$ if $X \subseteq Y$ and $Y \subseteq X$.

If two sets have no members in common, the sets are said to be **disjoint**. For example, $A = \{1, 2, a, b\}$ and $B = \{3, e, 5, c\}$ have no members in common; so they are disjoint sets. Set $X = \{1, 5, m, n\}$ and set $Y = \{1, 3, r, s\}$ are **not disjoint** because 1 is common to both sets.

In the discussion of particular sets, the assumption is always made that the sets under discussion are all subsets of some larger set, called the **universal set** U. The choice of the universal set depends upon the problem under consideration. For example, in discussing the set of all students and the set of all girl students, we may use the set of all humans as the universal set.

EXAMPLE 5

$A = \{1, 2, 3\}$, $B = \{2, 3, 4\}$, and $C = \{4, 5, 6\}$ are subsets of the universal set $U = \{1, 2, 3, 4, \ldots\}$. A and B are not disjoint; B and C are not disjoint; but A and C are disjoint.

We may use **Venn diagrams** to illustrate the relationships among sets. We shall use a rectangle to represent the universal set and closed figures inside the rectangle to represent the sets under consideration. Figure 1-1 shows such a Venn diagram.

Figure 1-1 shows that B is a subset of A; that is, $B \subseteq A$. In Figure 1-2, M and N are disjoint sets. In Figure 1-3, sets X and Y overlap; that is, they are not disjoint.

The shaded portion of the diagram indicates where the two sets overlap. The set containing the members that are common to two sets is said to be in the **intersection** of the two sets.

In general, we say that the intersection of **A** and **B**, written **A ∩ B**, is

$$A \cap B = \{x : x \in A \text{ and } x \in B\}$$

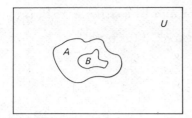

Figure 1-1

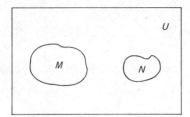

Figure 1-2

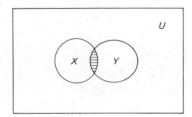

Figure 1-3

EXAMPLE 6

If $A = \{2, 3, 4, 5\}$ and $B = \{3, 5, 7, 9, 11\}$, find $A \cap B$.

Solution

$A \cap B = \{3, 5\}$ because 3 and 5 are in both A and B.

Figure 1-4 shows the sets and their intersection.

The **union** of two sets is the set that contains all members of the two sets.

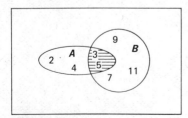

Figure 1-4

In general, the union of **A** *and* **B**, *written* **A** ∪ **B**, *is*

$$A \cup B = \{x: x \in A \text{ or } x \in B \text{ (or both)}\}$$

EXAMPLE 7
If $X = \{a, b, c, f\}$ and $Y = \{e, f, a, b\}$, find $X \cup Y$.

Solution

$X \cup Y = \{a, b, c, e, f\}$

EXAMPLE 8
Let $A = \{x: x \text{ is a natural number less than } 6\}$ and $B = \{1, 3, 5, 7, 9, 11\}$.

a. Find $A \cap B$.
b. Find $A \cup B$.

Solution

a. $A \cap B = \{1, 3, 5\}$
b. $A \cup B = \{1, 2, 3, 4, 5, 7, 9, 11\}$

We can illustrate the intersection and union of two sets by the use of Venn diagrams. The shaded region in Figure 1-5 represents $A \cap B$, the intersection of A and B, while the shaded region in Figure 1-6 represents $A \cup B$.

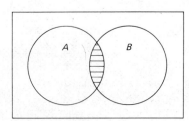

Figure 1-5

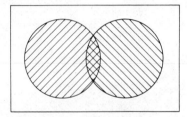

Figure 1-6

Exercise 1-1

Use "∈" or "∉" to indicate whether the given object is an element of the given set in the following problems.

1. x $\{x, y, z, a\}$
2.   2 $\{1, 2, 3, 4, 5\}$
3.   3 $\{1, 2, 4, 5, 6\}$
4. a $\{a, b, c, d, e\}$
5.   12 $\{1, 2, 3, 4, \ldots\}$
6.   6 $\{1, 2, 3, 4, \ldots\}$
7.   5 $\{x: x$ is a natural number greater than 5$\}$
8.   4 $\{x: x$ is a natural number less than 6$\}$
9.   3 $\emptyset$
10. -1 $\{x: x$ is a natural number$\}$

Write the following sets a second way.

11. $\{x: x$ is a natural number less than 8$\}$
12. $\{x: x$ is a natural number greater than 6 and less than 10$\}$
13. $\{x: x$ is a natural number greater than 10$\}$
14. $\{c: c$ is a natural number greater than 15$\}$
15. $\{w: w$ is a woman astronaut for the United States$\}$
16. $\{n: n$ is a natural number greater than 4 and less than 3$\}$
17. $\{1, 2, 3, 4, 5, 6, 7, 8, 9\}$
18. $\{3, 4, 5, 6, 7\}$
19. $\{7, 8, 9, 10, \ldots\}$
20. $\{18, 19, 20, 21, \ldots\}$
21. $\{0\}$
22. $\{1, 2, 3, 4\}$
23. Is $A \subseteq B$ if $A = \{1, 2, 3, 4\}$ and $B = \{1, 2, 3, 4, 5, 6\}$?
24. Is $A \subseteq B$ if $A = \{a, b, c, d\}$ and $B = \{c, d, a, b\}$?
25. Is $A \subseteq B$ if $A = \{a, b, c, d\}$ and $B = \{a, b, d\}$?
26. Is $A \subseteq B$ if $A = \{6, 8, 10, 12\}$ and $B = \{6, 8, 10, 14, 18\}$?

Use "$\subseteq$" notation to indicate which set is a subset of the other in the following problems.

27. $C = \{a, b, 1, 2, 3\}, D = \{a, b, 1\}$
28. $E = \{x, y, a, b\}, F = \{x, 1, a, y, b, 2\}$
29. $A = \{1, \pi, e, 3\}, D = \{1, \pi, 3\}$
30. $A = \{2, 3, 4, e\}, \emptyset$
31. $A = \{6, 8, 7, 4\}, B = \{8, 7, 6, 4\}$
32. $D = \{a, e, 1, 3, c\}, F = \{e, a, c, 1, 3\}$

Indicate if the following pairs of sets are equal.

33. $A = \{a, b, \pi, \sqrt{3}\}, B = \{a, \pi, \sqrt{3}, b\}$
34. $A = \{x, g, a, b\}, D = \{x, a, b, y\}$
35. $D = \{x: x \text{ is a natural number less than } 4\}, E = \{1, 2, 3, 4\}$
36. $F = \{x: x \text{ is a natural number greater than } 6\}, G = \{7, 8, 9, \ldots\}$
37. From the following list of sets, indicate which pairs of sets are disjoint.
    $A = \{1, 2, 3, 4\}$
    $B = \{x: x \text{ is a natural number greater than } 4\}$
    $C = \{4, 5, 6, \ldots\}$
    $D = \{1, 2, 3\}$

Find the intersection of sets A and B, $A \cap B$, if

38. $A = \{2, 3, 4, 5, 6\}$ and $B = \{4, 6, 8, 10, 12\}$
39. $A = \{a, b, c, d, e\}$ and $B = \{a, d, e, f, g, h\}$
40. $A = \{x, y, a, b\}$ and $B = \{a, b, y, x\}$
41. $A = \{a, b, c, d, e\}$ and $B = \{f, g, h, i, j\}$
42. $A = \{x: x \text{ is a natural number less than } 4\}$ and $B = \{3, 4, 5, 6\}$
43. $A = \emptyset$ and $B = \{x, y, a, b\}$

Find the union of sets A and B, $A \cup B$, if

44. $A = \{1, 2, 4, 5\}$ and $B = \{2, 3, 4, 5\}$
45. $A = \{2, 1, 3, 6\}$ and $B = \{4, 5, 7, 8\}$
46. $A = \{1, 3, 5, 7\}$ and $B = \{2, 4, 6, 8\}$
47. $A = \{a, e, i, o, u\}$ and $B = \{a, b, c, d\}$
48. $A = \emptyset$ and $B = \{1, 2, 3, 4\}$
49. $A = \{x: x \text{ is a natural number greater than } 6\}$ and $B = \{x: x \text{ is a natural number less than } 7\}$
50. $A = \{x: x \text{ is a natural number greater than } 5\}$ and $B = \{x: x \text{ is a natural number less than } 5\}$

1-2 Natural numbers and integers

OBJECTIVES: After completing this section, you should be able to:

a. Determine whether two sets are equivalent.
b. Establish a one-to-one correspondence between two equivalent sets.

1-2 Natural numbers and integers

c. Graph natural numbers and integers on a number line.
d. Use < or > notation to express inequalities.
e. Use chain inequalities to simplify inequalities.

Two sets may have the same number of elements even though the sets are not equal. Such sets are called **equivalent**. Any two **equivalent sets** can be put into one-to-one correspondence. For example, set $A = \{1, 2, 3, 4\}$ and set $B = \{x, y, z, a\}$ can be put in one-to-one correspondence as follows:

1 ⟷ x
2 ⟷ y
3 ⟷ z
4 ⟷ a

There are other ways to place these two sets in correspondence, such as

1 ⟷ a
2 ⟷ x
3 ⟷ z
4 ⟷ y

It is also possible to establish a one-to-one correspondence between certain pairs of sets with an infinite number of elements. For example, the set of natural numbers $N = \{1, 2, 3, 4, \ldots\}$ can be put into one-to-one correspondence with the set of even natural numbers $E = \{2, 4, 6, 8, \ldots\}$ as follows:

1 ⟷ 2
2 ⟷ 4
3 ⟷ 6
4 ⟷ 8
.
.
.
n ⟷ $2n$
.
.
.

Note that the set N of natural numbers is equivalent to a subset of itself, E.

There is a one-to-one correspondence between the real numbers and the points on a straight line; that is, there is exactly one point on the line for each number. Thus **the number line is a picture, or graph, of the real numbers.** Two numbers are said to be **equal** whenever they are represented by the same point on the number line. The **equation** $a = b$ (a equals b) means that the symbols a and b represent the same real number. Thus $3 + 4 = 7$ means that $3 + 4$ and 7 represent the same number. We can graph the natural numbers by choosing any point on the line to represent 0. We next choose any length to represent 1 unit, and using this unit we can graph the numbers 1, 2, 3, 4, . . . as follows:

As we have indicated earlier, the numbers 1, 2, 3, 4, 5, . . . are called the **natural numbers.**

Using the same unit length, we can graph numbers to the left of 0 and number them -1, -2, -3, and so on, as follows:

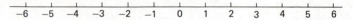

These numbers (the natural numbers, zero, and the negatives of the natural numbers) form the set of **integers.** Because the natural numbers form a subset of the integers, we may represent them in the following Venn diagram (Figure 1-7). In this Venn diagram, the universal set is chosen as the real numbers.

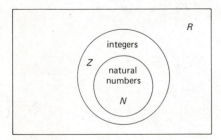

Figure 1-7

The point representing 0 divides the number line into two parts. All real numbers represented by points to the right of the 0 point are **positive numbers,** and those to the left are **negative numbers.** Because each real number is represented by a point on the number line, we have the following **tricotomy law**.

Trichotomy Law. If m is a real number, then one of the following is true:
  a. m is positive;
  b. m is zero; or
  c. m is negative.

We say that a is less than b (written $a < b$) if the point representing a is to the left of the point representing b on the real number line. For example, $4 < 7$ because 4 is to the left of 7 on the number line. We may also say that 7 is greater than 4 (written $7 > 4$). We may indicate that the number x is less than or equal to another y by writing $x \leq y$. We may also indicate that p is greater than or equal to 4 by writing $p \geq 4$.

EXAMPLE 1

Use "$<$" or "$>$" notation to write

  a. 6 is greater than 5;
  b. 10 is less than 15;
  c. 3 is to the left of 8 on the number line;
  d. x is less than or equal to 12.

Solution

  a. $6 > 5$
  b. $10 < 15$
  c. $3 < 8$
  d. $x \leq 12$

If we want to express the fact that -3 is less than -1 and greater than -5, we may use the **chain inequality** $-5 < -3 < -1$, which is a shorter way of writing $-5 < -3$ and $-3 < -1$.

EXAMPLE 2

Combine the following pairs of inequalities into one statement of inequality.

  a. $-6 < -2$ and $-7 < -6$
  b. $x < -3$ and $x > -7$

Solution

a. $-7 < -6 < -2$
b. $-7 < x < -3$

Exercise 1-2

Are the following pairs of sets equivalent?

1. $\{1, 2, 3, 4\}$ and $\{5, 6, 7, 8\}$
2. $\{a, b, 1, 2\}$ and $\{2, 3, a, c\}$
3. $\{1, 2, 3, 4, 5\}$ and $\{1, 4, 6, 8\}$
4. $\{1, 2, 3, 4, 5, \ldots\}$ and $\{3, 6, 9, 12, \ldots\}$

Establish a one-to-one correspondence between the following pairs of sets.

5. $\{2, 4, 6, 8\}$ and $\{6, 12, 14, 18\}$
6. $\{4, 5, 6, 8, 11\}$ and $\{a, d, c, f, g\}$
7. $\{1, 2, 3, 4, \ldots\}$ and $\{3, 6, 9, 12, \ldots\}$
8. $\{2, 4, 6, 8, \ldots\}$ and $\{3, 6, 9, 12, \ldots\}$
9. Graph the natural numbers from 2 to 9 on a number line.
10. Graph the natural numbers from 6 to 13 on a number line.
11. Graph the integers from -8 to $+8$ on a number line.
12. Graph the integers from -3 to 9 on a number line.

Insert the proper inequality symbol $<$ or $>$ between the following numbers.

13. $-8 \quad -2$
14. $-2 \quad -4$
15. $-6 \quad -1$
16. $0 \quad -4$
17. $-3 \quad 0$
18. $4 \quad -4$
19. $-2 \quad -6$
20. $-3 \quad -7$

Combine each of the following pairs of inequalities into one statement of inequality.

21. $-8 < -4, -4 < -2$
22. $-4 < -3, -2 > -3$
23. $-6 < 2, 2 < 4$
24. $-5 < 1, 1 < 5$
25. $-6 > -8, -6 < -2$
26. $-3 > -4, -3 < 0$
27. $-8 > -9, -5 > -8$
28. $-6 > -7, -3 > -6$

1-3 Rational numbers

OBJECTIVES: After completing this section, you should be able to:

a. Graph rational numbers on a number line.
b. Determine if two rational numbers are equal.
c. Write a rational number as a terminating or repeating decimal.
d. Write a terminating decimal as a fraction with integer numerator and denominator.

Clearly the integers do not account for all the points on the number line. We can plot points on the line to represent the numbers $\frac{1}{2}$, $\frac{1}{3}$, $\frac{2}{3}$, and so on. Any number that **can be** written as the quotient of two integers is called a **rational number**; that is, a rational number is a number that can be written as the ratio of two integers, where the denominator is not 0. We can plot the rational number r/n on the number line by dividing each unit interval into n parts and then counting r of these new intervals from 0. The following number line has $-\frac{3}{4}$ and $\frac{5}{4}$ plotted on it.

EXAMPLE 1

Graph $-\frac{2}{3}$ and $\frac{4}{3}$ on a number line.

Solution

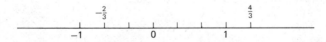

We can easily see that any integer is a rational number, because the integer a can be written as the ratio $a/1$. For example, $2 = \frac{2}{1}$, $-3 = -\frac{3}{1}$, and so on.

We may illustrate that integers are also rational numbers by the following Venn diagram (Figure 1–8).

Figure 1–9 shows that natural numbers are also rational numbers. Note that we may represent the same rational number in many different ways. For example, $3 = \frac{3}{1} = \frac{6}{2} = \frac{9}{3} = \frac{27}{9}$, and so on.

14   Sets and the Real-Number System

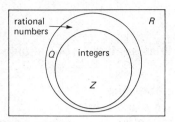

Figure 1-8

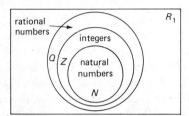

Figure 1-9

We say that two fractions are equal if they represent the same rational number. We can use the following rule to determine if two fractions are equal.

Two fractions **a/b** *and* **c/d** *are equal if and only if* **ad = bc.**

EXAMPLE 2

a. Show that $\frac{13}{52}$ and $\frac{1}{4}$ are equal.
b. Show that $\frac{9}{16}$ and $\frac{3}{5}$ are not equal.

Solution

a. $\frac{13}{52} = \frac{1}{4}$ because $13 \cdot 4 = 52 \cdot 1$
b. $\frac{9}{16} \neq \frac{3}{5}$ because $9 \cdot 5 \neq 16 \cdot 3$

We may represent rational numbers by decimals as well as by fractions. We convert a fraction to a decimal by dividing the numerator by the denominator. For example, $\frac{2}{5} = 2 \div 5 = 0.4$. To find the decimal equivalent to $\frac{2}{7}$, we divide as follows:

```
        0.285714285714 ...
    7 ) 2.00000000
        1 4
          60
          56
           40
           35
            50
            49
             10
              7
             30
             28
            ⎛20⎞
            ⎜14⎟
            ⎝ 6⎠
```

Note that the circled step in the division is the same as the first step and that each of the succeeding steps will be repetitions of previous steps. Thus the digits of the quotient will keep repeating. In dividing by 7, only seven remainders are possible (0, 1, 2, 3, 4, 5, or 6). Therefore the quotient must either terminate (remainder 0) or repeat if the division is performed more than seven times. We may write the decimal representation for $\frac{2}{7}$ as

$$\frac{2}{7} = 0.285714\overline{285714} \ldots$$

where the bar indicates the repeating block of digits.

In general, we may state that any rational number may be written as a terminating or repeating decimal.

Of course, we may also write a terminating decimal (such as 0.4) as a repeating decimal with the zero digit repeating ($0.4 = 0.4000\overline{0} \ldots$). Thus we can write every rational number as a repeating decimal. It can also be shown that **every repeating decimal represents a rational number.** (See Chapter 13.)

EXAMPLE 3

a. Show that $\frac{5}{8}$ can be written as a terminating decimal.
b. Show that $\frac{7}{11}$ can be written as a repeating decimal.

Solution

a. $\frac{5}{8} = 5 \div 8 = 0.625$

b. $\frac{7}{11} = 11 \overline{)7.0000}^{\,0.63\overline{63}\ldots}$

$$\underline{6\,6}$$
$$40$$
$$\underline{33}$$
$$70$$
$$\underline{66}$$
$$40$$
$$\underline{33}$$
$$7$$

EXAMPLE 4

Express 4.63 as the ratio of two integers.

Solution

$$4.63 = 4\frac{63}{100} = \frac{400}{100} + \frac{63}{100} = \frac{463}{100}$$

Exercise 1-3

1. Graph the rational numbers $1, \frac{1}{2}$, and $-\frac{3}{2}$ on a number line.
2. Graph the rational numbers $-\frac{1}{3}, 2$, and $\frac{5}{3}$ on a number line.
3. Graph the rational numbers $-\frac{7}{4}, \frac{3}{4}$, and $\frac{11}{4}$ on a number line.
4. Graph the rational numbers $-\frac{1}{2}, 2, \frac{3}{4}$, and $-\frac{1}{4}$ on a number line.

Are the following pairs of rational numbers equal?

5. $\frac{9}{12}$ and $\frac{27}{36}$ 6. $\frac{7}{21}$ and $\frac{12}{36}$
7. $\frac{8}{12}$ and $\frac{3}{4}$ 8. $\frac{36}{56}$ and $\frac{4}{7}$
9. $\frac{15}{52}$ and $\frac{5}{14}$ 10. $\frac{9}{21}$ and $\frac{21}{49}$

Change the following rational numbers from fractional form to repeating or terminating decimals.

11. $\frac{1}{2}$ 12. $\frac{3}{8}$ 13. $\frac{2}{5}$ 14. $\frac{9}{16}$
15. $\frac{2}{3}$ 16. $\frac{5}{6}$ 17. $\frac{5}{9}$ 18. $\frac{3}{11}$

Write the following terminating decimals as the ratio of two integers (fractions with integer numerators and denominators).

| 19. 0.125 | 20. 0.47 | 21. 0.615 | 22. 0.75 |
| 23. 4.69 | 24. 5.11 | 25. 0.33 | 26. 4.871 |

1-4 Irrational numbers

OBJECTIVES: After completing this section, you should be able to:

a. Recognize the relationships among the numbers in the real number system.
b. Use a table to find the decimal approximations of irrational numbers involving square roots.

Although it may appear that the rational numbers account for all of the points on the real number line, this is not the case. We can illustrate this fact by showing that there is at least one point on the number line which does not correspond to a rational number. By forming a right triangle with two legs of length 1, we have a hypotenuse of length $\sqrt{2}$ (see Figure 1-10).

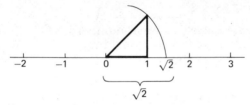

Figure 1-10

It can be shown (see Appendix I) that the $\sqrt{2}$ cannot be written as the ratio of two integers. But as Figure 1-10 shows, there is a point on the real number line that corresponds to $\sqrt{2}$.[1] The decimal representation for $\sqrt{2}$ is 1.4142135 . . . . This decimal does not terminate and does not repeat. We can approximate $\sqrt{2}$ by a decimal to any accuracy desired. For example, we may approximate $\sqrt{2}$ to three decimal places by 1.414 and to four decimal places by 1.4142. Other numbers, such as $\sqrt{3}$, $\sqrt{5}$, $2 + \sqrt{2}$, $4 - \sqrt{7}$, and so on, cannot be written as the ratio of two integers (nor as repeating decimals). These numbers are called **irrational numbers** because they are real

---

[1] The Pythagorean theorem states that the sum of the squares of the legs of a right triangle equals the square of the hypotenuse. Thus $1^2 + 1^2 = 2$; so the hypotenuse has length $\sqrt{2}$.

numbers that are not rational numbers. In addition to numbers involving square roots, there are some special irrational numbers. One very special irrational number is π. If the circumference of a circle is divided by its diameter, the quotient is π. π is used frequently in geometry and trigonometry and is frequently approximated by the decimal 3.14 (or more precisely by 3.1415926536).

Another special irrational number is e. The decimal representation for e is 2.71828 . . . . We can approximate e to three decimal places by the decimal 2.718. This number occurs frequently in economics, in biology, and in studies of radiation. We shall study functions involving e in Chapter 11.

The rational and irrational numbers together account for all points on the real number line. Clearly the set of rational numbers and the set of irrational numbers are disjoint sets. Their intersection is the empty set, and their union is the set of real numbers. Thus in Figure 1-9, the region inside the universal set of real numbers but outside the set of rational numbers represents the irrational numbers.

EXAMPLE 1

a. Is a natural number also a rational number?
b. Is a natural number also an irrational number?
c. Is an irrational number also a real number?

Solution

a. Yes; any natural number n can be written as $n/1$.
b. No; a natural number is rational and therefore cannot be irrational.
c. Yes; an irrational number is a real number that is not a rational number.

EXAMPLE 2

a. Is $3 + \sqrt{4}$ a rational or irrational number?
b. Is $\sqrt{7}$ a rational or irrational number?

Solution

a. $3 + \sqrt{4}$ is a rational number because it can be written as the ratio of two integers, $3 + 2 = \frac{5}{1}$.
b. $\sqrt{7}$ cannot be written as the ratio of two integers; so it is an irrational number.

EXAMPLE 3

Identify each of the following numbers as one or more of natural number, integer, rational number, irrational number, or real number.

a. $-\sqrt{9}$
b. $\sqrt{17}$
c. $\frac{4}{3}$
d. $\frac{6}{2}$

Solution

a. $-\sqrt{9} = -3$ is an integer, rational number, and a real number.
b. $\sqrt{17}$ is an irrational number and a real number.
c. $\frac{4}{3}$ is a rational number and a real number.
d. $\frac{6}{2} = 3$ is a natural number, an integer, a rational number, and a real number.

While we usually write a symbol for a particular irrational number, we shall occasionally want the decimal approximation of its value. We may, for example, use 3.14 to approximate the irrational number π (to two decimal places) in calculating the area of a circle. We may use Table I (in the Appendix) to find decimal approximations of the square roots of numbers. Thus the number $\sqrt{3}$ can be represented **approximately** by 1.732. This representation is correct to three decimal places.

EXAMPLE 4

Use Table I to write the decimal approximations of the following irrational numbers.

a. $\sqrt{60}$
b. $2\sqrt{7}$
c. $4 - \sqrt{70}$

Solution

a. $\sqrt{60} \cong$ (approximately equals) 7.7460   (use the column headed by $\sqrt{10\ n}$)
b. $2\sqrt{7} \cong 2(2.6458) = 5.2916$
c. $4 - \sqrt{70} \cong 4 - 8.3666 = -4.3666$

20 Sets and the Real-Number System

Exercise 1-4

Tell whether each of the following numbers is rational or irrational.

1. $\sqrt{16}$
2. $\sqrt{32}$
3. $\sqrt{36}$
4. $\sqrt{25}$
5. $\dfrac{\sqrt{144}}{5}$
6. $\dfrac{\sqrt{50}}{3}$
7. $\dfrac{3 - \sqrt{25}}{4}$
8. $\dfrac{6 + \sqrt{27}}{3}$
9. $\sqrt{\dfrac{9}{16}}$
10. $\sqrt{\dfrac{5}{12}}$
11. $\sqrt{\dfrac{8}{36}}$
12. $\sqrt{\dfrac{2}{72}}$
13. $3 + \sqrt{25}$
14. $4 + \sqrt{24}$
15. $2 - \sqrt{16}$
16. $3 - \sqrt{18}$
17. $2\sqrt{8}$
18. $3\sqrt{16}$
19. $\dfrac{8 - \sqrt{25}}{4}$
20. $\dfrac{6 + \sqrt{27}}{3}$

Use Table I to write the decimal approximations of the following, correct to three decimal places.

21. $\sqrt{24}$
22. $\sqrt{90}$
23. $3\sqrt{160}$
24. $2\sqrt{33}$
25. $5\sqrt{15}$
26. $8\sqrt{32}$
27. $4 + \sqrt{3}$
28. $2 - \sqrt{18}$
29. $4 - 2\sqrt{6}$
30. $4 + 3\sqrt{7}$
31. $\dfrac{2 - \sqrt{18}}{3}$
32. $\dfrac{4 + \sqrt{6}}{2}$
33. $\dfrac{17 - \sqrt{50}}{3}$
34. $\dfrac{5 - \sqrt{12}}{2}$

1-5 Properties of real numbers

OBJECTIVES: After completing this section, you should be able to recognize and use properties of real numbers.

We have seen that the real numbers can be represented geometrically on a number line. The properties of the real numbers are fundamental to the study of algebra. The properties and examples of their use follow.

I **Addition of real numbers is commutative.** For real numbers a and b,

$$a + b = b + a$$

Example: $5 + 6 = 6 + 5$

II **Multiplication of real numbers is commutative.** For real numbers a and b,

$$a \cdot b = b \cdot a$$

Example: $4 \cdot 7 = 7 \cdot 4$

III **Addition of real numbers is associative.** For real numbers a, b, and c,

$$a + (b + c) = (a + b) + c$$

Example: $3 + (2 + 5) = (3 + 2) + 5$

Note: The parentheses in $3 + (2 + 5)$ indicate that the sum of 2 and 5 should be added to 3, while $(3 + 2) + 5$ indicates the sum of 3 and 2 should be found before adding 5. But the associative property states that either procedure gives the same answer. Thus it is possible to indicate the sum of three numbers without any parentheses: $(a + b) + c = a + (b + c) = a + b + c$.

IV **Multiplication of real numbers is associative.** For real numbers a, b, and c,

$$a(b \cdot c) = (a \cdot b)c$$

Example: $4(3 \cdot 2) = (4 \cdot 3)2$

Note: The parentheses are used here to indicate multiplication.

V **0 is the additive identity.**

$$a + 0 = a$$

Example: $\frac{1}{2} + 0 = \frac{1}{2}$

VI **1 is the multiplicative identity.**

$$a \cdot 1 = a$$

Example: $-4 \cdot 1 = -4$

22 Sets and the Real-Number System

VII **Additive inverse exists.** For every real number a, there exists a real number b such that $a + b = 0$. We denote the additive inverse or negative of a as $-a$.

Example: $4 + (-4) = 0$

Note 1: -4 is the additive inverse of 4, and 4 is the additive inverse of -4. But the additive inverse of -4 is also written as $-(-4)$. Thus $-(-4) = 4$. In general, $-(-a) = a$.

Note 2: We do not define subtraction because we can use additive inverses and addition to perform subtractions. For example, $a - x = a + (-x)$; so we can use addition to subtract x from a.

VIII **Multiplicative inverse exists for nonzero reals.** For every real number $a \neq 0$, there exists a number b such that $ab = 1$. We denote the multiplicative inverse (or *reciprocal*) of a as $1/a$.

Note 1: We do not define division because we can use multiplicative inverses and multiplication to perform divisions. For example, if $x \neq 0$, $a \div x = a \cdot 1/x$; so we can use the multiplication to divide a by x.

Note 2: The number 0 has no multiplicative inverse. For if 0 has an inverse, say k, then $k = \frac{1}{0}$. But if $k = \frac{1}{0}$, then $k \cdot 0 = 1$, and there is no real number k such that $k \cdot 0 = 1$. Thus 0 has no inverse; so division by 0 is undefined.

Note 3: $1/a$ is the reciprocal of a, and a is the reciprocal of $1/a$. But the reciprocal of $1/a$ is also written $1/(1/a)$. Thus $1/(1/a) = a$.

IX **Multiplication is distributive over addition.** For real numbers a, b, and c,

$a(b + c) = ab + ac$

Example 1: $3(4 + 5) = 3 \cdot 4 + 3 \cdot 5$; that is, $3(9) = 12 + 15$

Example 2: $(8 + 3)x = 8x + 3x$

Note: The distributive property expresses the relationship that exists between multiplication and addition. This property will be used in factoring algebraic expressions.

EXAMPLE 1

Complete the following illustrations of the commutative laws of addition and multiplication.

a. $4 + 5 =$ _____
b. $10 \cdot 6 =$ _____

Solution

a. $4 + 5 = 5 + 4$
b. $10 \cdot 6 = 6 \cdot 10$

EXAMPLE 2

Complete the following illustrations of the distributive law.

a. $6(10 + 4) =$ _____
b. $\pi(x + 5) =$ _____

Solution

a. $6(10 + 4) = 6 \cdot 10 + 6 \cdot 4$
b. $\pi(x + 5) = \pi \cdot x + \pi \cdot 5$

Exercise 1-5

Tell which property of real numbers is being illustrated in the following.

1. $8 + 6 = 6 + 8$
2. $6 + (3 + 4) = (6 + 3) + 4$
3. $(5 + 3) + 6 = 5 + (3 + 6)$
4. $7 \cdot 3 = 3 \cdot 7$
5. $6(4 \cdot 5) = (6 \cdot 4)5$
6. $4 \cdot \frac{1}{4} = 1$
7. $5 + (-5) = 0$
8. $6(5 + 3) = 6 \cdot 5 + 6 \cdot 3$
9. $3 \cdot \frac{1}{3} = 1$
10. $6 + 0 = 6$
11. $\sqrt{3} + 0 = \sqrt{3}$
12. $5 - 5 = 0$
13. $-e \cdot 1 = -e$
14. $5(2 \cdot 3) = (5 \cdot 2)3$
15. $\sqrt{7} + 3 = 3 + \sqrt{7}$
16. $\pi \cdot \frac{1}{4} = \frac{1}{4} \cdot \pi$
17. $-3 + 0 = -3$
18. $\frac{4}{5} \cdot 1 = \frac{4}{5}$
19. $18 \div 18 = 1$
20. $(3 + 7)\pi = 3\pi + 7\pi$

21. Which property of real numbers permits the operation of subtraction?
22. Which property of real numbers permits the operation of division?
23. Which property expresses the relationship between multiplication and addition?
24. For which real number is there no multiplicative inverse?

Complete the following illustrations of the associative laws of addition and multiplication.

24 Sets and the Real-Number System

25. $5 + (6 + 7) =$ _____
26. $3 + (2 + 1) =$ _____
27. $(6 + 3) + 4 =$ _____
28. $(4 + 2) + 2 =$ _____
29. $7(3 \cdot 2) =$ _____
30. $6(7 \cdot 2) =$ _____
31. $(4 \cdot 2)3 =$ _____
32. $(6 \cdot 8)5 =$ _____

Use the distributive law to compute

33. $6(3 + 5)$
34. $5(8 + 10)$
35. $5(5 + 8)$
36. $8(6 + 4)$

1-6 Order of operations and symbols of grouping

OBJECTIVES: After completing this section, you should be able to:

a. Perform operations correctly in arithmetic expressions involving parentheses, brackets, braces, and bars.
b. Perform arithmetic operations in the proper order.
c. Evaluate the absolute values of real numbers.

We may use parentheses to indicate the order in which operations are to be performed. For example, $12 - (3 + 2)$ indicates we should add 3 and 2 and then subtract their sum from 12. Thus $12 - (3 + 2) = 12 - 5 = 7$. As mentioned in Section 1-5, we may also use parentheses to indicate multiplication. For example, $4(3 + 2)$ indicates we should add 3 and 2 and then multiply their sum by 4.

EXAMPLE 1

Compute

a. $16 + (3 - 1)$
b. $21 - (16 + 3)$
c. $3(4 - 2)$
d. $4(2 + 5)$

Solution

a. $16 + (3 - 1) = 16 + 2 = 18$
b. $21 - (16 + 3) = 21 - 19 = 2$
c. $3(4 - 2) = 3(2) = 6$
d. $4(2 + 5) = 4(7) = 28$

Brackets [ ] and braces { } are also symbols of inclusion and are treated the same as parentheses.

1-6 Order of operations and symbols of grouping

EXAMPLE 2

Compute

a. $3 + [5 - 1]$
b. $4\{4 + 2\}$
c. $3[6 - 2]$

Solution

a. $3 + [5 - 1] = 3 + 4 = 7$
b. $4\{4 + 2\} = 4\{6\} = 24$
c. $3[6 - 2] = 3[4] = 12$

Another symbol of grouping is the bar ———, which may be used to indicate division of one or more numbers by a number. For example, $\dfrac{4+6}{5}$ indicates that the sum of 4 and 6 is to be divided by 5. Thus $\dfrac{4+6}{5} = \dfrac{10}{5} = 2$.

EXAMPLE 3

Compute

a. $\dfrac{7-3}{2}$

b. $\dfrac{12+3}{3+2}$

c. $\dfrac{16+2}{3 \cdot 3}$

Solution

a. $\dfrac{7-3}{2} = \dfrac{4}{2} = 2$

b. $\dfrac{12+3}{3+2} = \dfrac{15}{5} = 3$

c. $\dfrac{16+2}{3 \cdot 3} = \dfrac{18}{9} = 2$

The operations of addition, subtraction, multiplication, and division are **binary operations.** This means that we can add only **two** numbers at a time, subtract only two numbers at a time, and so on. The use of parentheses helps us determine the order in which operations are to be performed, but we frequently see expressions such as $4 + 3 \cdot 5$, which contain three or more numbers with no parentheses. In order to avoid confusion about which operation should be performed first, we have the following rule.

If a problem contains several signs or symbols, the operations should be performed in the following order.

PROCEDURE	EXAMPLE
	Compute $4^2 \cdot (2 + 3) + 6 \div 2$.
1. Perform the operations inside the parentheses or other symbols of inclusion.	1. $4^2 \cdot (2 + 3) + 6 \div 2$ $= 4^2 \cdot 5 + 6 \div 2$
2. Raise numbers to powers and take roots.	2. $= 16 \cdot 5 + 6 \div 2$
3. Do all multiplications in the order in which they occur from left to right.	3. $= 80 + 6 \div 2$
4. Do all division in the order in which they occur from left to right.	4. $= 80 + 3$
5. Do all additions and subtractions in order from left to right.	5. $= 83$

The proper order of operations may be summarized as follows:
1. First perform all operations inside **parentheses**.
2. Then remove all **exponents** (or radicals).
3. Then perform all **multiplications**.
4. Then perform all **divisions**.
5. Then perform all **additions**.
6. Then perform all **subtractions**.

We can remember the proper order by recalling the initials P–E–M–D–A–S, or the phrase: **Please Entertain My Dear Aunt Sally.**

EXAMPLE 4

Compute $4 + 3 \cdot 5$.

Solution

$$4 + 3 \cdot 5 = 4 + 15 = 19$$

EXAMPLE 5

Compute $2^3 - 3 \cdot 2$.

Solution

$$2^3 - 3 \cdot 2 = 8 - 3 \cdot 2 = 8 - 6 = 2$$

EXAMPLE 6

Compute $12 - (4 - 1)^2 + 2$.

Solution

$$12 - (4 - 1)^2 + 2 = 12 - 3^2 + 2 = 12 - 9 + 2 = 14 - 9 = 5$$

EXAMPLE 7

Calculate $\dfrac{3 + (3 - 2)4}{3 + 4}$.

Solution

Because the bar is a symbol of inclusion, we shall calculate the numerator and denominator separately and then divide.

$$\frac{3 + (3 - 2)4}{3 + 4} = \frac{3 + (1)4}{3 + 4} = \frac{3 + 4}{3 + 4} = \frac{7}{7} = 1$$

There will be occasions when we are interested in the **distance** a number is from the origin (0) of the number line. The numbers 4 and −4 are both 4 units from the origin. We use the absolute value of a signed number to indicate its distance from 0 on the number line. The absolute value of any number ($\neq 0$) is positive. The absolute value of 0 is 0. We write the absolute value of the number a as $|a|$.

Note: If any number, *except 0*, has no attached sign, it is assumed to be a positive number.

EXAMPLE 8

Evaluate the following.

28 Sets and the Real-Number System

a. $|-4|$
b. $|+2|$
c. $|0|$
d. $|5|$

Solution

a. $|-4| = +4 = 4$
b. $|+2| = +2 = 2$
c. $|0| = 0$
d. $|5| = +5 = 5$

Exercise 1-6

Compute

1. $13 + (3 + 4)$
2. $12 + (7 + 1)$
3. $15 + (6 - 2)$
4. $17 + (16 - 12)$
5. $26 - (15 + 4)$
6. $37 - (16 + 12)$
7. $37 - (16 - 12)$
8. $15 - (19 - 6)$
9. $4(6 + 3)$
10. $5(7 + 6)$
11. $16[15 + 4]$
12. $12\{18 + 9\}$
13. $12[31 - 16]$
14. $8(16 - 12)$
15. $2 + (3 - 1)$
16. $15[18 - 3]$
17. $3 + \{16 + 2\}$
18. $4 + \{18 - 4\}$
19. $14 - [16 - 8]$
20. $12(12 - 8)$
21. $\dfrac{16 - 4}{3}$
22. $\dfrac{26 - 12}{7}$
23. $\dfrac{36 + 4}{10}$
24. $\dfrac{15 - 7}{4}$
25. $3 + 6 \cdot 8$
26. $15 - 2 \cdot 3$
27. $4^2 + 2 \cdot 5$
28. $3^2 + 5 \cdot 3$
29. $12 + 2^2 \cdot 3$
30. $26 - 3^2 \cdot 2$
31. $16 - 3 \cdot 4$
32. $\sqrt{36} + 4 \cdot 2$
33. $42 - 6 \div 2$
34. $12 + 12/3$
35. $(16 - 4)/4$
36. $(12 + 15)/9$
37. $\dfrac{19 - 5 \cdot 2}{3}$
38. $\dfrac{56 - 7 \cdot 2}{6}$
39. $\dfrac{3^2 - 4}{5}$
40. $\dfrac{2^3 + 6}{7}$
41. $\dfrac{17 - (2 + 3)}{2 + 4}$
42. $\dfrac{64 + 2 \cdot 6}{21 + 17}$

Write the absolute values of the following numbers.

43. −3
44. −6
45. 4
46. 5
47. 0
48. −1
49. Which is larger, |−3| or |2|?
50. Which is larger, |−2| or |3|?
51. Which is larger, |−6| or |−5|?

1-7 Addition and subtraction of signed numbers

OBJECTIVE: After completing this section, you should be able to compute the sum and differences of signed numbers.

Suppose you have stock in two different companies. Each of your stocks could either increase or decrease from day to day as follows:

	Day 1	*Day 2*	*Day 3*	*Day 4*
Company A	+$1 (gain)	−$1 (loss)	+$1 (gain)	−$1 (loss)
Company B	+$2 (gain)	−$2 (loss)	−$2 (loss)	+$2 (gain)
Net result	+$3 (gain)	−$3 (loss)	−$1 (loss)	+$1 (gain)

By looking at the examples, we see that if the signs of both numbers being added are the same, then the result has this common sign. This suggests the following procedure be used to add signed numbers that have the same sign.

PROCEDURE	EXAMPLES
To add signed numbers with same sign:	Compute $(-3) + (-4)$.
1. Add the absolute values (the numerical values, disregarding signs).	1. $3 + 4 = 7$
2. Affix their common sign to the sum.	2. $(-3) + (-4) = -7$

EXAMPLE 1

Compute $(+3) + (+5)$.

Solution

$$3 + 5 = 8$$
$$(+3) + (+5) = +8$$

We can extend the rule to the addition of three or more numbers.

EXAMPLE 2

Compute $(-2) + (-6) + (-7)$.

Solution

$$2 + 6 + 7 = 15$$
$$(-2) + (-6) + (-7) = -15$$

Referring to the stock examples, where the signs of the two numbers being added are different, suggests the following procedure.

PROCEDURE	EXAMPLE
To add two signed numbers with unlike signs:	Compute $(-4) + (+3)$.
1. Find the difference of their absolute values.	1. $4 - 3 = 1$
2. Affix the sign of the number with the larger absolute value.	2. $(-4) + (+3) = -1$

EXAMPLE 3

Compute $(+4) + (-5)$.

Solution

$$5 - 4 = 1$$
$$(+4) + (-5) = -1$$

EXAMPLE 4

Compute $5 + (-3)$.

Solution

$$5 - 3 = 2$$
$$5 + (-3) = +2$$

We can add three or more signed numbers by adding them two at a time, as the following examples show.

EXAMPLE 5

Compute $(-3) + 2 + (-5)$.

Solution

$$(-3) + 2 + (-5) = [(-3) + 2] + (-5) = (-1) + (-5) = -6$$

EXAMPLE 6

Compute $(-4) + 2 + 4 + (-1)$.

Solution

$$(-4) + 2 + 4 + (-1) = [(-4) + 2] + 4 + (-1) = (-2) + 4 + (-1)$$
$$= [(-2) + 4] + (-1) = 2 + (-1) = 1$$

We have seen (as a consequence of the existence of the additive inverse) that we can perform subtractions of signed numbers by adding the negative (additive inverse). Thus $4 - 3 = 4 + (-3) = 1$ and $-4 - (-2) = -4 + 2 = -2$.

In general, we can subtract one number from another by changing the sign of the number we are subtracting and proceeding as in addition.

EXAMPLE 7

Compute $(-7) - (-8)$.

Solution

$$(-7) - (-8) = (-7) + (+8) = +1$$

EXAMPLE 8

Subtract 8 from -16.

Sets and the Real-Number System

Solution

$$(-16) - 8 = (-16) - (+8) = (-16) + (-8) = -24$$

We may combine additions and subtractions in one problem.

EXAMPLE 9

Compute $3 + (-6) - (-4)$.

Solution

$$3 + (-6) - (-4) = 3 + (-6) + (+4) = (-3) + (+4) = 1$$

A problem like $5 - 3$ may be thought of as a subtraction problem or as an addition problem of the form $5 + (-3)$; that is, we may think of the negative sign as a sign on the 3 or as an indicator of the operation of subtraction. Fortunately, either approach will give the same result.

EXAMPLE 10

Compute $5 - 3$.

Solution

by subtraction: $\qquad 5 - 3 = 2$

by addition: $\qquad 5 + (-3) = 2$

EXAMPLE 11

Compute $-4 + 2 + 4 - 1$.

Solution

$$-4 + 2 + 4 - 1 = (-4 + 2) + 4 - 1$$
$$= -2 + 4 - 1 = (-2 + 4) - 1 = 2 - 1 = 1$$

Exercise 1-7

Compute the following sums.

1. $(-3) + (-4)$
2. $(+2) + (+6)$
3. $(-4) + (-7)$
4. $(-5) + (-2) + (-6)$

5. $2 + 4 + 5$
6. $(-2) + (-5) + (-7)$
7. $(-6) + (+7)$
8. $7 + (-8)$
9. $(-7) + 5$
10. $(-3) + (+4) + (2)$
11. $7 + (-6) + (-3)$
12. $(-7) + 6 + (-2)$
13. $(-2) + (-3) + (+2)$
14. $(-8) + (+6) + (-2)$
15. $(-6) + (+6)$
16. $(+3) + (-3) + 2$
17. $-6 + 5$
18. $3 + (-2) + 6$
19. $7 + (-3 + 2)$
20. $-3 + (2 + 1)$
21. $-4 + (-2 + 3)$
22. $-6 + (+8) + (-7)$
23. $-6 + (-3) + (-7)$
24. $6 + (-6) + (-5)$

Find the negatives (*additive inverses*) of the following numbers.

25. 2
26. 4
27. 6
28. 5
29. -6
30. -3
31. -2
32. -7
33. -4

Compute the following.

34. $(-3) - (+2)$
35. $(-2) - (-4)$
36. $(+6) - (-3)$
37. $(-7) - (-5)$
38. $(+6) - (-7)$
39. $(-7) - (-8)$
40. $(-3) - (-8)$
41. $4.6 - (+9.2)$
42. $(+7.3) - (+5.1)$
43. $(-3) + (-4) - (+6)$
44. $4 - 6 + (-7)$
45. $(-8) + (-6) - (-10)$
46. $(-4) - (-3) + (-6)$
47. $(8.61) - (6.2) + 4.62$
48. $(-7) - (-6) + (-8)$
49. $(7) - (-6) - (-8)$
50. $4 + (-4) + 6$
51. $15 - (-6) - (-8)$
52. $-6 - (-3) - (-4)$
53. $(-8) + (-6) - (+3)$

54. Ruth Sentz makes a deposit of $500 to a savings account which contains $265. What is the new balance in the account?
55. Raymond Carlton made a withdrawal of $465 from an account containing $805. What is the new balance?
56. Walter James wrote a check for $365. His checking account contained $298. By how much did he overdraw his account?
57. The highest point in the United States is Mount McKinley at 20,270 feet above sea level, and the lowest point is Death Valley at 280 feet below sea level. What is the difference in height between the two points?
58. The low temperature for January 15 was 10 degrees

34 Sets and the Real-Number System

below zero, and the high temperature was 35 degrees above zero. How many degrees did the temperature change during the day?

1-8 Products and quotients of signed numbers

OBJECTIVES: After completing this section, you should be able to:

a. Perform computations involving the multiplication and division of signed numbers.
b. Perform computations involving zero.

We can think of multiplication as repeated addition; that is, $3 \cdot 4 = 4 + 4 + 4 = 12$. We may also compute the product $4 \cdot 3$ by computing $3 + 3 + 3 + 3 = 12$ or by using the commutative law of multiplication to show that $4 \cdot 3 = 3 \cdot 4 = 12$.

We can compute the product $3\,(-4)$ by computing $(-4) + (-4) + (-4) = -12$. We can compute $(-4) \cdot 3$ by using the commutative law to write the product as $3 \cdot (-4) = (-4) + (-4) + (-4) = -12$.

The above discussion suggests that the product of two numbers with unlike signs is the negative of the product of their absolute values; that is, **the product of two numbers with unlike signs is always negative.**

To multiply two numbers with unlike signs, we proceed as follows:

PROCEDURE	EXAMPLE
	Multiply $(+4)$ by (-5).
1. Multiply their absolute values.	1. $\|4\| \cdot \|-5\| = 4 \cdot 5 = 20$
2. Attach the negative sign.	2. $4(-5) = -20$

EXAMPLE 1

Find the products.

   a. $(-3)(+6)$

b. 2(−8)
c. 5(−5)

Solution

a. (−3)(+6) = −18
b. 2(−8) = −16
c. 5(−5) = −25

EXAMPLE 2

Compute the product of (−1) and 4.

Solution

(−1)(4) = −4

As a result of the existence of an additive inverse (see Note 1 following Property VII), we know $-(-a) = a$. Thus to find the product (−4)(−3), we write

$$\begin{aligned}(-4)(-3) &= [(-1)(4)](-3) \\ &= (-1)[(4)(-3)] \\ &= (-1)(-12) \\ &= -(-12) \\ &= 12\end{aligned}$$

Thus the product (−4)(−3) = +12. We know that (+4)(+3) = +12. This illustrates the fact that the product of two numbers with like signs is the product of their absolute values; that is, **the product of two numbers with like signs will always be positive.**

EXAMPLE 3

Find the products.

a. (−4)(−2)
b. (+2)(+10)
c. (−1)(−6)

Solution

a. (−4)(−2) = +8
b. (+2)(+10) = +20
c. (−1)(−6) = 6

To find the product of more than two signed numbers, we can group them in pairs and compute their products two at a time.

EXAMPLE 4

Compute $(-3)(+2)(-4)$.

Solution

$(-3)(+2)(-4) = [(-3)(+2)](-4) = [-6](-4) = +24$

But a better way to determine the sign of the product of signed numbers is to use the following rule.

If there is an odd number of negative factors, the product will be negative; if there is an even number of negative factors, the product will be positive.

EXAMPLE 5

Compute $(-4)(+3)(-6)(-7)$.

Solution

$(-4)(+3)(-6)(-7)$ has three negative factors; so the product will be negative.

$(-4)(+3)(-6)(-7) = -504$

EXAMPLE 6

Compute $(+4)(-3)(2)(-1)$.

Solution

$(+4)(-3)(2)(-1)$ will be positive, because there are two negative factors.

$(+4)(-3)(2)(-1) = +24$ or 24

The same rules that apply to the products of signed numbers apply to the quotients of them.

The quotient of two signed numbers with like signs is positive.

1-8 Products and quotients of signed numbers

EXAMPLE 7

Compute $(-14) \div (-2)$.

Solution

$$(-14) \div (-2) = +(14 \div 2) = +7$$

Check: $(-2)(+7) = -14$

EXAMPLE 8

Compute $(+36) \div (+4)$.

Solution

$$(+36) \div (+4) = +(36 \div 4) = +9$$

The quotient of two numbers with unlike signs is negative.

EXAMPLE 9

Compute $(-28) \div (+4)$.

Solution

$$(-28) \div (+4) = -(28 \div 4) = -7$$

Check: $(+4)(-7) = -28$

EXAMPLE 10

Compute $(+45) \div (-5)$.

Solution

$$(+45) \div (-5) = -(45 \div 5) = -9$$

We may also combine the operations of multiplication and division of signed numbers.

EXAMPLE 11

Compute $\dfrac{(-42)(-6)}{(-21)}$.

Solution

$$\frac{(-42)(-6)}{(-21)} = \frac{+252}{-21} = -12$$

or

$$\frac{(-42)(-6)}{(-21)} = \frac{(-42)}{(-21)}(-6) = 2(-6) = -12$$

As we indicated in Property VII, Note 2, we cannot divide by 0. The following examples illustrate computations with 0.

EXAMPLE 12

Compute $(-4) + 0$.

Solution

$(-4) + 0 = -4$

EXAMPLE 13

Compute $0 - (-3)$.

Solution

$0 - (-3) = 0 + (+3) = 3$

Thus, $-(-3) = +3$, which reaffirms that $+3$ is the negative of -3.

EXAMPLE 14

Compute $144(-166)(0)(-366)$.

Solution

$144(-166)(0)(-366) = 0$

EXAMPLE 15

$\dfrac{147 \cdot (-66) + 177}{3 \cdot 114 \cdot 0}$ is undefined because $3 \cdot 114 \cdot 0 = 0$, and division by 0 is undefined.

Exercise 1-8

Compute.

1. $(-3)(-2)$
2. $(-4)(-3)$
3. $(4)(5)$
4. $(-5)(-6)$
5. $(-7)(+6)$
6. $(-2)(+3)$
7. $(2)(-2)(-1)$
8. $(3)(-2)(4)$
9. $(4)(-2)(-8)$
10. $-3 \cdot 4(-0.6)$
11. $-7(1.4)6$
12. $-4.8(-2)(-6)$
13. $-2(-3)(-1)$
14. $(-1)(2)(-4)$
15. $(-3)(-2)(-5)$
16. $2(-2)(-3)$
17. $(-6)(-2)(-1)$
18. $2(+2)(-1)$
19. $-8(-4)3(-1)$
20. $6(-2)5(-2)$
21. $7(-8)(-1)(-2)$
22. $-6(-3)(-2)(4)$
23. $4(-1)(-3)(2)$

Perform the following operations.

24. $(-14) \div (-7)$
25. $(-6) \div (-2)$
26. $(+14) \div (+2)$
27. $(+18)/(+9)$
28. $(-21)/(-7)$
29. $(-16)/(-8)$
30. $(-38.6) \div (19.3)$
31. $72.2/(-36.1)$
32. $43.2 \div (-8)$
33. $(-4.41) \div 21$
34. $144/(-16)$
35. $-176 \div 16$
36. $\dfrac{3(-18)}{-9}$
37. $\dfrac{21 \cdot 16}{-14}$
38. $\dfrac{(-48)(-16)}{64}$
39. $\dfrac{24(-6)}{36}$
40. $\dfrac{-14(-3)6}{(-4)(-7)}$
41. $\dfrac{2(-3)(-6)}{9(-2)}$
42. $\dfrac{6(-8)(-4)}{3(-16)}$
43. $\dfrac{(-2)(-3)(+4)}{6 \cdot 2}$
44. $-3.2 + 0 + 16$
45. $-42 + 0 + (-17)$
46. $16 - 0$
47. $0 - 4$
48. $0 - (-8)$
49. $0 - [16(-8)]$
50. $4.4 \cdot 0 \cdot (-3)$
51. $(-17)(0)(-16)$
52. $16 - 0(4.5)$
53. $32 + 0(14 + 3)$
54. $0 \div 6$
55. $0 \div (-37)$
56. $0 \cdot \dfrac{14}{17}$
57. $36 \div 0$
58. $\dfrac{(1.4)(-13)(0.6)}{3 \cdot 0}$
59. $\dfrac{17 \cdot 6 \cdot 14 \cdot 0}{14 \cdot 2}$

40 Sets and the Real-Number System

60. $\dfrac{3 \cdot 0 \cdot 6}{-6 + 14 - 8}$ 61. $\dfrac{(3 - 6 + 3) + 4}{16 - 8 + 2}$

CHAPTER TEST

1. Is $13 \in \{1, 3, 5, 7, \ldots\}$?
2. Is $6 \in \{x: x$ is a natural number less than $6\}$?
3. Write $\{x: x$ is a natural number less than $9\}$ a second way.
4. Write $\{1, 3, 5, 7, 9\}$ a second way.
5. Is $A \subseteq B$ if $A = \{1, 3, 8, 7, 6\}$ and $B = \{1, 2, 3, 4, 5, 6, 8\}$?
6. Does $A = B$ if $A = \{1, 3, 5, 7\}$ and $B = \{x: x$ is an odd natural number less than $9\}$?
7. Are A and B disjoint if $A = \{1, 2, 3, 4\}$ and $B = \{x: x$ is a natural number greater than $4\}$?
8. If $A = \{7, 8, \pi, 7, 0\}$ and $B = \{1, 2, 3, \pi, 6\}$, find $A \cap B$.
9. If $A = \{1, 3, 5, 7, 11, 6\}$ and $B = \{2, 3, 4, 7, 11\}$, find $A \cup B$.
10. Are $A = \{1, 2, 3, 4\}$ and $B = \{a, b, c, d\}$ equivalent sets?
11. Establish a one-to-one correspondence between $\{3, 6, 8, 11, \pi\}$ and $\{\pi, 6, 5, 4, 3\}$.
12. Graph the integers from -6 to 10 on a number line.
13. Insert the proper inequality ($<$ or $>$) between -6 and -8.
14. Combine $-1 > -4$ and $-1 < 3$ into one inequality.
15. Graph the rational numbers $-\frac{1}{4}$ and $\frac{7}{4}$ on a number line.
16. Are $\frac{17}{54}$ and $\frac{13}{39}$ equal?

Identify each of the following real numbers as one or more of natural number, integer, rational number, or irrational number.

17. $\sqrt{16}$
18. $\sqrt{15}$
19. $\frac{3}{4}$

Compute the following.

20. $36 + 71 - 23$
21. $16 \cdot 15 + 52 \div 13$
22. $3^3 - 6/2$
23. $6 + 7 = 7 + \underline{}$
24. $16 \cdot \underline{} = 4 \cdot 16$

Compute the following.

25. $3 \cdot 2 + 4$
26. $13 - (15 - 8)$
27. $0 \cdot 3 + 4^2 \cdot 2$
28. Complete the following illustration of the associative law of multiplication.

$$5(2 \cdot 7) = \underline{}$$

29. Complete the following illustration of the distributive law.

$$4(2 + 7) = 4 \cdot 2 + \underline{}$$

Write the absolute values of the following.

30. -6
31. 5

Compute the following.

32. $(-6) + (-7) + (+5)$
33. $(-4) - (+7) + (-8)$
34. $(-3) + (+6) - (-6)$
35. $(-2) \cdot (+6) \cdot (-5)$
36. $(+3)(-2)(-6)(-1)$
37. $\dfrac{(-33)(16)}{24}$
38. $\dfrac{6(-3)(-2)}{-9}$
39. $(-14)0 + 16 - 0$

operations on algebraic expressions

2-1 Combination of like terms

OBJECTIVES: After completing this section, you should be able to:

a. **Classify certain polynomials as monomials, binomials, or trinomials.**
b. **Write the numerical coefficients of algebraic terms.**
c. **Determine the degree of a polynomial.**
d. **Combine terms that differ only in their numerical coefficients.**

In algebra we are usually dealing with a mixture of numerals (such as 3, 6, −1) and letters (such as x, a, m). An expression containing one or more numerals or letters is called an algebraic expression. Examples of algebraic expressions are $3x - y$, $2x^3 + 4$, $15 - x$, $ax + by + c$, and $15x$. Note that the last algebraic expression ($15x$) is the product of a numeral (15) and a letter (x). Each numeral, letter, or product of numerals and letters is called a term. Thus $15x$ is an expression with one term, and $3x - y + 2z$ is an expression with three terms. **Note that each term is completely set off from the others by a plus (+) or minus (−) sign.** An expression with only one term (such as $15x$) is called a monomial. An algebraic expression with two or more terms is called a **multinomial**. If the multinomial has no letters under a radical sign ($\sqrt{\ }$) or in a denominator, it is called a **polynomial**. The expression $3x + 5by + 2y + 6$ is a polynomial with four terms. A polynomial with two terms is called a **binomial**, and one with three terms is called a **trinomial**.

In the term $15x$, we say that 15 is the **coefficient** of x, and x is the coefficient of 15. We shall frequently need to use the **numerical coefficient** in a term. The numerical coefficient in

15x is 15, and the numerical coefficient in $-5xy$ is -5. If no numeral appears as a coefficient in a term, the numerical coefficient is understood to be 1; that is, x means $1x$ and $-y$ means $-1y$. The letter in a term is frequently called a **literal factor** of the term. If a term contains no literal factors, it is called a **constant term**.

Because different values may be substituted in for a letter in a polynomial, the letter is frequently called a **variable**. Thus $x^3 + 2x^2 - 3x$ is a polynomial in one variable x and $x^2 - 3xy + y^2$ is a polynomial in two variables x and y.

The **degree of a term** containing one variable is the same as the exponent of that variable. For example, $14x$ is a first-degree term, and $6x^3$ is a third-degree term. If two or more variables are in a term, the degree of the term is the *sum* of the exponents of the variables. Thus the degree of $4x^2y$ is $2 + 1 = 3$, and the degree of $6xy$ is $1 + 1 = 2$. The **degree of a polynomial** is the degree of the term in the polynomial having the highest degree. Thus the polynomial $4x^3 - 2x + 4$ is a third-degree polynomial, and $2xy - 4x + 6$ is a second-degree polynomial.

EXAMPLE 1

Given the following polynomials, classify them as monomials, binomials, or trinomials; give the numerical coefficient in each term; and state the constant if there is one.

a. $3x - y$
b. $4x^2 + 3x + 2$
c. $4x^3$

Solution

a. Binomial; 3 is the numerical coefficient of x; -1 is the coefficient of y.
b. Trinomial; 4 is the numerical coefficient of x^2; 3 is the coefficient of x; 2 is a constant.
c. Monomial; 4 is the numerical coefficient of x^3.

EXAMPLE 2

Classify the following as monomials, binomials, or trinomials, and state the degree of the polynomial.

a. $4x^2 - 3$
b. $2x^3 - 3x + 2$
c. $4xy$

Solution

a. Binomial; second degree.
b. Trinomial; third degree.
c. Monomial; second degree.

Two or more terms that differ only in their numerical coefficients are called **similar terms**, or **like terms**. Thus $3x^2y$, $5x^2y$, and $16yx^2$ are like terms; $3x^2y$ and $3xy$ are not like terms because x is used twice as a factor in $3x^2y$ and only once in $3xy$. **We can add two like terms by adding their numerical coefficients and assigning their common literal factors.** For example, $3x + 2x = 5x$. Note that $3x = x + x + x$ and $2x = x + x$; so $3x + 2x = (x + x + x) + (x + x) = 5x$.

EXAMPLE 3

Add $5y$ to $3y$.

Solution

$3y$ and $5y$ are like terms; so their sum is $3y + 5y = (3 + 5)y = 8y$ **(note the use of the distributive property).**

EXAMPLE 4

Combine $3xy + 4x + 7xy$.

Solution

$3xy$ and $7xy$ are like terms; so their sum is $10xy$. $4x$ is not like any other term in this expression; so it cannot be combined with any of the other terms. We can only indicate the addition of $4x$. Thus $3xy + 4x + 7xy = 10xy + 4x$.

This procedure can easily be extended to the operation of subtraction. For example, $8x - 3x = 8x + (-3x) = 5x$.

EXAMPLE 5

Combine $4xy - 8xy$.

Solution

$$4xy - 8xy = 4xy + (-8xy) = -4xy$$

Performing the indicated additions and subtractions is called combining the terms. We can simplify a series of terms by combining the like terms.

EXAMPLE 6

Combine $-3x^2 + 4xy + 5x^2 + 7xy$.

Solution

$-3x^2$ and $5x^2$ are like terms; so their sum is $2x^2$. $4xy$ and $7xy$ are like terms; so their sum is $11xy$. Thus $-3x^2 + 4xy + 5x^2 + 7xy = 2x^2 + 11xy$.

Exercise 2-1

Classify the following as monomials, binomials, or trinomials; give the numerical coefficients of each term; and identify the constant if there is one.

1. $3x + 2y$
2. $4x^2 - 3x + 2$
3. $15x^2y$
4. $4x^3 + 2$
5. $6x^2 - 3x + 4$
6. $36x$
7. $46x^3 - 12$
8. $15x - 3y + 1$
9. State the degree of the polynomials in Problems 1 through 4.
10. State the degree of the polynomials in Problems 5 through 8.

Simplify the following by combining like terms.

11. $5x + 3x$
12. $16y + 4y$
13. $13y + 4y$
14. $10c + 11c$
15. $2y^3 + 4y^3 + 5y^3$
16. $5c + 6c + 2c$
17. $13z + 2z + 7z$
18. $36r^2 + 4r^2 + 5r^2$
19. $2x + 4x^2 + 3x$
20. $5y^2 + 6y + 7y^2$
21. $13x + 2y + 4x$
22. $7p + 3q + 5q$
23. $4x + 5a + 3x + 3a$
24. $2p + 4q + 5q + 7p$
25. $6t^2 + 4t^3 + 5t^2 + t^3$
26. $17r^2 + 4s + 6r^2 + 2s$
27. $3x^2y + 4xy - 7x^2y$
28. $4xy + 2yz + 3xy + 3yz$
29. $3a^2 + 2b^2 + 6a^2 + 4b^2$
30. $x^2y + xy + 4x^2y + 5xy$
31. $4x - 2x$
32. $15y - 4y$
33. $17xy - 12xy$
34. $37ab - 15ab$
35. $12ab - 6a - 2ab$
36. $16pq - 4p - 12pq$

37. $17mn + 4n - 4mn$
38. $36r^2 + 15r - 16r^2$
39. $5xy + 4xy - 7xy$
40. $12rs + 2rs - 8rs$
41. $15pq + 3pq - 21pq$
42. $6y^2z + 3y^2z - 11y^2z$
43. $4a - 3b + 2c - 4a + 2b$
44. $5r + 2s + u - 3r + 4s - 6u$
45. $15x^2y + 4xy - y^2 + 3x^2y - 5xy + 4y^2$
46. $3xy + 3y^2 + 6x - 4xy - 2y^2 + 3x$

2-2 Sums and differences of polynomials

OBJECTIVE: After completing this section, you should be able to find the sums and differences of polynomials.

The procedures for adding and subtracting polynomials are based on the procedures for combining like terms. If, for example, we want to add $5x + 3y$ and $7x - y$, we combine their like terms to find the sum. Thus the sum of $5x + 3y$ and $7x - y$ is $12x + 2y$. We may indicate the sum as $(5x + 3y) + (7x - y) = 12x + 2y$.

EXAMPLE 1

Combine $(3x^2 + 4xy + y^2) + (x^2 + 4x - 2xy - 2y^2)$.

Solution

Because the parentheses are preceded by no sign or a plus (+) sign, we may remove the parentheses without changing any signs. Thus we may write $(3x^2 + 4xy + y^2) + (x^2 + 4x - 2xy - 2y^2)$ as $3x^2 + 4xy + y^2 + x^2 + 4x - 2xy - 2y^2$. Combining like terms gives $4x^2 + 4x + 2xy - y^2$.

EXAMPLE 2

Compute $(4xy + 3x) + (5xy - 2x)$.

Solution

$(4xy + 3x) + (5xy - 2x) = 4xy + 3x + 5xy - 2x = 9xy + x$

We can subtract $4x - 2xy$ from $x^2 + 2x + 4xy$ by writing $(x^2 + 2x + 4xy) - (4x - 2xy)$

We may use the distributive law to remove the parentheses preceded by a minus sign; that is,
$$-(4x - 2xy) = (-1)(4x - 2xy) = -4x + 2xy.$$
Thus
$$(x^2 + 2x + 4xy) - (4x - 2xy) = x^2 + 2x + 4xy - 4x + 2xy$$
$$= x^2 - 2x + 6xy$$

The example above shows that **we may remove parentheses preceded by a minus sign if we change the sign of every term inside the parentheses.**

Thus we may subtract one polynomial from another by changing the sign of each term of the polynomial we are subtracting and proceeding as in addition.

EXAMPLE 3

Compute $(3x^2 + 4xy + 5y^2 + 1) - (6x^2 - 2xy + 4)$.

Solution

Removing the parentheses gives $3x^2 + 4xy + 5y^2 + 1 - 6x^2 + 2xy - 4$, which simplifies to $-3x^2 + 6xy + 5y^2 - 3$.

EXAMPLE 4

Compute $(14ab - 6a^2b + 4b) - (11ab + 4a^2b^2 - 4b + 2a^2b)$.

Solution

$$(14ab - 6a^2b + 4b) - (11ab + 4a^2b^2 - 4b + 2a^2b)$$
$$= 14ab - 6a^2b + 4b - 11ab - 4a^2b^2 + 4b - 2a^2b$$
$$= 3ab - 8a^2b - 4a^2b^2 + 8b$$

Exercise 2-2

Combine the following.

1. $(3x + 2y) + (2x + 5y)$
2. $(4x + 2y) + (5x + 3y)$
3. $(5m + 6n) + (3m + 4n)$
4. $(16p + 4q) + (11p + 5q)$
5. $(3x^2 + 4y) + (3x^2 + 2y)$

48 Operations on Algebraic Expressions

6. $(5a + 3c) + (2x + a)$
7. $(2x + 3y + 4m) + (2y + 3x + m)$
8. $(2x^2y + 4xy) + (x^2y - 4xy)$
9. $(2x - 3y + 4m) + (2y - 3x + m)$
10. $(4x^2y + 4mx + 2) + (3x^2y - mx + 4)$
11. $(12a + b) - (3 + 2b)$
12. $(15xy + 5y) - (6xy + 4y)$
13. $(11r + 5s) - (7r - 2s)$
14. $(4xy - 3x) - (3x - 4y)$
15. $(4x^2 - 3x^2y + 4) - (3x^2 + 2xy + 4x^2y)$
16. $(4xy - 3x + 2y) - (3x - 4y + 2xy)$
17. $(7x - 3y + 2) - (6x + 2y + 4)$
18. $(17x + 2xy + y) - (4x - 3xy - y)$
19. $(5a + 3c) + (2x - a) + (4x - 3c)$
20. $(6a - 3c) + (2x - 5a) + (4x - 3c)$
21. $(3x - 3y) + (3x - 4y) - (3x + 2y)$
22. $(2x - 3y) - (2y - x) + (3x + 2y)$
23. $(3a - 4b) - (6b - 2a) - (2a + 4b)$
24. $(2x - 2y) - (x + 2y) - (x - 4y)$
25. $(7 - 4ax^2y + 3axy + 4x^2y) - (7ax^2y - 2axy + 2x)$
26. $(17ar^2b + 6x + 6xr + 7) - (36ar^2b - 5x - 7xr - 7)$

2-3 Multiplication and division of monomials

OBJECTIVES: After completing this section, you should be able to:

a. Find the product of two or more monomials.
b. Find the quotient of two monomials.
c. Simplify expressions involving monomial products.

When we write a^5, we are writing the number a to the fifth power. The number a is called the **base**, and 5 is the **exponent**. Just as $3^5 = 3 \cdot 3 \cdot 3 \cdot 3 \cdot 3$, we have $a^5 = a \cdot a \cdot a \cdot a \cdot a$; that is, a^5 indicates that a is to be used as a **factor** 5 times. If we want to multiply a^3 times a^2, we can write $a^3 \cdot a^2 = (a \cdot a \cdot a)(a \cdot a) = a \cdot a \cdot a \cdot a \cdot a = a^5$; so $a^3 \cdot a^2 = a^5$. We could have obtained the same result by adding the exponents of a^3 and a^2, getting $a^3 \cdot a^2 = a^{3+2} = a^5$. This is an example of one of the basic laws of exponents.

To multiply two powers of the same base, add the exponents.

Written symbolically, this is $a^m \cdot a^n = a^{m+n}$.

EXAMPLE 1

Multiply

a. $x^4 \cdot x^3$
b. $y^2 \cdot y^5$
c. $3^2 \cdot 3^4$
d. $a^5 \cdot b^2$

Solution

a. $x^4 \cdot x^3 = x^7$
b. $y^2 \cdot y^5 = y^7$
c. $3^2 \cdot 3^4 = 3^6$
d. $a^5 \cdot b^2 = a^5 b^2$ (Bases are different.)

As Example 1d shows, we do not add the exponents if the numbers we are multiplying have different bases.

If we want to multiply the monomials $2x^2y$ and $3xy$, we can write $(2x^2y) \cdot (3xy) = 2 \cdot x \cdot x \cdot y \cdot 3 \cdot x \cdot y = 2 \cdot 3 \cdot x \cdot x \cdot x \cdot y \cdot y = 6x^3y^2$. **We multiply monomials by multiplying the numerical coefficients, finding the product of the corresponding literal factors, and writing these products together as the final product.**

EXAMPLE 2

Multiply the following.

a. $(8xy^3)(2x^3y)$
b. $(3x^2y^3)(4x^3yz)$
c. $(12m^2n)(-3mn^2)(4kn)$

Solution

a. $(8xy^3)(2x^3y) = 16x^4y^4$
b. $(3x^2y^3)(4x^3yz) = 12x^5y^4z$
c. $(12m^2n)(-3mn^2)(4kn) = -144m^3n^4k$

EXAMPLE 3

Simplify the following.

a. $(x^2y)(2xy) + (3x)(x^2y^2)$
b. $(3x^3y)(6y^2) - (2x^3y)(3y^2)$

Solution

The multiplications must be performed before the additions or subtractions.

a. $(x^2y)(2xy) + (3x)(x^2y^2) = 2x^3y^2 + 3x^3y^2 = 5x^3y^2$
b. $(3x^3y)(6y^2) - (2x^3y)(3y^2) = 18x^3y^3 - 6x^3y^3 = 12x^3y^3$

It is a fundamental fact of arithmetic that any number other than 0 divided by itself results in a quotient of 1. For example, $4 \div 4 = 1$, $3 \div 3 = 1$, and $\frac{1076}{1076} = 1$. By the same token, a literal number (not equal to 0) divided by itself results in a quotient of 1. Thus $y \div y = 1$, $x/x = 1$, and $xy/xy = 1$.

EXAMPLE 4

Compute $x^6 \div x^2$ if $x \neq 0$.

Solution

$$x^6 \div x^2 = \frac{x^6}{x^2} = \frac{x \cdot x \cdot x \cdot x \cdot x \cdot x}{x \cdot x} = \frac{x \cdot x}{x \cdot x} \cdot x \cdot x \cdot x \cdot x$$

$$= 1 \cdot x \cdot x \cdot x \cdot x = x^4$$

EXAMPLE 5

Compute $x^2 \div x^5$ if $x \neq 0$.

Solution

$$x^2 \div x^5 = \frac{x^2}{x^5} = \frac{x \cdot x}{x \cdot x \cdot x \cdot x \cdot x} = \frac{x \cdot x}{x \cdot x} \cdot \frac{1}{x \cdot x \cdot x}$$

$$= \frac{1}{x \cdot x \cdot x} = \frac{1}{x^3}$$

The rule we use to compute the quotient of two powers having the same base follows.

PROCEDURE	EXAMPLE
To compute the quotient of two powers having the same base:	a. Divide z^3 by z^3. b. Divide x^5 by x^3. c. Divide y^4 by y^6.
1. Determine if the exponent of the dividend (numerator) is greater than the exponent of the divisor (denominator).	1a. Exponents are equal. b. Dividend's exponent is greater. c. Dividend's exponent is less.

2a. If the exponents of the divisor and the dividend are equal, the quotient is one; that is,
$x^m \div x^n = 1$ if $m = n$

b. If the exponent of the dividend is greater, we subtract the exponent of the divisor from the exponent of the dividend and write the difference as the new exponent of the base; that is,
$x^m \div x^n = x^{m-n}$ if $m > n$

c. If the exponent of the dividend is less than that of the divisor, we subtract the exponent of the dividend from that of the divisor, and write the quotient as a fraction with numerator one and denominator the base to the new power; that is,
$x^m \div x^n = \dfrac{1}{x^{n-m}}$ if $m < n$

2a. $z^3 \div z^3 = 1$

b. $x^5 \div x^3 = x^2$
$\left(\text{Note that } \dfrac{x^5}{x^3} = \dfrac{x^3}{x^3} \cdot x^2 = x^2.\right)$

c. $y^4 \div y^6 = \dfrac{1}{y^2}$
$\left(\text{Note that } \dfrac{y^4}{y^6} = \dfrac{y^4}{y^4} \cdot \dfrac{1}{y^2} = \dfrac{1}{y^2}.\right)$

EXAMPLE 6

Find the quotients.

a. $a^5 \div a^4$
b. $m^7 \div m^9$
c. $x^5 \div x^5$
d. $y^{13} \div y^{10}$

Solution

a. $a^5 \div a^4 = a^{5-4} = a^1 = a$
b. $m^7 \div m^9 = \dfrac{1}{m^{9-7}} = \dfrac{1}{m^2}$
c. $x^5 \div x^5 = 1$
c. $y^{13} \div y^{10} = y^{13-10} = y^3$

To divide two monomials, we divide their numerical coefficients, find the quotients of the corresponding literal factors, and write the product of these quotients together as the final quotient.

EXAMPLE 7

Find the quotient $-15x^2y^3 \div 3xy^2$.

Solution

$$-15 \div 3 = -5$$
$$x^2 \div x = x$$
$$y^3 \div y^2 = y$$

Thus $-15x^2y^3 \div 3xy^2 = -5xy$. Note that we may write $-15x^2y^3 \div 3xy^2$ as $\dfrac{-15x^2y^3}{3xy^2}$.

The steps in the division may be combined into one step, as the following examples show.

EXAMPLE 8

Find the quotients.

a. $3x^3y^3 \div 6x^2y$
b. $14x^2y^3 \div 7xy^5$
c. $-8xy^2 \div 4x^2y^3$

Solution

a. $\dfrac{3x^3y^3}{6x^2y} = \dfrac{xy^2}{2}$

b. $\dfrac{14x^2y^3}{7xy^5} = \dfrac{2x}{y^2}$

c. $\dfrac{-8xy^2}{4x^2y^3} = \dfrac{-2}{xy}$

Exercise 2-3

Perform the following multiplications.

1. $x^2 \cdot x^4$
2. y^3y^5

3. $m^5 \cdot m^4$
5. $5^2 \cdot 5^4$
7. $4^2 \cdot 4^3$
9. $a^2b \cdot ab^2$
11. $lm^2 \cdot l^2m^3$
13. $2x^2 \cdot 5x^3$
15. $4m^2 \cdot 3m^4$
17. $2b^2 \cdot b^5$
19. $(4x^2)(6x^3)$
21. $(3x^2y) \cdot (6xy^3)$
23. $(-4xy)(3x^2t)$
25. $(4xy)(-3xy)(-2x^2)$
27. $(-2mn^2)(-3m^2n)(-5mn)$

4. $p^3 \cdot p^4$
6. $3^4 \cdot 3^3$
8. $7^2 \cdot 7^4$
10. $x^3y \cdot x^2y^2$
12. $r^3s \cdot rs^4$
14. $3y^3 \cdot 6y^2$
16. $6r^2 \cdot 5r^3$
18. $4c^3 \cdot c^4$
20. $(5y^3)(2y^4)$
22. $(16xy^2)(4xy^2)$
24. $(6pq)(-3p^2)$
26. $(-b^2c)(bc^3)(b^2c)$
28. $(-3x^2y)(2xy^3)(-10x^3y^3)$

Simplify the following.

29. $(4x^2y)(2xy^2) - (6x^3y)y^2$
30. $x^2(2x) - x(x^2)$
31. $(15xy)(2x^3y) + (3x^2y^2)(-x^2)$
32. $(2m^2n)(3mn) + (m^2n^2)(3m)$
33. $r^3 + (2r^2)(3r) - (6r)(2r^2)$
34. $(rs)(-s) + (3r)(s^2) - (6r)(-s)^2$

Perform the following divisions.

35. $24m^2n \div 8mn$
37. $76r^2s^4 \div 19rs^3$
39. $(-48m^2n) \div (6mn)$
41. $(-54r^3s^2) \div (-9rs^2)$
43. $x^3 \div x^7$
45. $r^2s \div r^3s$
47. $\dfrac{16x^5y^3}{8x^3y^4}$

49. $\dfrac{36x^5y^3}{72m^3y^3}$

51. $\dfrac{15x^2y}{5xy^2}$

53. $\dfrac{-18x^2y^3}{-36xy}$

55. $\dfrac{xy^2z}{xyz}$

57. $\dfrac{-12x^2y^3z}{4xy^2z}$

36. $34x^4 \div 17x^4$
38. $56x^5y^3 \div 8x^2y$
40. $(-36x^3y) \div (9xy)$
42. $(-86p^5q^2) \div (-43p^2)$
44. $r^5 \div r^8$
46. $m^4n^3 \div m^2n^4$
48. $\dfrac{32p^3q^4}{4p^4q}$

50. $\dfrac{48p^2q}{96p^3q}$

52. $\dfrac{-32r^2s}{8r^2s^2}$

54. $\dfrac{-48x^3y^2}{-16x^4y^2}$

56. $\dfrac{a^2b^3c^5}{ab^2c^4}$

58. $\dfrac{36x^4y^3z}{-9x^2y^4}$

54 Operations on Algebraic Expressions

59. $\dfrac{-6a^2b^3c^4}{-2abc^2}$

60. $\dfrac{14a^2b^3r}{-7a^3br^2}$

Simplify each of the following.

61. $\dfrac{4x^3}{2x} + \dfrac{8x^5}{x^3}$

62. $\dfrac{15x^3y^2}{5xy} + \dfrac{6x^5y^3}{x^3y^2}$

63. $\dfrac{44m^3n}{4m^2n}$

64. $\dfrac{56r^2s}{6r} - 16rs$

2-4 Symbols of grouping

OBJECTIVE: After completing this section, you should be able to simplify algebraic expressions involving symbols of grouping.

Because the letters used in algebra represent real numbers, the operations of arithmetic apply to algebraic expressions. We have seen how additions, subtractions, multiplications, and divisions are performed with certain algebraic expressions. The symbols of grouping are used in algebra the same way as they are in the arithmetic of real numbers. The following examples illustrate how to simplify expressions involving symbols of grouping.

EXAMPLE 1

Simplify $-(3y - 2x) + (3y - 4x) - (6y + 4x)$.

Solution

$-(3y - 2x) + (3y - 4x) - (6y + 4x)$
$= -3y + 2x + 3y - 4x - 6y - 4x$
$= -6y - 6x$

When there are two or more symbols of grouping involved, begin with the innermost and work outward.

EXAMPLE 2

Simplify $3x^2 - [2x - (3x^2 - 2x)]$.

Solution

$$3x^2 - [2x - (3x^2 - 2x)] = 3x^2 - [2x - 3x^2 + 2x]$$
$$= 3x^2 - [4x - 3x^2]$$
$$= 3x^2 - 4x + 3x^2$$
$$= 6x^2 - 4x$$

EXAMPLE 3

Simplify $4x^2 - [(3x^2 - 1) - (2x + 1) + x^2]$.

Solution

$$4x^2 - [(3x^2 - 1) - (2x + 1) + x^2]$$
$$= 4x^2 - [3x^2 - 1 - 2x - 1 + x^2]$$
$$= 4x^2 - [4x^2 - 2x - 2]$$
$$= 4x^2 - 4x^2 + 2x + 2$$
$$= 2x + 2$$

There will be occasions when we have to insert parentheses in an algebraic expression. To include several terms in parentheses (or other symbols of grouping) preceded by a plus (+) sign without changing the value of the expression, we simply insert the parentheses.

EXAMPLE 4

Insert parentheses, preceded by a plus sign, to group the last two terms of $3x^2 - 4y - 2x + 3y^2$.

Solution

$$3x^2 - 4y + (-2x + 3y^2)$$

To include several terms in parentheses (or other symbols of grouping) preceded by a minus (−) sign without changing the value of the expression, we insert the parentheses and change the sign of every term included in the parentheses.

EXAMPLE 5

Insert parentheses preceded by a minus sign to group the last two terms of $3x^2 - 4y - 2x + 3y^2$.

Solution

$$3x^2 - 4y - (2x - 3y^2)$$

We frequently use parentheses to indicate multiplication. For example, $2(x + 3)$ represents "2 times the sum of x and 3." By the distributive law, $2(x + 3) = 2 \cdot x + 2 \cdot 3 = 2x + 6$.

EXAMPLE 6

Multiply the following.

a. $3(x + 4)$
b. $5(2x - y)$

Solution

a. $3(x + 4) = 3 \cdot x + 3 \cdot 4 = 3x + 12$
b. $5(2x - y) = 5 \cdot 2x - 5 \cdot y = 10x - 5y$

EXAMPLE 7

Compute the following.

a. $\dfrac{5x - x}{2} + 6x$

b. $\dfrac{9y^2 - 6y^2}{3y} + 2y$

Solution

Recall that the bar is a symbol of inclusion; so the numerator should be simplified before dividing.

a. $\dfrac{5x - x}{2} + 6x = \dfrac{(5x - x)}{2} + 6x = \dfrac{4x}{2} + 6x = 2x + 6x = 8x$

b. $\dfrac{9y^2 - 6y^2}{3y} + 2y = \dfrac{(9y^2 - 6y^2)}{3y} + 2y = \dfrac{3y^2}{3y} + 2y = y + 2y = 3y$

Exercise 2-4

Simplify the following.

1. $4x - (3x + 2) + (4x - 6)$
2. $4m - (2m - 1) + (3 - m)$

3. $15x - (2x + 1) + (x - 4)$
4. $(2p + 6) - (p - 3) + (2p + 6)$
5. $7p - [6p - (3p + 2)]$
6. $2a - [2a - (a + 1)]$
7. $x^2 - [2x^2 + (2x - x^2)]$
8. $xy - [x + (2xy - 4x)]$
9. $x^2 - [3x - (x^2 + 2x)]$
10. $xy - [x^2 - (xy + x^2)]$
11. $2x^3 - [(x^2 - x^3) - (x^2 - 3) + x^3]$
12. $16x + [x^2 - (3x - 2) + (4x^2 - 1)]$

Insert parentheses, preceded by a minus sign, to include the last three terms of the following.

13. $2x + 3y + 2z - 6$
14. $7x^2 - 3xy + 4x + 6$
15. $4ab + 3abx - 3x^2 + 2ax$
16. $ad - bc + 4ac - 2bd$

Multiply and simplify the following.

17. $2(5 + 6)$
18. $4(7 + 13)$
19. $4(x + 3)$
20. $5(x - y)$
21. $6(x + 2y)$
22. $2(1 + 3x)$

Simplify the following.

23. $\dfrac{5x + 3x}{2x}$
24. $\dfrac{7y^2 - 4y^2}{3y}$
25. $\dfrac{5y^2 + 2y^2}{y} + 3y^2$
26. $\dfrac{8x^2 - 2x^2}{3x} + 2x^2$
27. $\dfrac{12z^2 - 6z^2}{2z} + 4z$
28. $\dfrac{14x^2 + 6x^2}{5x} + 3x$
29. $\dfrac{x^2 + 3x^2}{4x - 2x}$
30. $\dfrac{y^3 + 2y^3}{5y - 2y} + 2y^2$

2-5 Products and quotients of polynomials

OBJECTIVES: After completing this section, you should be able to:
a. Find the product of two polynomials.
b. Find the quotient of two polynomials.

In Chapter 1 we discussed the distributive law, which was stated symbolically as $a(b + c) = ab + ac$. By the use of this law, we can multiply a polynomial by a monomial. For example, $x(2x + 3) = x \cdot 2x + x \cdot 3 = 2x^2 + 3x$.

EXAMPLE 1

Find the following products.

a. $4(5 + 3)$ (use the distributive law)
b. $-4(x^2 - 3)$
c. $x(5x^2 + 7x)$
d. $2y(x + y)$

Solution

a. $4(5 + 3) = 4 \cdot 5 + 4 \cdot 3 = 20 + 12 = 32$
(Note also that $4(5 + 3) = 4(8) = 32$.)
b. $-4(x^2 - 3) = -4 \cdot x^2 - (-4)3 = -4x^2 + 12$
(Note that we can distribute -4 over a difference.)
c. $x(5x^2 + 7x) = x \cdot 5x^2 + x \cdot 7x = 5x^3 + 7x^2$
d. $2y(x + y) = 2y \cdot x + 2y \cdot y = 2xy + 2y^2$

We can extend the distributive law to cover polynomials with more than two terms.

EXAMPLE 2

Find the product $5(x + y + 2)$.

Solution

$$5(x + y + 2) = 5x + 5y + 10$$

Thus we multiply a polynomial by a monomial by multiplying each term of the polynomial by the monomial.

EXAMPLE 3

Find the following products.

a. $-4ab(3a^2b + 4ab^2 - 1)$
b. $(4a + 5b + c)ac$

Solution

a. $-4ab(3a^2b + 4ab^2 - 1) = -12a^3b^2 - 16a^2b^3 + 4ab$
b. $(4a + 5b + c)ac = 4a \cdot ac + 5b \cdot ac + c \cdot ac$
  $= 4a^2c + 5abc + ac^2$

The distributive law can be used to show us how to multiply two polynomials. Consider the indicated multiplication $(a + b)(c + d)$. If we treat the sum $(a + b)$ as a single quantity, the distributive law gives

$$(a + b)(c + d) = (a + b) \cdot c + (a + b) \cdot d = ac + bc + ad + bd$$

Thus we see that the product can be found by multiplying $(a + b)$ by c, $(a + b)$ by d, and then adding the products. This is frequently set up as follows:

$$
\begin{array}{l}
a + b \\
\underline{c + d} \\
ac + bc \qquad\qquad [c \text{ times } (a + b)] \\
\underline{ad + bd} \quad [d \text{ times } (a + b)] \\
ac + bc + ad + bd \quad [\text{sum of products}]
\end{array}
$$

The procedure used above to multiply two binomials can be used to multiply any two polynomials. The procedure follows.

PROCEDURE	EXAMPLE
To multiply two polynomials:	Multiply $(3x + 4xy + 3y)$ by $(x - 2y)$.
1. Write one of the polynomials above the other.	1. $\begin{array}{l} 3x + 4xy + 3y \\ \underline{x - 2y} \end{array}$
2. Multiply each term of the top polynomial by each term of the bottom one, and write the similar terms of the product under each other.	2. $\begin{array}{l} 3x^2 + 4x^2y + 3xy \\ \underline{- 6xy - 8xy^2 - 6y^2} \end{array}$
3. Add like terms to form the product.	3. $3x^2 + 4x^2y - 3xy - 8xy^2 - 6y^2$

EXAMPLE 4

Multiply $(4x^2 + 3xy + 4x)(2x - 3y)$.

Solution

$$
\begin{array}{l}
4x^2 + 3xy + 4x \\
\underline{2x - 3y} \\
8x^3 + 6x^2y + 8x^2 \\
 \underline{- 12x^2y - 9xy^2 - 12xy} \\
8x^3 - 6x^2y + 8x^2 - 9xy^2 - 12xy
\end{array}
$$

EXAMPLE 5

Multiply $(x^2 + 3xy + 2y^2)(x^2 - y^2)$.

Solution

$$\begin{array}{r} x^2 + 3xy + 2y^2 \\ \underline{x^2 - y^2\phantom{{}+2y^2}} \\ x^4 + 3x^3y + 2x^2y^2\phantom{{}-2y^4} \\ \underline{\phantom{x^4+3x^3y+{}}-x^2y^2 - 3xy^3 - 2y^4} \\ x^4 + 3x^3y + \phantom{{}-{}}x^2y^2 - 3xy^3 - 2y^4 \end{array}$$

We have found quotients of two monomials in Section 2-3. We can divide a polynomial by a monomial by dividing the monomial into each term of the polynomial.

EXAMPLE 6

Divide $4x^3y^2 + 8x^2y^3 - 12x^4y^5$ by $4xy^2$.

Solution

$$(4x^3y^2 + 8x^2y^3 - 12x^4y^5) \div 4xy^2 = \frac{4x^3y^2 + 8x^2y^3 - 12x^4y^5}{4xy^2}$$

$$= \frac{4x^3y^2}{4xy^2} + \frac{8x^2y^3}{4xy^2} - \frac{12x^4y^5}{4xy^2}$$

$$= x^2 + 2xy - 3x^3y^3$$

In some cases, the monomial will not be a factor of every term of the polynomial. In these cases, we reduce the resulting fractions to their simplest terms.

EXAMPLE 7

Divide $6x^5 - 4x^4 + 2$ by $2x$.

Solution

$$(6x^5 - 4x^4 + 2) \div 2x = \frac{6x^5 - 4x^4 + 2}{2x}$$

$$= \frac{6x^5}{2x} - \frac{4x^4}{2x} + \frac{2}{2x} = 3x^4 - 2x^3 + \frac{1}{x}$$

Note that in reducing the fraction $(6x^5 - 4x^4 + 2)/2x$, the **entire** numerator must be divided by $2x$; that is, each term must be divided.

The division of one polynomial by another is done in a manner similar to long division in arithmetic.

The procedure follows.

PROCEDURE	EXAMPLE
To divide one polynomial into another:	Divide $(4x^2 + 3x + 2)$ by $(x + 2)$.
1. Write the division problem with both polynomials in descending powers of a variable.	1. $x + 2 \overline{)4x^2 + 3x + 2}$
2. Divide the highest power of the divisor into the highest power of the dividend to obtain the first partial quotient.	2. x divided into $4x^2$ is $4x$.
3. Write this partial quotient above the highest power in the dividend. Multiply the divisor by this quotient; write the product under the dividend; and subtract like terms.	3. $\begin{array}{r} 4x \phantom{{}+3x+2} \\ x+2\overline{)4x^2 + 3x + 2} \\ 4x^2 + 8x \\ \hline -5x \end{array}$
4. To the remainder bring down the next term of the dividend to form a new partial dividend. Divide the highest power of the divisor into the highest power of the partial dividend and write this partial quotient above the dividend. Multiply the partial quotient times the divisor; write the product under the partial dividend; and subtract.	4. $\begin{array}{r} 4x - 5 \\ x+2\overline{)4x^2 + 3x + 2} \\ 4x^2 + 8x \\ \hline -5x + 2 \\ -5x - 10 \\ \hline 12 \end{array}$
5. Repeat step 4 until all terms of the dividend have been used. Any remainder is written over the divisor.	5. All terms have been used. The quotient is $$4x - 5 + \frac{12}{x + 2}$$

EXAMPLE 8

Divide $(4x^3 - 13x - 22)$ by $(x - 3)$.

Solution

$$
\begin{array}{r}
4x^2 + 12x + 23 \\
x-3\overline{)4x^3 - 0x^2 - 13x - 22} \\
\underline{4x^3 - 12x^2} \\
12x^2 - 13x \\
\underline{12x^2 - 36x} \\
23x - 22 \\
\underline{23x - 69} \\
47
\end{array}
$$

($0x^2$ is inserted so each power of x is present.)

The quotient is $4x^2 + 12x + 23$, with remainder 47.

Note: If the remainder is 0, the divisor is a factor of the dividend.

EXAMPLE 9

Divide $(3x^2 + 3x - 6)$ by $(x - 1)$.

Solution

$$
\begin{array}{r}
3x + 6 \\
x-1\overline{)3x^2 + 3x - 6} \\
\underline{3x^2 - 3x} \\
6x - 6 \\
\underline{6x - 6} \\
0
\end{array}
$$

Thus $x - 1$ is a factor of $3x^2 + 3x - 6$.

Exercise 2-5

Multiply the following.

1. $5(x + y)$
2. $3(2x - 6)$
3. $ab(ax + by)$
4. $cd(cm + dn)$
5. $2ax(x^2 + 2xy)$
6. $3by(x^2 + 2xy)$
7. $-x(a + bx + cx^2)$
8. $-y(2x + 2xy + 3y^2)$
9. $ax^2y(4xy^2 + 3x^2y + 4y)$
10. $(3x^3 + 4x^2y + 5xy^2 + 2y^3)2xy$

11. $(x - y)(x^2 + 2x + y^2)$
12. $(x + 2)(2x^2 + x + 4)$
13. $(x - y)(x + y + xy)$
14. $(y + 1)(y^2 - 3y + 2)$
15. $(y - 2z)(y^2 - 2yz + x^2)$
16. $(y - 3)(y^2 - 4y + 2)$
17. $(z - b)(a^2 + ab + ab^2)$
18. $(c + d)(c + 2cd - d)$
19. $(2a - b)(a^2 - 3ab + b^2)$
20. $(3c - d)(c^2 - 4d^2)$
21. $(a^2 - b^2)(a^2 + 2ab + ab^2)$
22. $(c^2 + m^2)(c - cm + m^2)$
23. $(a - 3b)(a^2 + ab + ab^2)$
24. $(x^2 - y^2)(x + xy + y)$

Divide the following.

25. $(3x^2y + 6x^2 + 9x^3y^2) \div 3x^2$
26. $(5a^3b + 10a^2b^2 - 15ab^3) \div 5ab$
27. $(16m^3n - 8m^2n^2 - 32m^2n^3) \div 8m^2n$
28. $(9x^2y^2 - 27xy^2 - 3xy^3) \div 3xy^2$
29. $(4x^3y^2 - 12x^2y^2 - 8x^2y) \div 4x^2y^2$
30. $(5a^3b^2 - 10a^4b^2 + 15a^3b) \div 5a^2b^2$
31. $(24a^2c - 6a^3c^2 - 12a^2c) \div 6a^2c$
32. $(144n^2 - 60n^3 - 6n^4) \div 12n^2$
33. $(6x^2 - 5x - 6) \div (2x - 3)$
34. $(m^2 - 7m + 12) \div (m - 4)$
35. $(2a^2 - a - 15) \div (2a + 5)$
36. $(28x^2 - 13x - 6) \div (4x - 3)$
37. $(2y^3 + 4y + y^2 + 2) \div (2y + 1)$
38. $(x^3 - 7x - 6) \div (x - 3)$
39. $(x^2 - 4x + 3) \div (x - 2)$
40. $(3x^2 - 5x - 6) \div (x - 3)$
41. $(4x^3 - 5x - 7) \div (x - 2)$
42. $(15x^3 - 6x + 6) \div (x + 3)$
43. $(x^2 + 4xy - y^2) \div (x - y)$
44. $(2x^2 + 5xy - y^2) \div (x + y)$

CHAPTER TEST

Classify the following as monomials, binomials, or trinomials.

1. $3x^2 - 14x$
2. $5x^2y^3$
3. $6x^2 - 3xy + y^2$

Write each term of the following and state the numerical coefficient of each term.

4. $6ab + 4a^2$
5. $3a^2 + 4ax + 13x^2$

Simplify the following expressions by combining like terms.

6. $7x^2 + 8x + 4x^2 + 11x - 6$
7. $(16pq - 7p^2) + (5pq + 5p^2)$
8. $(3x^3 + 4x^2 + 6x) + (4x^3 + 17x)$
9. $5x^3 - 4x^2y^2 + 3x^2y^2 - 7x^3$
10. $(4m^2 - 3n^2 + 5) - (3m^2 + 4n^2 + 8)$
11. $(4a + 2b) - (3a + 3c) + (6b + 2c)$

Multiply the following.

12. $(5x^3)(7x^2)$
13. $(-3x^2y)(2xy^3)(4x^2y^2)$

Divide the following.

14. $39r^3s^2 \div 13r^2s$
15. $-15m^3n \div 5mn^4$

Simplify the following.

16. $(3mx)(2mx^2) - (4xm^2)x^2$
17. $\dfrac{15y^3z^2}{3yz^2} - \dfrac{8y^4z}{y^2z}$
18. $8 - [4 - (q + 5)]$
19. $x^3 + [3x - (x^3 - 3x)]$
20. $4(x + 2y) - (4x + 3y)$

Multiply the following.

21. $2(x + y)$
22. $ax^2(2x^2 + ax + ab)$
23. $(3y + 4)(2y - 3)$
24. $(x + 1)(x^2 + 2x + 4)$
25. $(a - 2b)(a^2 - 3ab + b^2)$

Perform the following divisions.

26. $(18m^2n + 6m^3n + 12m^4n^2) \div 6m^2n$
27. $(16x^2y + 4xy^2 + 8x) \div 4xy$
28. $(3x^4 + 4x^3 + 2x^2 - 4x - 5) \div (x^2 - 1)$
29. $(a^3 + 3a^2 + 6a + 8) \div (a + 2)$

factoring

3-1 Factoring—type I: factoring out common monomials

OBJECTIVE: After completing this section, you should be able to remove monomial factors from polynomials.

In Chapter 1, we discussed the distributive law, which was stated symbolically as $a(b + c) = ab + ac$. By use of this law, we can multiply a polynomial by a monomial. For example, $x(2x + 3) = x \cdot 2x + x \cdot 3 = 2x^2 + 3x$. We can also extend the distributive law to cover polynomials with more than two terms. For example, $5(x + y + 2) = 5x + 5y + 10$, and $4ab(3a^2b + 4ab^2 - 1) = 12a^3b^2 + 16a^2b^3 - 4ab$.

We can "factor" monomial factors out of a polynomial by using the distributive law "backward." $ab + ac = a(b + c)$ is an example showing that a is a monomial factor of the polynomial $ab + ac$. But it is also a statement of the distributive law (with the sides of the equation interchanged). The monomial factor of a polynomial must be a factor of each term of the polynomial; so it is frequently called a common monomial factor. We can factor common monomials from a polynomial as follows.

66 Factoring

PROCEDURE	EXAMPLE
To factor common monomials from a polynomial:	Factor $4x^2y + 4xy^2 + 4xy$.
1. Inspect the factors of each term and identify the factors common to all terms.	1. $4xy$ is a factor common to all terms.
2. Divide the common monomial factor into each term of the polynomial.	2. $(4x^2y + 4xy^2 + 4xy) \div 4xy$ $= x + y + 1$
3. Write the polynomial in factored form as the product of the monomial and the quotient from Step 2.	3. $4x^2y + 4xy^2 + 4xy$ $= 4xy(x + y + 1)$
4. Examine the polynomial to make sure you have not overlooked some other monomial factor.	
5. Mentally multiply the polynomial by the monomial to check your factorization.	

EXAMPLE 1

Factor the common monomials from $2x^2y + 4xy + 8xy^2$.

Solution

2 is a factor of all three terms.
x is a factor of all three terms.
y is a factor of all three terms.
There are no other factors of all three terms.
Thus $2x^2y + 4xy + 8xy^2 = 2xy(x + 2 + 4y)$.

EXAMPLE 2

Factor $-3x^2t - 3x + 9xt^2$.

Solution

$3x$ can be factored out, giving

$$-3x^2t - 3x + 9xt^2 = 3x(-xt - 1 + 3t^2)$$

or $-3x$ can be factored out (factoring out the negative will make the first term of the polynomial positive), giving

$$-3x^2t - 3x + 9xt^2 = -3x(xt + 1 - 3t^2)$$

Note: Factoring $-3x$ out of the middle term $(-3x)$ of the polynomial gives 1. This is because factoring is accomplished by **dividing** each term of the polynomial by the monomial.

Exercise 3-1

Factor the common monomials from the following polynomials.

1. $4x^2 + 8$
2. $6y^2 + 9$
3. $x^2 + 2xy$
4. $xy + 2y^2$
5. $-ax^2 + ay$
6. $-bm^2 + bcn$
7. $6xy + 12x$
8. $4mn - 8m$
9. $9ab - 3a$
10. $5x^3y + 5x$
11. $6ab - 12a^2b + 18b^2$
12. $4a^3 + 6a^2b + 4ab^2$
13. $8a^2b - 16ax + 4bx^2$
14. $15mx^2 - 3m^2x - 9ab$
15. $16x^2y^3 - 8x^2y - 9x^3y$
16. $12y^3z + 4yz^2 - 8y^2z^3$
17. $36x^2 - 4y^2$
18. $18x^2 - 2m^2$
19. $4x^2 + 8xy + 4y^2$
20. $3x^3 - 21x^2y + 18xy^2$

3-2 Products of two binomials

OBJECTIVE: After completing this section, you should be able to find the product of two binomials.

We have seen how to multiply two polynomials in Chapter 2. The following example illustrates this method.

EXAMPLE 1

Multiply $(4x + y)(2x - y)$.

Solution

$$\begin{array}{r} 4x + y \\ 2x - y \\ \hline 8x^2 + 2xy \\ -4xy - y^2 \\ \hline 8x^2 - 2xy - y^2 \end{array}$$

68 Factoring

A more efficient way of multiplying two binomials is the FOIL technique.

PROCEDURE	EXAMPLE
To multiply two binomials:	Multiply $(a + b)(c + d)$.
1. **F**—Multiply the **First** terms.	1. First terms: a and c Product: ac
2. **O**—Multiply the **Outside** terms.	2. Outside terms: a and d Product: ad
3. **I**—Multiply the **Inside** terms.	3. Inside terms: b and c Product: bc
4. **L**—Multiply the **Last** terms.	4. Last terms: b and d Product: bd
5. Add these products; simplify if possible.	5. $(a + b)(c + d) = ac + ad + bc + bd$

EXAMPLE 2

Multiply $(2x + 4)(3x - 1)$.

Solution

$$(2x + 4)(3x - 1) = 2x \cdot 3x + 2x \cdot (-1) + 4 \cdot 3x + 4 \cdot (-1)$$
$$= 6x^2 - 2x + 12x - 4$$
$$= 6x^2 + 10x - 4$$

EXAMPLE 3

Multiply $(x - y)(2x + y)$.

Solution

$$(x - y)(2x + y) = x(2x) + xy + (-y) \cdot 2x + (-y)y$$
$$= 2x^2 + xy - 2xy - y^2$$
$$= 2x^2 - xy - y^2$$

3-3 Factoring—type II: factoring trinomials

Exercise 3-2

Find the following products.

1. $(x + 1)(x + 1)$
2. $(x - 1)(x - 1)$
3. $(x - 2)(x - 2)$
4. $(x + 3)(x + 3)$
5. $(y + 2)(y + 3)$
6. $(y + 1)(y + 2)$
7. $(x + 4)(x + 2)$
8. $(x - 3)(x - 2)$
9. $(x + 2)(x - 1)$
10. $(x - 6)(x + 3)$
11. $(x - 3)(x + 1)$
12. $(x - 2)(x + 4)$
13. $(-a - 2)(a + 3)$
14. $(b + 6)(-b - 3)$
15. $(x - 2)(x + 2)$
16. $(x + 1)(x - 1)$
17. $(a - 3)(a + 3)$
18. $(b + 4)(b - 4)$
19. $(x - 3)^2$
20. $(y + 2)^2$
21. $(a + 1)^2$
22. $(b - 6)^2$
23. $(y - 8)^2$
24. $(z - 5)^2$
25. $(x + y)(x - y)$
26. $(x + y)(x + y)$
27. $(x + y)(2x + y)$
28. $(x + 2y)(x + y)$
29. $(2x + 1)(2x - 2)$
30. $(4x - 3)(5x + 2)$
31. $(3x - 1)(4x - 3)$
32. $(4y - 1)(3y + 2)$
33. $(5x - 2)(-3x + 2)$
34. $(-2a - 1)(2b + 3)$
35. $(3x + 2)^2$
36. $(4y - 1)^2$
37. $(2x + 3y)(3x + 1)$
38. $(5x - 1)(2x + y)$
39. $(3x + 4y)(5x + 2y)$
40. $(2x + 3y)(3x + 2y)$
41. $(2x + y)(2x - y)$
42. $(2x + y)(2x + 3y)$

3-3 Factoring—type II: factoring trinomials

OBJECTIVE: You should be able to factor trinomials that are products of two binomials.

We have seen in the previous section that the product of two binomials can result in a trinomial. Thus certain trinomials can be factored into the product of two binomials. If we write one of these trinomials with the powers of a variable in descending order, we can make the following observations.

1. The first term of the trinomial is the product of the two first terms of the binomial factors.
2. The third term of the trinomial is the product of the last terms of the binomial factors.
3. The middle term of the trinomial is the sum of the products of the inside and outside terms of the binomial factors.

EXAMPLE 1

Factor $x^2 + 5x + 6$.

Solution

The possible first-degree factors for the first terms of the binomial factors are x and x. Thus we can begin by writing

$(x \quad)(x \quad)$

The possible positive factors of the last term of the trinomial are 6 and 1 or 2 and 3. If we try 6 and 1, we get

$$(x + 6)(x + 1) = x^2 + x + 6x + 6$$
$$= x^2 + (1 + 6)x + 6$$

But we see that the sum we need for the middle term is 5. Because the sum of 2 and 3 is 5, we have our factors.

$$(x + 2)(x + 3) = x^2 + 3x + 2x + 6$$
$$= x^2 + (3 + 2)x + 6$$
$$= x^2 + 5x + 6$$

All the terms of the example above are positive signs. Of course, this is not the case with all trinomials. The signs in the trinomial we are factoring can be most helpful in determining the signs in the factors, as the following analysis shows.

4. From observation 1 above, the last terms of the binomial factors will have the same sign if the last term of the trinomial is positive. From observation 3 above, if the sign on the last term of the trinomial is positive, the signs on the last terms of both binomials will be the same as the sign on the trinomial's middle term.

For example, the last term of $x^2 + 5x + 6$ is positive, so the last terms of its binomial factors will be the same. The sign on the trinomial's middle term is positive, so both signs are positive (see Example 1 above).

5. The signs on the last terms of the binomials will be different if the sign on the last term of the trinomial is negative. The sign of the middle term of the trinomial will be the same as the sign of the larger of the two products of inside and outside terms.

For example, the last term of $x^2 + x - 6$ is negative, so the last terms of its binomial factors will be different. The sign on the trinomial's middle term is positive, so the larger of the two products of inside and outside terms is positive. The factorization of $x^2 + x - 6$ is $(x + 3)(x - 2)$.

EXAMPLE 2

Use sign analysis to factor $x^2 - 7x + 6$.

Solution

1. Since this sign is positive, the signs will be the same on binomial factors.

$x^2 \ominus 7x \oplus 6$

2. Since this sign is negative, both the signs will be negative on the binomial factors.

Because the x^2 term has coefficient 1, we can begin by writing the factors as

$(x -)(x -)$

Because the signs are the same, 6 represents the **product** of the last two terms, and 7 represents the **sum** of these two terms. The factors of 6 are 2 and 3 or 6 and 1. Clearly the required factors are 6 and 1, because $6 + 1 = 7$. Thus

$x^2 - 7x + 6 = (x - 6)(x - 1) \quad [\text{or } (x - 1)(x - 6)]$

Be sure to check your factorization by multiplying.

EXAMPLE 3

Use sign analysis to factor $x^2 - x - 6$.

Solution

1. Signs are different on factors.

$x^2 \ominus x \ominus 6$

2. Sign of largest product is negative.

Because the x^2 term has coefficient 1, we begin as follows:

$(x -)(x +)$

Because the signs of the last terms are opposite, we need two factors of 6 whose **difference** is 1. Choosing from factors 6 and 1 or 2 and 3, we see 2 and 3 will work. Placing the 3 after the minus sign will give the required negative middle term on the trinomial. Thus

$x^2 - x - 6 = (x - 3)(x + 2)$

72 Factoring

We can write a general procedure for factoring trinomials that are products of two binomials.

PROCEDURE	EXAMPLE
To factor trinomials into binomials:	Factor $-7x + 4 + 3x^2$.
1. Write the trinomials in descending powers of a variable.	1. $3x^2 - 7x + 4$
2. Determine two first-degree factors of the first term of the trinomial.	2. Factors of $3x^2$ are $3x$ and x.
3. Use sign analysis; begin to write factors.	3. Signs are the same, both are negative: $(3x -)(x -)$
4. Find all possible factors of last term of trinomial.	4. Possible factors are 1 and 4, 2 and 2.
5. Select a pair of factors that you think will give the correct middle term for the trinomial. Multiply the binomials to see if you get the correct middle term. If you do not, try them in another order, then try another pair of factors.	5. If the factors are 1 and 4, we get $(3x - 1)(x - 4)$ or $(3x - 4)(x - 1)$: $(3x - 1)(x - 4)$ $= 3x^2 - 13x + 4$ no $(3x - 4)(x - 1)$ $= 3x^2 - 7x + 4$ yes
6. If no factors appear to work, begin again with Step 2.	The factorization is $(3x - 4)(x - 1)$
7. Check by multiplying.	7. $(3x - 4)(x - 1) = 3x^2 - 7x + 4$

EXAMPLE 4

Factor $5x^2 - 8x - 4$.

Solution

Factors of $5x^2$ are $5x$ and x.
Signs are different.
Factors of the third term are 1 and 4 or 2 and 2.
Thus the possibilities are

Possible Factors		Middle Term
1. $(5x + 1)(x - 4)$		$-19x$
2. $(5x - 1)(x + 4)$	(Changes sign of no. **1**.)	$+19x$
3. $(5x + 4)(x - 1)$		$-x$
4. $(5x - 4)(x + 1)$	(Changes sign of no. **3**.)	$+x$
5. $(5x - 2)(x + 2)$		$+8x$ (close, but wrong sign)
6. $(5x + 2)(x - 2)$	(Changes sign of no. **5**.)	$-8x$ BINGO!

The correct factorization is $(5x + 2)(x - 2)$. (Note that nos. **2, 4,** and **6** differ from nos. **1, 3,** and **5**, respectively, only in the sign of the middle term.)

In the previous examples, we factored trinomials of the form $x^2 + bx + c$ or $ax^2 + bx + c$, where a was a prime. Thus we had only one possible set of first-degree factors for the first term of the trinomials. Now we shall factor trinomials where there are two or more possible factors of the first term. The procedures for factoring are the same as those described in the previous section, except that more patience in finding the correct factors may be required.

EXAMPLE 5

Factor $6x^2 + 11x + 3$.

Solution

The possible first-degree factors of the first term are x and $6x$ or $2x$ and $3x$.
The signs are alike and positive.
The factors of the third term are 1 and 3.
Thus the possibilities are

Possible Factors	Middle Term
$(x + 1)(6x + 3)$	$9x$
$(x + 3)(6x + 1)$	$19x$
$(2x + 1)(3x + 3)$	$9x$
$(2x + 3)(3x + 1)$	$11x$

Of these possibilities, we need the one that gives $11x$ as a middle term. Thus $6x^2 + 11x + 3 = (2x + 3)(3x + 1)$.

Note: The possible factors that are underlined each have a common monomial factor. Since the original polynomial does not have a monomial factor, these possibilities can be eliminated immediately.

EXAMPLE 6

Factor $6x^2 - 37x + 35$.

Solution

Factors of $6x^2$ are x and $6x$ or $2x$ and $3x$.
Sign of third term is positive, of second is negative; so signs in both factors are negative.
Factors of third term are 1 and 35 or 5 and 7.
Thus we have

Possible Factors	Middle Term	
$(x - 1)(6x - 35)$	$-41x$	
$(x - 35)(6x - 1)$	$-211x$	
$(x - 5)(6x - 7)$	$-37x$	BINGO!
$(x - 7)(6x - 5)$	$-47x$	
$(2x - 1)(3x - 35)$	$-73x$	
$(2x - 35)(3x - 1)$	$-107x$	
$(2x - 5)(3x - 7)$	$-29x$	
$(2x - 7)(3x - 5)$	$-31x$	

The correct factorization is $(x - 5)(6x - 7)$.

Note: In the previous example, there is no need to check all factors after the correct one has been found. Visual inspection would quickly eliminate those factors that would give very large middle terms.

EXAMPLE 7

Factor $12 - 31x + 9x^2$.

Solution

First we write the trinomial in descending powers of x:
$9x^2 - 31x + 12$.
The possible factors of $9x^2$ are x and $9x$ or $3x$ and $3x$.
The signs on both binomial factors are negative.

There are no monomial factors of the polynomial; so no binomial factor will have monomial factor.

The factors of the last term are 1 and 12, 2 and 6, or 3 and 4.

Possible Factors	Middle Term	(Close?)
$(x - 1)\underline{(9x - 12)}$		NO (monomial factor)
$(x - 12)(9x - 1)$		NO
$(x - 2)\underline{(9x - 6)}$	$-24x$	(monomial factor)
$(x - 6)(9x - 2)$		NO
$(x - 3)(9x - 4)$	$-31x$	BINGO!
$(x - 4)\underline{(9x - 3)}$	$-39x$	(monomial factor)
$(3x - 1)\underline{(3x - 12)}$	$-39x$	(monomial factor)
$(3x - 2)\underline{(3x - 6)}$	$-24x$	(monomial factor)
$\underline{(3x - 3)}(3x - 4)$		NO (monomial factor)

The correct factorization is $(x - 3)(9x - 4)$.

Note: There is no need to continue this process after the correct factors are found. The process was continued in this example to show all the possibilities.

EXAMPLE 8

Factor $x^2 - 11xy + 10y^2$.

Solution

In this trinomial, the powers of y increase as the powers of x decrease. When this happens, we factor exactly as before except that both terms of the binomial factors will contain literal numbers. Factors of the first term are x and x. Signs on binomials are alike, both positive. Factors of the last term are y and $10y$ or $2y$ and $5y$. The sum of xy and $10xy$ gives $11xy$; so the correct factorization is

$$x^2 - 11xy + 10y^2 = (x - y)(x - 10y)$$

Remember always to check by multiplication.

Not all trinomials can be factored using the methods discussed above. The trinomial $x^2 + 3x + 11$ cannot be factored into the form $(x - a)(x - b)$ with a and b both real numbers.

Exercise 3-3

Factor the following trinomials.

1. $x^2 + 3x + 2$
2. $x^2 - 3x + 2$
3. $x^2 - x - 2$
4. $x^2 + x - 2$
5. $x^2 + 6x + 5$
6. $x^2 - 6x + 5$
7. $x^2 - 4x - 5$
8. $x^2 + 4x - 5$
9. $x^2 + 6x + 8$
10. $x^2 - 6x + 8$
11. $x^2 - x - 12$
12. $x^2 + 4x - 12$
13. $y^2 - 8y - 9$
14. $y^2 + 9y + 20$
15. $y^2 - 9y - 36$
16. $z^2 + 6z - 27$
17. $2x^2 - x - 1$
18. $x^2 + x - 6$
19. $7x^2 + 15x + 2$
20. $3y^2 - 10y + 3$
21. $3y^2 - 7y - 6$
22. $5y^2 + 4y - 12$
23. $7a^2 - 10a - 8$
24. $11b^2 + 5b - 16$
25. $3x^2 + 4x + 1$
26. $12x^2 + 11x + 2$
27. $15x^2 + 26x + 7$
28. $12x^2 + 16x + 5$
29. $8x^2 - 10x + 3$
30. $8x^2 - 17x + 2$
31. $6x^2 - 7x + 1$
32. $10x^2 - 27x + 5$
33. $10x^2 + 13x - 3$
34. $6x^2 - x - 2$
35. $12x^2 - 5x - 2$
36. $12x^2 + 7x - 5$
37. $6x^2 + 25x + 14$
38. $6x^2 - 17x + 12$
39. $12x^2 - 28x + 15$
40. $10x^2 + 19x + 6$
41. $6x^2 - 37x + 35$
42. $6x^2 + 67x - 35$
43. $4x^2 - 4x - 15$
44. $9x^2 - 9x - 10$
45. $4 + 11x + 6x^2$
46. $9 - 47x + 10x^2$
47. $21 - 26x + 8x^2$
48. $3 + 10x + 3x^2$
49. $x^2 + xy - 6y^2$
50. $2x^2 + 11xy + 15y^2$
51. $4x^2 - 8xy - 5y^2$
52. $3a^2 - 7ab + 2b^2$
53. $10y^2 + 3yz - z^2$
54. $6m^2 - mn - n^2$

3-4 Factoring—type III: perfect trinomial squares

OBJECTIVES: After completing this section, you should be able to:

a. **Square a binomial.**
b. **Factor a trinomial that is the square of a binomial.**

The square of a binomial is found by multiplying the binomial times itself; that is, $(a + b)^2 = (a + b)(a + b)$. Multiplying gives $a^2 + 2ab + b^2$. Note that the resulting trinomial has as its terms the square of the first term of the binomial, the square of the last term of the binomial, and twice the product of the two terms of the binomial.

EXAMPLE 1

Expand $(6x - 3)^2$.

Solution

The first term is $6x$; the second term is -3.

$$(6x - 3)^2 = (6x)^2 + 2(6x)(-3) + (-3)^2 = 36x^2 - 36x + 9$$

EXAMPLE 2

Expand $(2x + 3y)^2$.

Solution

$$(2x + 3y)^2 = 4x^2 + 12xy + 9y^2$$

A trinomial is called a perfect square if it can be factored into the square of a binomial. We can recognize a perfect square by:

1. Two of the terms will be perfect squares.
2. The remaining term will be twice the product of the square roots (disregarding signs) of the perfect-square terms.

EXAMPLE 3

Is $x^2 - 2xy + y^2$ a perfect trinomial square?

Solution

x^2 and y^2 are perfect squares; $2xy$ is twice the product of their square roots. Therefore $x^2 - 2xy + y^2$ is a perfect trinomial square.

Perfect trinomial squares can be factored in the same manner as other trinomials, but they can be factored quickly and easily if they are recognized. A perfect trinomial square may be factored as follows.

PROCEDURE	EXAMPLE
To factor a perfect trinomial square:	Factor $9x^2 - 12xy + 4y^2$.
1. Write a binomial squared, with the square roots of the perfect squares as the terms.	1. $(3x \quad 2y)^2$
2. The sign between the terms of the binomial is the same as the sign on the remaining (middle) term of the trinomial.	2. $9x^2 - 12xy + 4y^2 = (3x - 2y)^2$

EXAMPLE 4

Factor $4x^2 - 12x + 9$.

Solution

$4x^2 - 12x + 9$ is a perfect trinomial square because $4x^2$ is a square, 9 is a square, and $12x$ is twice the product of their square roots. The sign of $12x$ is negative; so the factorization is

$$4x^2 - 12x + 9 = (2x - 3)^2$$

EXAMPLE 5

Factor $16x^2 + 40xy + 25y^2$.

Solution

$16x^2 + 40xy + 25y^2$ is a perfect trinomial square because $16x^2$ and $25y^2$ are squares, and $40xy$ is twice the product of their square roots. The sign on $40xy$ is positive; so the factorization is $(4x + 5y)^2$. Multiplying to check, we get

$$(4x + 5y)^2 = 16x^2 + 40xy + 25y^2$$

Exercise 3-4

Expand the following.

1. $(x + 2)^2$
2. $(x - 3)^2$
3. $(x + 4)^2$
4. $(x - 2)^2$

5. $(2x + 3)^2$
6. $(3x + 2)^2$
7. $(6x - 5)^2$
8. $(4x - 9)^2$
9. $(3x - 2y)^2$
10. $(2a + b)^2$
11. $(4x + 3y)^2$
12. $(2y - 5)^2$

Are the following polynomials perfect trinomial squares?

13. $x^2 + 4x + 4$
14. $x^2 + 6x + 9$
15. $x^2 + 16xy + 16y^2$
16. $x^2 + 10xy + 25y^2$
17. $4x^2 + 8x + 16$
18. $4x^2 + 12x + 9$
19. $9x^2 + 24x + 16$
20. $25y^2 + 30y + 9$

Factor the following.

21. $x^2 - 4x + 4$
22. $x^2 - 6x + 9$
23. $x^2 - 10xy + 25y^2$
24. $x^2 + 12x + 36$
25. $4x^2 - 8x + 4$
26. $9x^2 + 12x + 4$
27. $4x^2 + 48x + 144$
28. $16x^2 + 16x + 4$
29. $4x^2 + 20xy + 25y^2$
30. $25x^2 + 30xy + 9y^2$

3-5 Factoring—type IV: difference of two squares

OBJECTIVES: After completing this section, you should be able to:

a. Find products of conjugate binomials.
b. Factor a polynomial that is the difference of two squares.

The product of the binomials $x + y$ and $x - y$ is $(x + y)(x - y) = x^2 - xy + xy - y^2 = x^2 - y^2$. Note that the two binomial factors are identical except that their second terms have opposite signs. Any time two binomials differ only in one sign, we say that each one is the **conjugate** of the other. **The product of two binomial conjugates is always the difference of two squares** ($x^2 - y^2$ in the example above).

EXAMPLE 1

Multiply $(2x - 3)(2x + 3)$.

Solution

Because the two binomials are conjugates, the middle term will be 0 $(6x - 6x)$; so the product is $4x^2 - 9$. Note that this product is $(2x)^2 - 3^2$, the difference of two squares.

Factoring

EXAMPLE 2

Multiply $(3x + 5y)(3x - 5y)$.

Solution

The factors are binomial conjugates; so the product is the difference of two squares:

$$(3x + 5y)(3x - 5y) = 9x^2 - 25y^2$$

To factor a polynomial that is the difference of two squares, we observe that such a polynomial results from the product of two conjugate binomials, whose terms are the square roots of the squares.

EXAMPLE 3

Factor $x^2 - y^2$.

Solution

$x^2 - y^2$ is the difference of two squares; so the factorization is $(x+y)(x-y)$. Checking, we see $(x+y)(x-y) = x^2 - y^2$.

EXAMPLE 4

Factor $25x^2 - 36y^2$.

Solution

$25x^2 - 36y^2$ is the difference of two squares; so the factorization is $(5x + 6y)(5x - 6y)$. If we check by multiplication, we get $(5x + 6y)/(5x - 6y) = 25x^2 - 36y^2$.

Exercise 3-5

Multiply the following.

1. $(a - b)(a + b)$
2. $(y - z)(y + z)$
3. $(a - 3)(a + 3)$
4. $(b - 4)(b + 4)$
5. $(x - 5)(x + 5)$
6. $(y + 6)(y - 6)$
7. $(x - y)(x + y)$
8. $(x - b)(x + b)$
9. $(3x + 2)(3x - 2)$
10. $(4x - 3)(4x + 3)$
11. $(5x + 4)(5x - 4)$
12. $(2x + 5)(2x - 5)$

13. $(3x - 2y)(3x + 2y)$ 14. $(x - 3y)(x + 3y)$
15. $(7x - 2y)(7x + 2y)$ 16. $(3x - y)(3x + y)$

Factor the following.

17. $a^2 - b^2$ 18. $a^2 - 36$
19. $x^2 - y^2$ 20. $x^2 - 9$
21. $4x^2 - 1$ 22. $9x^2 - 4$
23. $36a^2 - 25$ 24. $25x^2 - 16$
25. $a^2 - 49$ 26. $y^2 - 64$
27. $4a^2 - y^2$ 28. $16x^2 - b^2$
29. $16x^2 - 25y^2$ 30. $49a^2 - 144b^2$

3-6 Products and factors involving cubes

OBJECTIVES: After completing this section, you should be able to:

a. Mentally compute the cube of a binomial.
b. Factor perfect binomial cubes.
c. Factor the sum and difference of two cubes.

If we raise a number to the third power, we say we are computing the **cube** of the number. Thus the cube of the binomial $a + b$ is $(a + b)^3$. The following formulas give the special products called **perfect binomial cubes**.

$$(a + b)^3 = a^3 + 3a^2b + 3ab^2 + b^3$$
$$(a - b)^3 = a^3 - 3a^2b + 3ab^2 - b^3$$

EXAMPLE 1

Compute $(2x + 3)^3$.

Solution

This is of the form $(a + b)^3$, with $a = 2x$ and $b = 3$.

$$(2x + 3)^3 = (2x)^3 + 3(2x)^2(3) + 3(2x)(3^2) + 3^3$$
$$= 8x^3 + 36x^2 + 54x + 27$$

EXAMPLE 2

Factor $x^3 - 6x^2 + 12x - 8$.

Solution

$x^3 - 6x^2 + 12x - 8$ is the cube of $(a - b)$, where $a = x$ and $b = 2$. Thus $x^3 - 6x^2 + 12x - 8 = (x - 2)^3$.

Two additional formulas involving cubes are important in algebra. They are the formulas for factoring the **sum of two cubes** and the **difference of two cubes**:

$$a^3 + b^3 = (a + b)(a^2 - ab + b^2)$$
$$a^3 - b^3 = (a - b)(a^2 + ab + b^2)$$

Note: One factor of $a^3 + b^3$ is found by taking the cube roots of the two terms and adding them. The remaining factor can be gotten from the squares and product of the terms of the first factor. The same process works for $a^3 - b^3$.

EXAMPLE 3

Factor $x^3 - 27$.

Solution

Using the formula for the difference of two cubes, with $a = x$ and $b = 3$, gives

$$x^3 - 27 = (x - 3)(x^2 + 3x + 9)$$

EXAMPLE 4

Factor $64x^3 + 8y^3$.

Solution

This is the sum of two cubes, $(4x)^3 + (2y)^3$; so

$$64x^3 + 8y^3 = (4x + 2y)(16x^2 - 8xy + 4y^2)$$

Exercise 3-6

Multiply the following.

1. $(x + 1)^3$
2. $(y + 4)^3$
3. $(a - 5)^3$
4. $(b - 2)^3$
5. $(2x + y)^3$
6. $(3x - 5)^3$

7. $(2x - 4)^3$
8. $(5y - b)^3$
9. $(x + 1)(x^2 - x + 1)$
10. $(x + 3)(x^2 - 3x + 9)$
11. $(x - 2)(x^2 + 2x + 4)$
12. $(x - 5)(x^2 + 5x + 25)$

Factor the following.

13. $x^3 + 3x^2 + 3x + 1$
14. $x^3 + 6x^2 + 12x + 8$
15. $x^3 - 12x^2 + 48x - 64$
16. $y^3 - 9y^2 + 27y - 27$
17. $x^3 - 64$
18. $x^3 - 8$
19. $y^3 + 27$
20. $y^3 + 125$
21. $x^3 + 27y^3$
22. $y^3 + 8z^3$
23. $m^3 - 64n^3$
24. $p^3 - 27q^3$
25. $8x^3 + 27y^3$
26. $27x^3 + 8y^3$
27. $8x^3 - 64m^3$
28. $125p^3 - 64q^3$

3-7 Factoring completely

OBJECTIVE: After completing this section, you should be able to factor certain polynomials completely.

A polynomial is said to be factored completely if all possible factorizations have been completed. For example, $(2x - 4)(x + 3)$ is not factored completely because a 2 can still be factored out of $(2x - 4)$. Confining our attention to factors with integer coefficients, we can factor a number of polynomials completely using the following procedure.

PROCEDURE	EXAMPLE
To factor completely:	Factor $4x^2 + 28x + 48$.
1. Look for a common monomial factor in each term.	1. $4(x^2 + 7x + 12)$
2. Look at binomial factors for the difference of two squares.	2. not the difference of two squares
3. Look at trinomial factors for perfect squares.	3. not perfect trinomial square
4. Try other methods of factoring trinomial factors.	4. $4(x + 3)(x + 4)$ (factored completely)

This procedure can be indicated as follows:

Look for: monomials first.
Then for: difference of two squares.
Then for: trinomial squares.
Then for: other methods of factoring trinomials.

EXAMPLE 1

Factor completely $12x^2 - 36x + 27$.

Solution

$$12x^2 - 36x + 27 = 3(4x^2 - 12x + 9) \quad \text{(monomial)}$$
$$= 3(2x - 3)^2 \quad \text{(perfect square)}$$

EXAMPLE 2

Factor completely $4x^2y^2 - 6xy^2 - 4y^2$.

Solution

$$4x^2y^2 - 6xy^2 - 4y^2 = 2y^2(2x^2 - 3x - 2)$$
$$= 2y^2(2x + 1)(x - 2)$$

EXAMPLE 3

Factor completely $16x^2 - 64y^2$.

Solution

$$16x^2 - 64y^2 = 16(x^2 - 4y^2)$$
$$= 16(x + 2y)(x - 2y)$$

Note: Factoring the difference of two squares immediately would give $(4x + 8y)(4x - 8y)$, which is not factored completely (we could still factor 4 from $4x + 8y$ and 4 from $4x - 8y$).

EXAMPLE 4

Factor $6 + 5x - x^2$.

Solution

We first write the polynomial in descending powers of x:

$$6 + 5x - x^2 = -x^2 + 5x + 6$$

Next we factor -1 out to make the second degree term (x^2) positive:
$$-x^2 + 5x + 6 = -1(x^2 - 5x - 6)$$
Now we can factor the new polynomial using the usual methods:
$$-1(x^2 - 5x - 6) = -1(x + 1)(x - 6)$$
Multiplying to check gives
$$-1(x + 1)(x - 6) = -1(x^2 - 5x - 6) = -x^2 + 5x + 6$$
$$= 6 + 5x - x^2$$

Exercise 3-7

Factor completely the following.

1. $5x + 5y$
2. $4b - 12$
3. $5x^2 - 5y^2$
4. $6x^2 - 6y^2$
5. $9x^2 - 36y^2$
6. $4x^2 - 100y^2$
7. $4x^2 + 24x + 32$
8. $4b^2 - 4b - 160$
9. $3a^2 - 12a - 63$
10. $4b^2 - 4b - 120$
11. $a^3 - 8a^2 + 15a$
12. $x^3 + 4x^2 - 21x$
13. $ax^2 + 4ax - 12a$
14. $ax^2 - 14ax - 51a$
15. $12 + 16x + 4x^2$
16. $3 - 6m - 45m^2$
17. $100a - 15ay - ay^2$
18. $12x^3 - x^3y - 63x^3y^2$
19. $a^3 + 16a^2 + 64a$
20. $12b^2 - 12b + 3$
21. $3x^2 + 6x + 9$
22. $a^2b^2 + 6a^2b + 9a^2$
23. $100 + 160m + 64m^2$
24. $12c^2 + 12c^2a + c^2a^2$
25. $4a^2x^2 + 4a^2x$
26. $5a^2x^2 - 5a^2$

3-8 Trinomials in quadratic form; factoring by grouping

OBJECTIVES: After completing this section, you should be able to:

a. Factor trinomials that are in quadratic forms.
b. Use grouping to factor certain polynomials.

Most of the polynomials we have factored have been second-degree polynomials or **quadratic polynomials**. Some polynomials that are not quadratic are in a form which can be factored in the same manner as quadratics. For example, x^4+4x^2+4 can be written as a^2+4a+4, where $a=x^2$.

EXAMPLE 1

Factor $x^4 + 4x^2 + 4$ completely.

Solution

The trinomial is in the form of a perfect square; so letting $a = x^2$ gives

$$x^4 + 4x^2 + 4 = a^2 + 4a + 4 = (a + 2)^2$$

so

$$x^4 + 4x^2 + 4 = (x^2 + 2)^2$$

EXAMPLE 2

Factor $x^4 - 16$ completely.

Solution

$x^4 - 16$ can be treated as the difference of two squares, $(x^2)^2 - 4^2$; so

$$x^4 - 16 = (x^2 - 4)(x^2 + 4)$$

But $x^2 - 4$ can be factored into $(x-2)(x+2)$; so

$$x^4 - 16 = (x - 2)(x + 2)(x^2 + 4)$$

Some problems in factoring, which cannot be factored by the methods learned thus far, can be factored by the use of grouping. The grouping is done so that common factors (frequently binomial factors) can be removed.

EXAMPLE 3

Factor $ax + bx + ay + by$ completely.

Solution

The first two terms have the factor x in common, and the last two terms have the factor y in common. Factoring the x from the first two terms and y from the last two terms gives

$$(a + b)x + (a + b)y$$

Now the factor $(a + b)$ can be factored from the two terms, giving

$$(a + b)(x + y)$$

If this step bothers you, let $u = a + b$. Then $(a + b)x + (a + b)y = ux + uy = u(x + y)$. By substituting $a + b$ back in for u we get $(a + b)(x + y)$.

EXAMPLE 4

Factor $2ax - 3ay + 2bx - 3by$ completely.

Solution

$2x$ can be factored from the first and third terms, and $-3y$ can be factored from the second and fourth terms. Thus we can rewrite $2ax - 3ay + 2bx - 3by$ as

$$2x(a + b) - 3y(a + b)$$

Again we can factor $(a + b)$ from the two resulting terms, which gives

$$(a + b)(2x - 3y)$$

Note that we could group the terms differently, but the final result would be the same:

a is a factor of the first and second terms.
b is a factor of the third and fourth terms.

Rewriting the terms gives

$$a(2x - 3y) + b(2x - 3y)$$

Factoring $(2x - 3y)$ from the two resulting terms gives

$$(2x - 3y)(a + b)$$

EXAMPLE 5

Factor $x^3 - y^3 + 3x - 3y$ completely.

Solution

The first two terms can be treated as the difference of two cubes.

$$\begin{aligned} x^3 - y^3 + 3x - 3y &= (x - y)(x^2 + xy + y^2) + 3(x - y) \\ &= (x - y)(x^2 + xy + y^2 + 3) \end{aligned}$$

Multiplying the factors provides a check of the factorization.

Exercise 3-8

Factor the following completely.

1. $x^4 + 6x^2 + 9$
2. $x^6 + 10x^3 + 25$
3. $p^4 - 18p^2 + 81$
4. $m^4 - 8m^2 + 16$
5. $p^4 - q^2$
6. $y^4 - z^2$
7. $x^4 - 81$
8. $y^4 - 16x^4$
9. $x^8 - 256$
10. $x^6 - y^4$
11. $4x + 4y + Bx + By$
12. $2a + 2y + Ca + Cy$
13. $ax + 4y + 4x + ay$
14. $6x + by + 6y + bx$
15. $3x - 3y - ax + ay$
16. $xy - xz + my - mz$
17. $5p + pq - 5m - mq$
18. $mn - an + mx - ax$
19. $3am - xm + 12ax - 4x^2$
20. $2ab + 4b^2 - 6ax - 12bx$
21. $x^3 - 4x^2 + 5x - 20$
22. $4m^3 + 8m - m^2p - 2p$
23. $x^2 - y^2 + 4x - 4y$
24. $x^2 - y^2 + mx + my$
25. $x^2 + 4x + 4 - 3x - 6$
26. $5m - 15 + m^2 - 6m + 9$
27. $x^2 + 8x + 16 - y^2$
28. $m^2 - 2my + y^2 - 25$
29. $x^2 + 2xy - 16 + y^2$
30. $y^2 - x^2 + 6x - 9$
31. $x^3 - y^3 + 4x - 4y$
32. $p^3 + q^3 + mp + mq$

CHAPTER TEST

Find the following products.

1. $(x - 4)(x - 2)$
2. $x(3x + 1)(2x - 3)$
3. $(x - 4)^2$
4. $(2a + 3b)^2$
5. $(b - 3)(b + 3)$
6. $(5y^2 - 2)(5y^2 + 2)$
7. $(x^2 + y^3)^2$

Factor the following completely.

8. $5x^2 - 10$
9. $6ab^2 - 12a^2b + 24a^2b^2$
10. $x^2 - 5x - 6$
11. $y^2 - 7y + 10$
12. $5x^2 + 6x - 27$
13. $6x^2 - 19x + 15$
14. $8x^2 - 2xy - 15y^2$
15. $x^2 - 10x + 25$
16. $4y^2 + 12y + 9$
17. $16x^2 - 1$
18. $4x^2 - 25y^2$
19. $4x^2 - 16x + 16$
20. $15a^2y^2 - 15a^2b^2$
21. $x^4 + 6x^2y^2 + 9y^4$
22. $x^4 - 81$
23. $x^6 - p^4$
24. $6x - 6m + xy - my$
25. $x^2 + 4x + 4 - y^2$

4

algebraic fractions

4-1 Simplifying algebraic fractions

OBJECTIVES: You should be able to:
a. **Simplify algebraic fractions correctly;**
b. **Use factoring to divide one polynomial by another.**

The fraction $\frac{6}{8}$ can be reduced to $\frac{3}{4}$ by dividing both the numerator and denominator by 2. In the same manner, the algebraic fraction $\frac{(x+2)(x+1)}{(x+1)(x+3)}$ can be reduced to $\frac{(x+2)}{(x+3)}$ by dividing both the numerator and denominator by $x+1$.

EXAMPLE 1

Simplify $\frac{(x-1)(x+3)}{(x+2)(x-1)}$ (write in lowest terms).

Solution

$x-1$ is a factor of both the numerator and denominator. Dividing both by $x-1$ gives the equivalent (reduced) fraction $\frac{(x+3)}{(x+2)}$.

We simplify algebraic fractions by factoring the numerator and denominator and then dividing both the numerator and denominator by the common factors.

EXAMPLE 2

Simplify $\frac{6a^2bc}{9ab^2c}$.

Solution

$$\frac{6a^2bc}{9ab^2c} = \frac{2 \cdot \overset{1\cdot a \cdot 1 \cdot 1}{\cancel{3}\cancel{a^2}\cancel{b}\cancel{c}}}{3 \cdot \underset{1 \cdot 1 \cdot b \cdot 1}{\cancel{3}\cancel{a}\cancel{b^2}\cancel{c}}}$$

$$= \frac{2a}{3b}$$

EXAMPLE 3

Simplify $\dfrac{3x^2 - 14x + 8}{x^2 - 16}$.

Solution

$$\frac{3x^2 - 14x + 8}{x^2 - 16} = \frac{(3x - 2)(x - 4)}{(x - 4)(x + 4)}$$

$$= \frac{(3x - 2)\cancel{(x - 4)}}{\cancel{(x - 4)}(x + 4)}$$

$$= \frac{3x - 2}{x + 4}$$

We can only reduce fractions by dividing both the numerator and denominator by the same quantity. If the numerator or denominator has more than one term and division by a monomial is being used to reduce the fraction, the monomial must be divided into **every** term.

EXAMPLE 4

Reduce the fraction $\dfrac{3x + 6}{3x}$.

Solution

We cannot "cancel" the $3x$'s. If we divide numerator and denominator by $3x$, we must divide $3x$ into both terms of the numerator; that is, $3x$ must be divided into both $3x$ **and** 6. To reduce the fraction, we should first factor the numerator. Thus

$$\frac{3x + 6}{3x} = \frac{3(x + 2)}{3x} = \frac{x + 2}{x}$$

Because x cannot be divided into both terms of the numerator, the fraction is in lowest terms.

If one polynomial is a factor of another, we can write their quotient as a fraction, factor, and simplify the fraction.

EXAMPLE 5

Divide $(4x^2 - 9y^2)$ by $(2x - 3y)$.

Solution

We may write the quotient as the fraction $\dfrac{(4x^2 - 9y^2)}{(2x - 3y)}$. Simplifying gives the quotient:

$$\frac{4x^2 - 9y^2}{2x - 3y} = \frac{(2x - 3y)(2x + 3y)}{2x - 3y}$$
$$= 2x + 3y$$

EXAMPLE 6

Divide $(3x^2 + 3x - 6)$ by $(x - 1)$.

Solution

$$\frac{3x^2 + 3x - 6}{x - 1} = \frac{3(x - 1)(x + 2)}{x - 1} = 3(x + 2) = 3x + 6$$

Note that this quotient was found by "long division" in Example 9 of Section 2–5.

There are three signs associated with a fraction: one on the fraction, one on the numerator, and one on the denominator. Any two of these signs can be changed without changing the value of the fraction. For example,

$$+\frac{-a}{+b} = +\frac{+a}{-b} = -\frac{+a}{+b}$$

EXAMPLE 7

Simplify $\dfrac{2x - 4}{4 - x^2}$.

Solution

$$\frac{2x-4}{4-x^2} = \frac{2(x-2)}{-(x^2-4)} = \frac{2(x-2)}{-(x-2)(x+2)}$$

$$= \frac{2}{-(x+2)} = \frac{-2}{x+2}$$

Exercise 4-1

Simplify the following fractions.

1. $\dfrac{(x-2)(x-3)}{(x+2)(x-2)}$

2. $\dfrac{(x-5)(x-2)}{(x+3)(x-5)}$

3. $\dfrac{(2x+1)(3x-1)}{(2x+1)(3x+1)}$

4. $\dfrac{(3x-1)(2x-1)}{(5x+1)(3x-1)}$

5. $\dfrac{5x^2y^3}{15xy^2}$

6. $\dfrac{8x^3y^6}{14x^2y^2}$

7. $\dfrac{18x^3z^3}{9x^3z}$

8. $\dfrac{15a^4b^5}{30a^3b}$

9. $\dfrac{4m^3n^3}{8m^4n^3}$

10. $\dfrac{12p^4q^8}{6pq^9}$

11. $\dfrac{x-3y}{3x-9y}$

12. $\dfrac{3c-bc}{3a-ab}$

13. $\dfrac{x^2-6x+8}{x^2-16}$

14. $\dfrac{x^2-2x-3}{x^2-9}$

15. $\dfrac{x^2-2x+1}{x^2-4x+3}$

16. $\dfrac{y^2-8y+15}{y^2+4y-21}$

17. $\dfrac{3x-3y}{x^2-y^2}$

18. $\dfrac{3a^3-3ab^2}{a^2-2ab+b^2}$

19. $\dfrac{a^2+3a+2}{4-a^2}$

20. $\dfrac{x^2-5x+6}{9-x^2}$

21. $\dfrac{5x+10}{5x}$

22. $\dfrac{4x+8}{4x}$

23. $\dfrac{4x^2+4x}{4x}$

24. $\dfrac{6x^2+6x}{6x}$

25. $\dfrac{15x^2+5x}{5x}$

26. $\dfrac{12x^2+6x}{6x}$

Use factoring to perform the following divisions.

27. $(x^2 + 7x + 6) \div (x + 6)$
28. $(6x^2 + 15x + 9) \div (2x + 3)$
29. $(15x^2 + x - 6) \div (5x - 3)$
30. $(10x^2 - 17x + 3) \div (2x - 3)$
31. $(x^2 - x - 6) \div (x - 3)$
32. $(3x^2 + 3x - 6) \div (x + 2)$

4-2 Products and quotients of fractions

OBJECTIVES: After completing this section, you should be able to perform computations involving products and quotients of algebraic fractions.

In arithmetic we learned to multiply fractions by writing the product as the product of the numerators divided by the product of the denominators. For example, $\frac{4}{5} \cdot \frac{10}{12} \cdot \frac{2}{5} = \frac{80}{300}$, which reduces to $\frac{4}{15}$.

We can also find the product by reducing the fractions before we indicate the multiplication in the numerator and denominator. For example, in $\frac{4}{5} \cdot \frac{10}{12} \cdot \frac{2}{5}$, we can divide the numerator and denominator by 5 and then by 4, giving $\frac{1}{1} \cdot \frac{2}{3} \cdot \frac{2}{5} = \frac{4}{15}$.

We multiply algebraic fractions by the same procedure used to multiply the fractions of arithmetic.

EXAMPLE 1

Multiply $\dfrac{a^2}{b} \cdot \dfrac{c}{a} \cdot \dfrac{b^2}{a^2}$.

Solution

$$\frac{a^2}{b} \cdot \frac{c}{a} \cdot \frac{b^2}{a^2} = \frac{\overset{1}{\cancel{a^2}}}{b} \cdot \frac{c}{a} \cdot \frac{b^2}{\underset{1}{\cancel{a^2}}}$$

$$= \frac{1}{\underset{1}{\cancel{b}}} \cdot \frac{c}{a} \cdot \frac{\overset{b}{\cancel{b^2}}}{1}$$

$$= \frac{cb}{a}$$

EXAMPLE 2

Multiply $\dfrac{4x^2}{5y} \cdot \dfrac{10x}{y^2} \cdot \dfrac{y}{8x^2}$.

Solution

$$\dfrac{4x^2}{5y} \cdot \dfrac{10x}{y^2} \cdot \dfrac{y}{8x^2} = \dfrac{\overset{1}{\cancel{4x^2}}}{\underset{1\cdot 1}{\cancel{5y}}} \cdot \dfrac{\overset{2}{\cancel{10x}}}{y^2} \cdot \dfrac{\cancel{y}}{\underset{2}{\cancel{8x^2}}}$$

$$= \dfrac{1}{1} \cdot \dfrac{\overset{1}{\cancel{2x}}}{y^2} \cdot \dfrac{1}{\underset{1}{\cancel{2}}} = \dfrac{x}{y^2}$$

We frequently have to factor the numerators and denominators of algebraic fractions to use this method of multiplication.

EXAMPLE 3

Multiply $\dfrac{-4x+8}{3x+6} \cdot \dfrac{2x+4}{4x+12}$.

Solution

$$\dfrac{-4x+8}{3x+6} \cdot \dfrac{2x+4}{4x+12} = \dfrac{-4(x-2)}{3(x+2)} \cdot \dfrac{2(x+2)}{4(x+3)}$$

$$= \dfrac{\overset{1}{\cancel{-4}}(x-2)}{3\cancel{(x+2)}} \cdot \dfrac{2\cancel{(x+2)}}{\underset{1}{\cancel{4}}(x+3)}$$

$$= \dfrac{-2(x-2)}{3(x+3)}$$

EXAMPLE 4

Multiply $\dfrac{x-2}{x^2-9} \cdot \dfrac{x^2-6x+9}{4-x^2}$.

Solution

$$\dfrac{x-2}{x^2-9} \cdot \dfrac{x^2-6x+9}{4-x^2} = \dfrac{x-2}{(x+3)(x-3)} \cdot \dfrac{(x-3)(x-3)}{(2-x)(2+x)}$$

$$= \dfrac{x-2}{(x+3)\cancel{(x-3)}} \cdot \dfrac{\cancel{(x-3)}(x-3)}{(2-x)(2+x)}$$

$$= \dfrac{(x-2)(x-3)}{(x+3)(2-x)(2+x)}$$

Note that $(x-2)$ is a factor of the numerator and $(2-x)$ is a factor of the denominator. Rewriting $(2-x)$ as $-(x-2)$ will permit us to reduce the product further.

$$\frac{(x-2)(x-3)}{(x+3)(2-x)(2+x)} = \frac{\overset{1}{(x-2)}(x-3)}{(x+3)[\underset{-1}{-(x-2)}](2+x)}$$

$$= \frac{x-3}{(x+3)(-1)(2+x)}$$

$$= \frac{x-3}{-(x+3)(2+x)}$$

$$= -\frac{x-3}{(x+3)(2+x)}$$

Note this can be written as

$$-\frac{x-3}{(x+3)(x+2)} \quad \text{or} \quad \frac{3-x}{(x+3)(x+2)}$$

In arithmetic we learned to divide one fraction by another by inverting the divisor and multiplying. The same rule applies to division of algebraic fractions. The reason this rule works can be seen in the following example.

EXAMPLE 5

Divide $\dfrac{a}{b} \div \dfrac{c}{d}$.

Solution

$$\frac{a}{b} \div \frac{c}{d} = \frac{\dfrac{a}{b}}{\dfrac{c}{d}}$$

Multiplying numerator and denominator by d/c (the divisor inverted) gives

$$\frac{\dfrac{a}{b} \cdot \dfrac{d}{c}}{\dfrac{c}{d} \cdot \dfrac{d}{c}} = \frac{\dfrac{a}{b} \cdot \dfrac{d}{c}}{1} = \frac{a}{b} \cdot \frac{d}{c}$$

96 Algebraic Fractions

Thus we see dividing by c/d is the same as multiplying by d/c.

The rule for division may be stated as follows:

To divide one fraction by another, invert the divisor and multiply the fractions.

EXAMPLE 6

Divide $\dfrac{a^2b}{c}$ by $\dfrac{ab}{c^2}$.

Solution

$$\frac{a^2b}{c} \div \frac{ab}{c^2} = \frac{a^2b}{c} \cdot \frac{c^2}{ab}$$

$$= \frac{\overset{a \cdot 1}{\cancel{a^2b}}}{\underset{1}{\cancel{c}}} \cdot \frac{\overset{c}{\cancel{c^2}}}{\underset{1 \cdot 1}{\cancel{ab}}}$$

$$= \frac{ac}{1}$$

$$= ac$$

Exercise 4-2

Multiply the following fractions.

1. $\dfrac{3}{5} \cdot \dfrac{5}{6} \cdot \dfrac{2}{7}$

2. $\dfrac{5}{8} \cdot \dfrac{4}{15} \cdot \dfrac{3}{2}$

3. $\dfrac{24}{35} \cdot \dfrac{7}{8} \cdot \dfrac{2}{3}$

4. $\dfrac{6}{7} \cdot \dfrac{21}{5} \cdot \dfrac{15}{26}$

5. $\dfrac{x^2}{y^2} \cdot \dfrac{x}{y} \cdot \dfrac{y^3}{x^4}$

6. $\dfrac{m^4}{n^3} \cdot \dfrac{n^2}{m} \cdot \dfrac{n^5}{m^5}$

7. $\dfrac{a^2}{b^5} \cdot \dfrac{b^2}{a^2} \cdot \dfrac{b^4}{a^2}$

8. $\dfrac{p^3}{q} \cdot \dfrac{q^3}{p} \cdot \dfrac{p^3}{q}$

9. $\dfrac{10x^2}{3y^2} \cdot \dfrac{27y}{15x} \cdot \dfrac{y}{2x}$

10. $\dfrac{6x^3}{8y^3} \cdot \dfrac{16x}{9y^2} \cdot \dfrac{15y^4}{x^3}$

Exercise 4-2

11. $\dfrac{3x-6}{4x+8} \cdot \dfrac{5x+10}{x-2}$

12. $\dfrac{8x-16}{x-3} \cdot \dfrac{4x-12}{3x-6}$

13. $\dfrac{7x+21}{3x-15} \cdot \dfrac{x+5}{2x+6}$

14. $\dfrac{4x+12}{6x-12} \cdot \dfrac{4x-8}{5x+10}$

15. $\dfrac{x^2+7x+12}{3x^2+13x+4} \cdot \dfrac{3x+1}{x+3}$

16. $\dfrac{2b^2-b-15}{b^2-4b+4} \cdot \dfrac{b-2}{b-3}$

17. $\dfrac{x^2+7x+12}{3x^2+13x+4} \cdot \dfrac{3x^2+4x+1}{x^2+5x+6}$

18. $\dfrac{2b^2-b-15}{b^2-25} \cdot \dfrac{b^2+10b+25}{b^2-4b+3}$

19. $\dfrac{x^2+7x+10}{x^2-3x-10} \cdot \dfrac{x^2-4x+4}{x^2+5x+6}$

20. $\dfrac{x^2-16}{x^2+x-12} \cdot \dfrac{x^2-9}{x^2+8x+16}$

21. $\dfrac{x^2-2x-3}{2x^2+3x+1} \cdot \dfrac{2x^2+7x+3}{x^2-5x+6}$

22. $\dfrac{9x^2-12x+4}{9x^2-15x+6} \cdot \dfrac{x^2-3x+2}{9x^2-4}$

23. $\dfrac{x-4}{x^2-25} \cdot \dfrac{x^2+8x+15}{16-x^2}$

24. $\dfrac{x^2-8x+7}{x^2-2x+1} \cdot \dfrac{x^2-1}{49-x^2}$

25. $\dfrac{x-5}{x^2-5x+4} \cdot \dfrac{x-3}{5-x}$

26. $\dfrac{x^2-3x-4}{x^2-7x+6} \cdot \dfrac{x^2-1}{4-x}$

Perform the following divisions.

27. $\dfrac{5}{9} \div \dfrac{10}{27}$

28. $\dfrac{18}{25} \div \dfrac{9}{5}$

29. $\dfrac{7}{10} \div \dfrac{21}{28}$

30. $\dfrac{4}{9} \div \dfrac{8}{27}$

31. $\dfrac{4a^2b}{9cd} \div \dfrac{2ab}{3c}$

32. $\dfrac{15ac^2}{7bd} \div \dfrac{5a}{14b^2d}$

33. $\dfrac{4x^2y}{3y^2} \div \dfrac{2xy}{y^2}$

34. $\dfrac{6ab}{5c} \div \dfrac{3ac}{2d}$

35. $\dfrac{24}{3x-6} \div \dfrac{4}{x^2-4}$

36. $\dfrac{a^2-b^2}{b^2} \div \dfrac{a+b}{ab}$

37. $\dfrac{15}{4x-12} \div \dfrac{5}{6x-18}$ 38. $\dfrac{16}{x-2} \div \dfrac{4}{3x-6}$

39. $\dfrac{y^2-2y+1}{7y^2-7y} \div \dfrac{y^2-4y+3}{35y^2}$

40. $\dfrac{z^2+6z+9}{z^2-9} \div \dfrac{z^2-1}{z^2-2z-3}$

41. $\dfrac{x^2-1}{x^2-5x+6} \div \dfrac{x^2-4x-5}{x^2-6x+9}$

42. $\dfrac{b^2+2b}{c^2-5c} \div \dfrac{b^2-6b-16}{c^2-4c-5}$

43. $\dfrac{x^2-x-2}{x^3-x} \div \dfrac{x^2-4x+4}{x^3+6x^2-7x}$ 44. $\dfrac{2c+2d}{x^2-4y^2} \div \dfrac{c^2-d^2}{x-2y}$

45. $\dfrac{r^2-2rs+s^2}{r+2s} \div \dfrac{r^2-s^2}{r^2+3rs+2s^2}$ 46. $\dfrac{x^2-y^2}{a^2-b^2} \div \dfrac{4x-4y}{9a-9b}$

4-3 Addition and subtraction of algebraic fractions

OBJECTIVE: After completing this section, you should be able to perform computations involving addition and subtraction of fractions.

If two fractions are to be added, it is convenient that each be expressed with the same denominator. If the denominators are not the same, we can write the equivalents of each of the fractions with a common denominator.

EXAMPLE 1

We can add $\frac{1}{4}$ and $\frac{3}{4}$, because they have a common denominator.
 To add $\frac{1}{4}$ and $\frac{1}{3}$, it is convenient to write them as $\frac{3}{12}$ and $\frac{4}{12}$; so they have the same denominators.

 To add $\frac{1}{4}$ and $\frac{1}{3}$, we could also use equivalent fractions whose denominators are 24, 36, or 120. However, we usually use the *lowest common denominator* (L.C.D.) when we write the equivalent fractions. The *lowest common denominator* is the smallest number into which all denominators will divide. We can find the lowest common denominator as follows.

4-3 Addition and subtraction of algebraic fractions

PROCEDURE	EXAMPLE
To find the L.C.D. of a set of fractions:	Find the L.C.D. of $\dfrac{1}{x^2-x}, \dfrac{1}{x^2-1}, \dfrac{1}{x^2}$
1. Completely factor each denominator.	1. $x^2 - x = x(x-1)$ $x^2 - 1 = (x-1)(x+1)$ $x^2 = x \cdot x$
2. Write the L.C.D. as the product of each of these factors, used the maximum number of times it occurs in any one denominator.	2. The factor x occurs a maximum of two times in one denominator; $x-1$ occurs once; $x+1$ occurs once. Thus L.C.D. $= x \cdot x(x-1)(x+1) = x^2(x-1)(x+1)$.

EXAMPLE 2

Find the L.C.D. of $\dfrac{3}{4x^2}, \dfrac{5}{6x}$, and $\dfrac{1}{2y}$.

Solution

1. Factor the denominators: $2 \cdot 2xx, \; 2 \cdot 3x, \; 2y$.
2. Write the product of factors the maximum number of times they occur in any one denominator:

   L.C.D. $= 2 \cdot 2 \cdot 3 \cdot xxy = 12x^2y$

EXAMPLE 3

Find the L.C.D. of $\dfrac{x+2}{x^2+2x+1}, \dfrac{4}{x^2+x}, \dfrac{x}{x^3-x^2}$.

Solution

1. The factors are $(x+1)(x+1), \; x(x+1), \; xx(x-1)$.
2. The L.C.D. is

   $xx(x+1)(x+1)(x-1) = x^2(x+1)^2(x-1)$

To add or subtract fractions that have different denominators, we can write the equivalents of the fractions, with each new fraction having the L.C.D. as its denominator.

100 Algebraic Fractions

To write the equivalent fractions for a set of fractions, using their L.C.D. as the denominator of each new fraction, we proceed as follows.

PROCEDURE	EXAMPLE
To write the equivalent fractions with the L.C.D. as denominator of each:	Write the equivalents of $\frac{1}{x^2}$, $\frac{x}{4y}$, $\frac{3}{xy^2}$, using their L.C.D. as the denominator of each.
1. Find the L.C.D. of the fractions.	1. L.C.D. is $4x^2y^2$.
2. Divide each denominator into the L.C.D. and multiply the quotient times the corresponding numerator and denominator to get the equivalent fraction.	2. $\frac{1}{x^2} = \frac{1(4y^2)}{x^2(4y^2)} = \frac{4y^2}{4x^2y^2}$ $\frac{x}{4y} = \frac{x(x^2y)}{4y(x^2y)} = \frac{x^3y}{4x^2y^2}$ $\frac{3}{xy^2} = \frac{3(4x)}{xy^2(4x)} = \frac{12x}{4x^2y^2}$

EXAMPLE 4
Write the equivalents of $\frac{1}{6x^2}$, $\frac{x}{2y}$, and $\frac{4}{xy^2}$ using their L.C.D. as the denominator of each.

Solution
1. The L.C.D. is $2 \cdot 3 \cdot x^2y^2 = 6x^2y^2$.
2. $\frac{1}{6x^2} = \frac{1(y^2)}{6x^2(y^2)} = \frac{y^2}{6x^2y^2}$

$\frac{x}{2y} = \frac{x(3x^2y)}{2y(3x^2y)} = \frac{3x^3y}{6x^2y^2}$

$\frac{4}{xy^2} = \frac{4(6x)}{xy^2(6x)} = \frac{24x}{6x^2y^2}$

EXAMPLE 5
Write the equivalents of

$$\frac{1}{a^2 - ab}, \quad \frac{a}{a^2 - b^2}, \quad \frac{b}{a^2 + 3ab + 2b^2}$$

using their L.C.D. as the denominator of each.

Solution

The factors of the denominators are $a(a - b)$, $(a + b)(a - b)$, $(a + b)(a + 2b)$.
The L.C.D. is $a(a - b)(a + b)(a + 2b)$.

$$\frac{1}{a(a-b)} = \frac{1(a+b)(a+2b)}{a(a-b)(a+b)(a+2b)}$$

$$= \frac{(a+b)(a+2b)}{a(a-b)(a+b)(a+2b)}$$

$$\frac{a}{(a+b)(a-b)} = \frac{a(a)(a+2b)}{(a+b)(a-b)(a)(a+2b)}$$

$$= \frac{a^2(a+2b)}{a(a-b)(a+b)(a+2b)}$$

$$\frac{b}{(a+b)(a+2b)} = \frac{b(a)(a-b)}{(a+b)(a+2b)(a)(a-b)}$$

$$= \frac{ab(a-b)}{a(a-b)(a+b)(a+2b)}$$

The sum of two or more fractions with like denominators is a fraction with the common denominator and a numerator equal to the sum of fractions' numerators.

EXAMPLE 6

Add $\dfrac{x}{x+2} + \dfrac{4}{x+2} + \dfrac{x}{x+2}$.

Solution

The common denominator is $x + 2$.
The numerator of the sum is $x + 4 + x = 2x + 4$.
Thus

$$\frac{x}{x+2} + \frac{4}{x+2} + \frac{x}{x+2} = \frac{2x+4}{x+2}$$

This sum will reduce to

$$\frac{2(x+2)}{x+2} = 2$$

The difference of two fractions with like denominators is the difference of the numerators over the common denominator.

EXAMPLE 7

Subtract $\dfrac{3x-1}{x^2+1} - \dfrac{2x-3}{x^2+1}$.

Solution

$$\dfrac{3x-1}{x^2+1} - \dfrac{2x-3}{x^2+1} = \dfrac{(3x-1)-\overline{(2x-3)}}{x^2+1}$$ (Recall the bar is a symbol of inclusion.)

$$= \dfrac{3x-1-2x+3}{x^2+1}$$ [Note that $-(2x-3) = -2x+3$.]

$$= \dfrac{x+2}{x^2+1}$$

The difference is in lowest terms.

The process of adding or subtracting fractions is called combining the fractions. A problem may involve both addition and subtraction.

To combine two or more fractions with unlike denominators, write the equivalents of the fractions with their L.C.D. as the denominator of each, and add or subtract as indicated.

EXAMPLE 8

Add $\dfrac{3x}{a^2} + \dfrac{4}{ax}$.

Solution

1. The L.C.D. is a^2x.
2. $\dfrac{3x}{a^2} = \dfrac{3x^2}{a^2x}, \quad \dfrac{4}{ax} = \dfrac{4a}{a^2x}$
3. $\dfrac{3x^2}{a^2x} + \dfrac{4a}{a^2x} = \dfrac{3x^2 + 4a}{a^2x}$
4. The sum is in lowest terms.

EXAMPLE 9

Combine $\dfrac{y-3}{y-5} + \dfrac{y-23}{y^2-y-20}$.

Solution

$y^2 - y - 20 = (y-5)(y+4)$; so the L.C.D. is $(y-5)(y+4)$. Writing the equivalent fractions gives

$$\dfrac{y-3}{y-5} + \dfrac{y-23}{(y-5)(y+4)} = \dfrac{(y-3)(y+4)}{(y-5)(y+4)} + \dfrac{y-23}{(y-5)(y-4)}$$

$$= \dfrac{(y^2+y-12)+(y-23)}{(y-5)(y+4)}$$

$$= \dfrac{y^2+2y-35}{(y-5)(y+4)}$$

But $y-5$ is a factor of the numerator, as its factorization shows.

$$\dfrac{y^2+2y-35}{(y-5)(y+4)} = \dfrac{(y-5)(y+7)}{(y-5)(y+4)}$$

$$= \dfrac{y+7}{y+4}$$

EXAMPLE 10

Combine $\dfrac{a}{a+1} + \dfrac{4}{a^2-1} - \dfrac{a-1}{a^2+a}$.

Solution

$a+1 = a+1$, $a^2-1 = (a-1)(a+1)$, and $a^2+a = a(a+1)$; so the L.C.D. of the fractions is $a(a+1)(a-1)$. Writing the equivalent fractions gives

$$\dfrac{a(a)(a-1)}{(a+1)(a)(a-1)} + \dfrac{4(a)}{(a-1)(a+1)(a)} - \dfrac{(a-1)(a-1)}{a(a+1)(a-1)}$$

$$= \dfrac{(a^3-a^2)+(4a)-(a^2-2a+1)}{a(a+1)(a-1)}$$

$$= \dfrac{a^3-a^2+4a-a^2+2a-1}{a(a+1)(a-1)}$$

$$= \dfrac{a^3-2a^2+6a-1}{a(a+1)(a-1)}$$

Algebraic Fractions

Exercise 4-3

Combine the following fractions.

1. $\frac{1}{3} + \frac{2}{3} + \frac{1}{3}$
2. $\frac{2}{7} + \frac{3}{7} + \frac{1}{7}$
3. $\frac{1}{9} + \frac{2}{9} + \frac{4}{9}$
4. $\frac{5}{8} + \frac{1}{8} + \frac{3}{8}$
5. $\frac{x}{x-2} + \frac{3}{x-2} + \frac{x}{x-2}$
6. $\frac{y}{y+3} + \frac{2}{y+3} + \frac{y}{y+3}$
7. $\frac{x}{x+5} + \frac{3}{x+5} + \frac{2}{x+5}$
8. $\frac{x}{x+1} + \frac{2}{x+1} + \frac{x}{x+1}$
9. $\frac{1}{6} + \frac{5}{15}$
10. $\frac{1}{9} + \frac{5}{12}$
11. $\frac{4}{5x^2} + \frac{3}{4x}$
12. $\frac{3a}{xy} + \frac{2a}{x^2}$
13. $\frac{a}{b^2} + \frac{b}{abc}$
14. $\frac{3}{m^3n} + \frac{2}{mn^2}$
15. $\frac{x+6}{x-5} + \frac{x-6}{x-4}$
16. $\frac{8}{x-1} + \frac{3}{x+1}$
17. $\frac{5a}{3a-9} + \frac{a-1}{a-3}$
18. $\frac{7}{4x+4y} + \frac{5}{x+y}$
19. $\frac{x-4}{x^2+3x} + \frac{x+2}{2x+6}$
20. $\frac{b-1}{b^2+2b} + \frac{b}{3b+6}$
21. $\frac{1}{m-2} + \frac{m+1}{m^2-4}$
22. $\frac{x}{x^2+x} + \frac{x+1}{x^2-1}$
23. $\frac{x-1}{x^2-x-2} + \frac{x-3}{x^2+x-6}$
24. $\frac{x-7}{x^2-9x+20} + \frac{x+2}{x^2-5x+4}$
25. $\frac{3}{2x-1} + \frac{2}{1-2x}$
26. $\frac{4x}{3x-2} + \frac{3x-1}{2-3x}$
27. $\frac{5}{7} - \frac{3}{7}$
28. $\frac{8}{9} - \frac{5}{9}$

29. $\dfrac{4x}{5a} - \dfrac{2x}{5a}$

30. $\dfrac{7a}{3x} - \dfrac{2b}{3x}$

31. $\dfrac{n+1}{6n^2} - \dfrac{4}{3n}$

32. $\dfrac{x+4}{5x^2} - \dfrac{1}{3x}$

33. $\dfrac{15}{2x-8} - \dfrac{3}{x-4}$

34. $\dfrac{12x}{x-5} - \dfrac{16x}{3x-15}$

35. $\dfrac{4a}{3x+6} - \dfrac{5a^2}{4x+8}$

36. $\dfrac{4b}{4a+10} - \dfrac{3b}{6a+12}$

37. $\dfrac{5}{a-4} - \dfrac{3}{a+3}$

38. $\dfrac{7}{a-6} - \dfrac{3}{a-2}$

39. $\dfrac{x+2}{x^2-3x+2} - \dfrac{3}{x-1}$

40. $\dfrac{4x-1}{x^2-2x-8} - \dfrac{2}{x+2}$

41. $\dfrac{4y+3}{y^2-y-2} - \dfrac{y+3}{2-y}$

42. $\dfrac{x+1}{x^2+x-6} - \dfrac{2x-1}{2-x}$

43. $\dfrac{2y-1}{y^2-3y-10} - \dfrac{y-3}{y^2-25}$

44. $\dfrac{x-2}{x^2+x-6} - \dfrac{3x+5}{2x^2-x-6}$

45. $\dfrac{3x-1}{2x-4} + \dfrac{4x}{3x-6} - \dfrac{4}{5x-10}$

46. $\dfrac{x-1}{5x+5} + \dfrac{2x}{3x+3} - \dfrac{6}{7x+7}$

47. $\dfrac{2x+1}{4x-2} + \dfrac{5}{2x} - \dfrac{x+1}{2x^2-x}$

48. $\dfrac{4}{5x+15} + \dfrac{6}{5x} - \dfrac{x-2}{x^2+3x}$

49. $\dfrac{2x-1}{4x^2-9} + \dfrac{3x+1}{2x^2-5x+3} - \dfrac{2x}{x^2-1}$

50. $\dfrac{x-1}{x^2+x-6} + \dfrac{3x+2}{x^2-x-2} - \dfrac{2x}{x^2-9}$

4-4 Complex fractions

OBJECTIVE: After completing this section, you should be able to simplify complex fractions.

A complex fraction is a fractional expression that contains one or more fractions in its numerator or denominator or in both its numerator and denominator. The fractions

$$\frac{\frac{2}{3}}{\frac{3}{5}}, \quad \frac{1+\frac{x}{2}}{2}, \quad \text{and} \quad \frac{3-\frac{2}{a}}{1+\frac{4}{a}}$$

are all complex fractions. We can simplify complex fractions as follows.

PROCEDURE	EXAMPLE
To simplify complex fractions:	Simplify $\dfrac{1+\dfrac{1}{x}}{2-\dfrac{3}{y}}$.
1. Find the L.C.D. of all the fractions in the numerator and denominator.	1. The L.C.D. of $\dfrac{1}{x}$ and $\dfrac{3}{y}$ is xy.
2. Multiply both the numerator and denominator of the complex fraction by the L.C.D.	2. $\dfrac{xy\left(1+\dfrac{1}{x}\right)}{xy\left(2-\dfrac{3}{y}\right)} = \dfrac{xy+y}{2xy-3x}$
3. Simplify the resulting fraction if possible.	3. $\dfrac{xy+y}{2xy-3x} = \dfrac{y(x+1)}{x(2y-3)}$

EXAMPLE 1

Simplify

$$\frac{\frac{2a^2}{3b^2}}{\frac{5a^2}{6b}}$$

Solution

The L.C.D. is $6b^2$.

$$\frac{6b^2\left(\frac{2a^2}{3b^2}\right)}{6b^2\left(\frac{5a^2}{6b}\right)} = \frac{4a^2}{5a^2b}$$

Reducing this fraction gives $\frac{4}{5b}$.

EXAMPLE 2

Simplify

$$\frac{\frac{1}{r} - \frac{1}{s}}{\frac{1}{r} + \frac{1}{s}}$$

Solution

The L.C.D. is rs.

$$\frac{rs\left(\frac{1}{r} - \frac{1}{s}\right)}{rs\left(\frac{1}{r} + \frac{1}{s}\right)} = \frac{s - r}{s + r}$$

EXAMPLE 3

Simplify

$$\frac{3x + \frac{y+2}{x+3}}{x + \frac{y}{3}}$$

Solution

The L.C.D. is $3(x + 3)$.

$$\frac{3(x+3)\left(3x + \frac{y+2}{x+3}\right)}{3(x+3)\left(x + \frac{y}{3}\right)} = \frac{3(x+3)(3x) + 3(y+2)}{3(x+3)x + (x+3)y}$$

$$= \frac{9x^2 + 27x + 3y + 6}{3x^2 + 9x + xy + 3y}$$

Exercise 4-4

Simplify the following complex fractions.

1. $\dfrac{\frac{2}{5}}{\frac{4}{10}}$

2. $\dfrac{\frac{9}{10}}{\frac{3}{5}}$

3. $\dfrac{\frac{4x^2}{5y^3}}{\frac{2xy}{3y}}$

4. $\dfrac{\frac{6a^2b}{c^2}}{\frac{3ab^2}{2c}}$

5. $\dfrac{1+\frac{1}{a}}{2+\frac{2}{b}}$

6. $\dfrac{2-\frac{1}{a}}{3-\frac{2}{a}}$

7. $\dfrac{1+\frac{a}{b}}{1+\frac{b}{a}}$

8. $\dfrac{1+\frac{2}{x}}{2-\frac{3}{x}}$

9. $\dfrac{\frac{1}{x}+\frac{1}{y}}{\frac{1}{x}-\frac{1}{y}}$

10. $\dfrac{\frac{1}{2}+\frac{1}{x}}{\frac{1}{3}-\frac{1}{x}}$

11. $\dfrac{a+b}{\frac{1}{a}+\frac{1}{b}}$

12. $\dfrac{\frac{1}{x}-\left(\frac{1}{2}\right)x}{x+2}$

13. $\dfrac{\frac{1}{2x}+\frac{2}{3x}}{\frac{2}{x}-\frac{4}{5x}}$

14. $\dfrac{\frac{2x}{3}-\frac{x}{4}}{\frac{3x}{4}-\frac{5x}{2}}$

15. $\dfrac{\frac{3}{2a}+\frac{4}{3b}}{\frac{2}{3a}+\frac{1}{6b}}$

16. $\dfrac{\frac{5}{2a}-\frac{3}{5b}}{\frac{1}{ab}+\frac{2}{3a}}$

CHAPTER TEST

1. Reduce $\dfrac{5x + 10}{5x}$.

2. Simplify $\dfrac{10x^3y^2}{15xy^3}$.

3. Simplify $\dfrac{5x^2 - 10xy}{6x - 12y}$.

4. Simplify $\dfrac{x^2 + 5x + 6}{x^2 + 3x + 2}$.

5. Use factoring to divide $x^2 + 8x + 12$ by $x + 6$.

6. Are the fractions $-\dfrac{5}{x-y}$ and $\dfrac{5}{y-x}$ equal?

7. Simplify $\dfrac{x^2 - 8x + 16}{x^2 - 16}$.

8. Multiply $\dfrac{2x^2}{y} \cdot \dfrac{4y}{x^2} \cdot \dfrac{x}{6y^2}$.

9. Multiply $\dfrac{x^2 - 7x + 12}{x^2 - 1} \cdot \dfrac{x^2 + 4x + 3}{x^2 - 9}$.

10. Divide $\dfrac{18a^2b}{14cd^2} \div \dfrac{6ab^2}{7c^2d}$.

11. Divide $\dfrac{z^2 - 6z + 5}{4z^2 - 4} \div \dfrac{z^2 - 7z + 10}{z^2 - 4z + 4}$.

12. Add $\dfrac{5x}{3x - 2} + \dfrac{4x}{2 - 3x}$.

13. Add $\dfrac{x + 1}{x^2 - x - 12} + \dfrac{2x + 1}{x^2 - 2x - 8}$.

14. Subtract $\dfrac{y + 2}{2y^2 - y - 6} - \dfrac{y - 3}{y^2 - 4}$.

15. Combine $\dfrac{x - 3}{6x - 4} + \dfrac{x + 2}{2x - 6} - \dfrac{5}{3x^2 - 11x + 6}$.

16. Simplify $\dfrac{1 + \dfrac{3}{x}}{\dfrac{1}{2} + \dfrac{1}{3x}}$.

17. Simplify $\dfrac{\dfrac{3}{a} - \dfrac{2}{b}}{\dfrac{1}{ab} + \dfrac{2}{3b}}$.

linear equations and inequalities in one variable

5-1 Equations

OBJECTIVES: After completing this section, you should be able to:

a. **Distinguish between identities and conditional equations.**
b. **Determine if equations are true for stated values of the variable.**
c. **Write number sentences as equations.**
d. **Evaluate formulas, given sufficient information.**

An equation is a mathematical statement that two quantities are equal. For example, $2 + 2 = 4$ is an equation, which states that $2 + 2$ and 4 represent the same number. The equation $2x + 4y = 7$ is an equation, which states that $2x + 4y$ and 7 represent the same number. Some equations involving variables are true only for certain values of the variables. For example, $3x = 6$ is true only if $x = 2$. Equations that are true only for certain values of the variable(s) are called **conditional equations** or simply **equations**. The values of the variables for which the equation is true are called the **solutions** of the equation. Thus 2 is a solution of the equation $3x = 6$. The set containing the solution(s) of an equation is called the **solution set** of the equation.

Certain equations are true for all values of the variable(s). These are called **identities**. The equation $2(x - 1) = 2x - 2$ is an example of an identity. We say an equation is an identity if both sides of the equation represent different ways of writing the same expression or number. Thus $2 + 2 = 4$ is an identity, and $2x + 4y = 7$ is not.

Identities were used in Chapter 1 to discuss the properties

of the real numbers. For example, the distributive property was stated by the identity

$$a(b + c) = ab + ac$$

EXAMPLE 1

Which of the following are identities and which are (conditional) equations?

a. $2 + 2 = 4$
b. $2 + x = 4$
c. $2x + 1 = 3x + 1 - x$
d. $2(1 + x) = 4$

Solution

a. identity
b. conditional equation
c. identity
d. conditional equation

EXAMPLE 2

Determine if the following equations are true for the given values of the variables.

a. $x + 4 = 12$, $x = 8$
b. $3y - 12 = 6$, $y = 6$
c. $8x + 4 = 32$, $x = 4$

Solution

a. $(8) + 4 = 12$; equation is true for $x = 8$.
b. $3(6) - 12 = 6$; equation is true for $y = 6$.
c. $8(4) + 4 = 36 \neq 32$; equation is not true for $x = 4$.

We can write equations to express sentences symbolically. For example, we may write "the sum of 4 and 6 is 10" as $4 + 6 = 10$, or "the product of 5 and x is 20" as $5x = 20$.

EXAMPLE 3

Write the following statements as equations.

a. The sum of a number and 8 is 12.
b. The product of a number and 4 is 36.
c. Twice a number added to 3 is 15.

Solution

a. $x + 8 = 12$
b. $x \cdot 4 = 36$, or $4x = 36$
c. $2x + 3 = 15$

Equations are frequently used to write rules in a shorthand fashion. When a rule is written as an equation, the equation is called a formula. Consider the following examples.

EXAMPLE 4

The perimeter of a square is four times the length of one of its sides. Write this rule as a formula.

Solution

Let s represent the length of a side and P represent the perimeter.

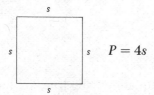

$P = 4s$

We can evaluate formulas by substituting in for the given variables.

EXAMPLE 5

The selling price of an item is equal to the cost of the item plus the markup. This may be written as the formula $S = C + M$. Find the selling price of an item which has a cost of $39.50 and a markup of $8.75.

Solution

$S = \$39.50 + \$8.75$
or
$S = \$48.25$

The selling price is $48.25.

EXAMPLE 6

The formula for the area of a trapezoid is

$$A = \frac{(b_1 + b_2)h}{2}$$

where b_1 and b_2 are the parallel bases and h is the altitude of the trapezoid. Find the area of a trapezoid with $b_1 = 10$ inches, $b_2 = 12$ inches, and $h = 6$ inches.

Solution

$$A = \frac{(10 + 12)6}{2} = \frac{22 \cdot 6}{2} = 66$$

The area is 66 square inches.

Exercise 5-1

Which of the following are identities and which are conditional equations?

1. $x + 5 = 2$
2. $x - 4 = 3$
3. $x - 1 = 7$
4. $3 - x = 2$
5. $2x - 1 = 4$
6. $3x + 1 = 6$
7. $2x = x + x$
8. $4x = 5x - x$
9. $5x + 1 = 8x + 1 - 3x$
10. $6x - 2 = 6x - 2 + 2x$
11. $2x - 3 = 3x + 4$
12. $5x + 4 = 3x + 4 - 2x$
13. $2(x - 1) = 2x - 2$
14. $3(2x + 1) = 6x + 3$

Determine if the following equations are true for the given values of the variables.

15. $x + 4 = 6, x = 2$
16. $x - 5 = 7, x = 2$
17. $x - 5 = 12, x = 7$
18. $y + 7 = 8, y = 1$
19. $3y + 2 = -4, y = -2$
20. $2x - 3 = 5, x = 1$

Write the following statements as equations.

21. The sum of 4 and 3 is 7.
22. The sum of 5 and 6 is 11.
23. The product of 3 and 4 is 12.
24. The product of 8 and 12 is 96.
25. The sum of x and 3 is 8.
26. The sum of y and 5 is 12.
27. The product of 3 and x is 6.

28. Twice x plus 4 is 12.
29. Three times x minus 4 is 8.
30. Twice y added to 7 is 12.
31. The sum of a number and 7 is 16.
32. The sum of a number and 8 is 12.
33. The product of a number and 3 is 15.
34. The product of a number and 5 is 20.
35. Twice a number added to 4 is 24.
36. Twice a number added to 7 is 13.
37. Twice a number minus 6 is 0.
38. Twice a number minus 5 is 8.
39. Three times a number added to 4 is 6.
40. Six times a number plus 6 is 15.
41. Four times a number subtracted from 19 is 7.
42. Three times a number minus 5 is 4.

Write the following rules as formulas.

43. The circumference C of a circle is π times the diameter d of the circle.
44. The area A of a circle is π times the square of the radius r.
45. The perimeter P of a triangle with sides of length a, b, and c is the sum of the lengths of the sides.
46. The area A of a square is the square of the length of one of the sides s.
47. The interest I on a loan equals the principal P of the loan times the rate of interest r times the time t.
48. The discount d on a sale is the price P times the discount rate r.
49. The perimeter P of a rectangle is twice the length l added to twice the width w.
50. The distance traveled d is equal to the rate r times the time t.

Evaluate the following formulas.

51. The velocity of a body falling freely is $V = 32t$, where t represents the number of seconds that have passed since the body was dropped. Find V if $t = 5$.
52. The formula for the perimeter of a square is $P = 4s$, where s is the length of one side. Find P if $s = 4$ inches.
53. The formula for the area of a triangle is $A = \frac{1}{2} bh$. Find A if $b = 4$ inches and $h = 6$ inches.
54. The formula for the area of a rectangle is $A = lw$. Find A if $l = 6$ inches and $w = 5$ inches.

55. The fomula for interest on a loan is $I = Prt$. Find I if $P = \$300$, $r = 6$ percent (per year), and $t = 3$ years.

56. The formula for distance traveled is $d = rt$, where r is the rate and t is the time. Find d if r is 45 miles per hour and t is 4 hours.

5-2 Solution of linear equations in one variable

OBJECTIVE: After completing this section, you should be able to solve linear equations in one variable.

As we discussed in Section 5-1, some equations are true for all values of the variable, and others are true only for certain values of the variable. Certain equations can be solved easily by inspection. For example, $x + 3 = 4$ has $x = 1$ as its solution. However, we need a general procedure to solve more complicated equations. We will begin by studying linear equations in one variable. $4x - 12 = 16$ is an example of a linear equation in one variable. The variables in a linear equation have only the exponent 1 and do not appear in the denominator of any fraction. For this reason, linear equations are also called **first-degree equations.**

Two equations are said to be **equivalent** if they have exactly the same solutions. For example,

$$4x - 12 = 16$$
$$4x = 28$$

and

$$x = 7$$

are all equivalent equations because they each have one solution, 7. We can solve a more complicated linear equation by finding an equivalent equation whose solution is easily found. We use the following properties of equality to reduce an equation to a simple equivalent equation.

Substitution Property. The equation formed by substituting one expression for an equal expression in an equation is equivalent to the original equation.

Example: $5x - 4x = 6$ is equivalent to $x = 6$ because $5x - 4x = x$. We may write the **solution set** of $5x - 4x = 6$ as $\{6\}$.

Note: This property permits us to simplify an equation by replacing it with an equivalent equation with arithmetic operations performed. For example, $3(x-3) + \frac{1}{2}(2x-6) = 4$ can be simplified to the equation $3x - 9 + x - 3 = 4$, and further to $4x - 12 = 4$.

Addition Property. The equation formed by adding the same quantity to both sides of an equation is equivalent to the original equation.

Example: $x - 4 = 6$ is equivalent to $x = 10$ because 4 was added to both sides of $x - 4 = 6$. Thus 10 is the solution to $x - 4 = 6$.

Note: This property permits us to use subtraction as well as addition to find equivalent equations. For example, by adding -5 to both sides (or *subtracting* 5 from both sides), we see that $x + 5 = 12$ is equivalent to $x = 7$.

Multiplication Property. The equation formed by multiplying the same quantity times both sides of an equation is equivalent to the original equation.

Example: $\frac{1}{3}x = 6$ is equivalent to $x = 18$ because 3 was multiplied times both sides of $\frac{1}{3}x = 6$. Thus the solution set for $\frac{1}{3}x = 6$ contains 18.

Note: This property permits us to use division as well as multiplication to find equivalent equations. For example, multiplying both sides by $\frac{1}{5}$ (or dividing both sides by 5) of $5x = 20$ gives the equivalent equation $x = 4$. Thus 4 is the solution to $5x = 20$.

The correct use of the above properties permits us to reduce any linear equation in one unknown to an equivalent equation whose solution is obvious. We may solve linear equations in one unknown using the following procedure.

1. Determine what has been done to the variable. (Has a number been added to it, subtracted from it, etc.?)
2. Perform the inverse operation that will "undo" what has been done to the variable. This will "isolate" the variable and give the solution of the equation.

EXAMPLE 1

Solve the equation $x + 4 = 5$ for x.

Solution

We note that 4 has been added to x. The inverse of addition is subtraction. If we subtract 4 from both sides of the equation, we have an *equivalent* equation: $x + 4 - 4 = 5 - 4$ or $x = 1$. Thus the solution of $x + 4 = 5$ is 1.

Check: $(1) + 4 = 5$

EXAMPLE 2

Solve $17x = 34$.

Solution

We note that x has been *multiplied* by 17. The inverse of multiplication is division. If we *divide both sides* by 17, we get the equivalent equation

$$\frac{17x}{17} = \frac{34}{17}, \quad \text{or} \quad x = 2$$

Thus the solution of $17x = 34$ is 2.

Check: $17(2) = 34$

As the previous examples illustrate, a linear equation in one unknown has only one possible solution. Linear equations frequently are more complicated than those discussed above. They may contain fractions, parentheses, and variables on both sides of the equation. The steps necessary to reduce an equation may be taken in many different orders, but following a general procedure will make our work easier. The procedure follows.

PROCEDURE	EXAMPLE
To solve a linear equation:	Solve $\dfrac{3x}{4} + 3 = \dfrac{2(x-1)}{6}$.
1. If the equation contains fractions, multiply both sides by the L.C.D. of the fractions.	1. L.C.D. is 12. The equivalent equation is $$12\left(\frac{3x}{4} + 3\right) = 12\left[\frac{2(x-1)}{6}\right]$$ or $9x + 36 = 4(x - 1)$

118 Linear Equations and Inequalities in One Variable

2. Remove any parentheses in the equation.

3. Perform any additions or subtractions to get all terms containing the variable on one side and all other terms on the other side.

4. Divide both sides of the equation by the coefficient of the variable.

5. Check the solution by substitution.

2. $9x + 36 = 4x - 4$

3. $5x + 36 = -4$ (Subtract $4x$ from both sides.)
$5x = -40$ (Subtract 36 from both sides.)

4. $x = \dfrac{-40}{5} = -8$ (Divide both sides by 5.)

5. $\dfrac{3(-8)}{4} + 3 = \dfrac{2(-8-1)}{6}$ since $\dfrac{-24}{4} + 3 = \dfrac{-18}{6}$

===

EXAMPLE 3

Solve $2(3y - 1) = 4(y - 5)$ for y.

Solution

1. No fractions are involved.
2. $6y - 2 = 4y - 20$ (removing parentheses)
3. $2y - 2 = -20$ (subtracting $4y$ from both sides)
   $2y = -18$ (adding 2 to both sides)
4. $y = -9$ (dividing both sides by 2)

Check: $2[3(-9) - 1] = 4(-9 - 5)$ since $-56 = -56$

EXAMPLE 4

Solve $\dfrac{2x}{4} = \dfrac{3(x - 4)}{2}$.

Solution

1. $4\left(\dfrac{2x}{4}\right) = 4\left[\dfrac{3(x-4)}{2}\right]$ (multiplying both sides by 4)
   or $2x = 6(x - 4)$
2. $2x = 6x - 24$ (removing parentheses)
3. $-4x = -24$ (subtracting $6x$ from both sides)
4. $x = 6$ (dividing both sides by -4)

Check: $\dfrac{2(6)}{4} = \dfrac{3(6-4)}{2}$ since $3 = 3$

EXAMPLE 5

Solve $\dfrac{3x}{5} + 2 = x - 3$.

Solution

$$3x + 5 \cdot 2 = 5(x - 3)$$
$$3x + 10 = 5x - 15$$
$$-2x + 10 = -15$$
$$-2x = -25$$
$$x = \tfrac{25}{2}$$

Check: $\dfrac{3(\tfrac{25}{2})}{5} + 2 = \tfrac{25}{2} - 3$ since $\tfrac{15}{2} + 2 = \tfrac{19}{2}$

Some linear equations contain decimal coefficients. We may write a decimal as a fraction with denominator 10, 100, or some higher power of 10. For example, 0.08 can be written as $\tfrac{8}{100}$. To solve an equation with one decimal coefficient, the first step is to multiply both sides of the equation by the appropriate power of 10 to remove the decimal.

EXAMPLE 6

Solve $2x + 0.45 = -3x$.

Solution

$$100(2x + 0.45) = 100(-3x)$$
$$200x + 45 = -300x$$
$$500x + 45 = 0$$
$$500x = -45$$
$$x = -\tfrac{45}{500} \text{ or } x = -0.09$$

Check: $2(-0.09) + 0.45 = -3(-0.09)$ since $0.27 = 0.27$

Linear Equations and Inequalities in One Variable

If more than one decimal is contained in an equation, we multiply both sides of the equation by the power of 10 that will remove all decimals.

EXAMPLE 7

Solve $0.4x + 0.37 = 0.21x - 4$.

Solution

$$100(0.4x + 0.37) = 100(0.21x - 4)$$
$$40x + 37 = 21x - 400$$
$$19x = -437$$
$$x = -23$$

Check: $0.4(-23) + 0.37 = 0.21(-23) - 4$ since $-9.2 + 0.37$
$= -4.83 - 4$

Exercise 5-2

Solve the following equations.

1. $x + 3 = 12$
2. $16x = -32$
3. $x + 8 = 7$
4. $-5x = 30$
5. $x + 7 = 12$
6. $-2x = -8$
7. $x + 10 = 6$
8. $16y = 4$
9. $y - 6 = 4$
10. $25x = 15$
11. $y - 7 = 7$
12. $\frac{y}{2} = 6$
13. $z - 2 = -3$
14. $\frac{x}{4} = 7$
15. $x - 4 = -2$
16. $\frac{y}{3} = -7$
17. $x + 5 = -2$
18. $\frac{x}{-4} = -6$
19. $x + 8 = -9$
20. $4x = 12$
21. $2x - 3 = 5$
22. $3x + 5 = 8$
23. $4x + 2 = 5$
24. $7x - 3 = 2$
25. $3y - 4 = 8$
26. $4y + 5 = 13$
27. $2z + 5 = 3$
28. $5y + 6 = 1$
29. $2x - 3 = 4$
30. $3y - 2 = 5$
31. $3x + 2 = 2x + 4$
32. $5z + 6 = 3z - 4$

33. $2x - 4 = x + 1$
34. $x + 1 = 4 - x$
35. $5x - 12 = 12x$
36. $16x - 1 = 12x + 3$
37. $4x - 7 = 8x - 2$
38. $3x + 2 = 3 - x$
39. $4x - 1 = 3(x - 1)$
40. $3x + 4 = 4(x - 1)$
41. $15 - x = 3(x - 1)$
42. $2(x - 1) = 3x + 4$
43. $3(x - 2) = 2x - 4$
44. $4x - 3 = 3(x - 4)$
45. $3(x - 2) = 4(3 - x)$
46. $2(3x + 2) = 3(3x - 1)$
47. $3x + \frac{1}{2} = 8$
48. $\frac{x}{2} + 3 = 4$
49. $\frac{x}{2} + 4 = \frac{1}{3}$
50. $y + \frac{1}{3} = \frac{2}{5}$
51. $\frac{5x}{2} - 4 = 2x - 7$
52. $5 - \frac{x}{4} = x - 1$
53. $\frac{2x}{3} - 1 = \frac{x - 2}{2}$
54. $\frac{4y}{3} + 1 = \frac{y - 2}{5}$
55. $\frac{2 - y}{3} = y + \frac{1}{2}$
56. $6z + 1 = \frac{3}{4} - \frac{z}{3}$
57. $\frac{6y - 1}{2} = \frac{5y - 10}{3}$
58. $\frac{3x + 2}{4} = \frac{5x + 4}{3}$
59. $\frac{5x - 1}{2} = 3(x - 1)$
60. $\frac{4(x + 2)}{3} = x + 3$
61. $\frac{6x + 5}{2} = \frac{5(2 - x)}{3}$
62. $\frac{7x - 1}{3} = \frac{2(x + 3)}{5}$
63. $0.4x = 16$
64. $0.8x = 24$
65. $0.4x + 5 = 9$
66. $0.3x + 2 = 11$
67. $2x + 0.6 = 4$
68. $3x - 5 = 0.4$
69. $0.5x = 0.6$
70. $0.7x = 1.4$
71. $0.3x + 0.4 = 1$
72. $0.4x + 0.5 = 2.1$
73. $0.2x + 0.16 = 4$
74. $5x + 0.4 = 0.45$
75. $0.7x + 0.36 = 0.2x + 0.06$
76. $0.8x + 0.07 = x - 0.1$
77. $2.1x + 0.3 = 0.3x - 3.3$
78. $0.76x - 0.002 = x + 0.01$

5-3 Literal equations

OBJECTIVE: After completing this section, you should be able to solve linear equations involving more than one literal number.

If an equation has more than one variable, we may solve for any one variable in terms of the others. The procedures used to solve these equations are identical to those for other linear equations.

EXAMPLE 1

Solve $4c + 2x = 5b$ for x.

Solution

$2x = 5b - 4c$ (subtracting $4c$ from both sides)

$x = \dfrac{5b - 4c}{2}$ (dividing both sides by 2)

EXAMPLE 2

Solve $4(x - b) = 3c$ for x.

Solution

$4x - 4b = 3c$ (multiplying $x - b$ by 4)

$4x = 4b + 3c$ (adding $4b$ to both sides)

$x = \dfrac{4b + 3c}{4}$ (dividing both sides by 4)

Note: Some students may be inclined to "cancel" the 4's in the solution above. But this is not correct. To eliminate the 4 in the denominator, we must divide *both* terms of the numerator by 4. Since 4 will not divide into $3c$, we cannot simplify the solution any further.

EXAMPLE 3

Solve $2(2x - b) = \dfrac{2c}{3}$ for x.

Solution

$6(2x - b) = 2c$ (multiplying both sides of the equation by 3)

$12x - 6b = 2c$ (removing parentheses)

$12x = 6b + 2c$ (adding $6b$ to both sides)

$x = \dfrac{6b + 2c}{12}$ (dividing both sides by 12)

$= \dfrac{2(3b + c)}{2 \cdot 6}$

$= \dfrac{3b + c}{6}$

Note: Here both the numerator and denominator were divisible by 2; therefore we could simplify the solution to $x = (3b + c)/6$. We cannot reduce it further because no factor of 6 divides both $3b$ and c.

Formulas are literal equations, and we may solve for any letter in the formula by assuming that it is a variable in the equation.

EXAMPLE 4

A formula for selling price is $S = C + O + P$, where S is selling price, C is cost, O is overhead, and P is profit. Solve this formula for O, the overhead.

Solution

Subtracting C and P from both sides, we get

$$S - C - P = O, \quad \text{so} \quad O = S - C - P$$

EXAMPLE 5

The formula for the perimeter of a rectangle is $P = 2l + 2w$. Solve the formula for l.

Solution

$$P - 2w = 2l$$

$$\frac{P - 2w}{2} = l$$

Thus

$$l = \frac{P - 2w}{2}$$

EXAMPLE 6

A formula for velocity is $V = K + 32t$. Solve the formula for t.

Solution

$$V - K = 32t$$

$$\frac{V - K}{32} = t$$

Thus
$$t = \frac{V - K}{32}$$

Exercise 5-3

Solve the following equations for x.

1. $5x - b = 2$
2. $2x - 4c = 4$
3. $3x - a = 4$
4. $3x + a = b$
5. $3x - b = 2x - 3b$
6. $5x + 2a = 2x - a$
7. $4x + b = 2b - x$
8. $3x - a = a - 2x$
9. $5m - x = 3m + 3x$
10. $x - 2b = b - 3x$
11. $4x - 8 = 2x + b$
12. $2x + 4a = 3x - 2b$

Solve the following equations for y.

13. $3y - 2n = 4y - n$
14. $4y + 6b = 2y - 7b$
15. $\frac{3y}{n} = y - n$
16. $2y - \frac{a}{b} = y + b$
17. $3(y - a) = 2y + a$
18. $3y + b = 4(y - 2b)$
19. $ay + b = 2b$
20. $2c - by = 2by + 4c$
21. Solve the formula $d = rt$ for t.
22. Solve the formula $d = rt$ for r.
23. Solve the formula $I = Prt$ for P.
24. Solve the formula $I = Prt$ for t.
25. Solve the formula $F = \frac{9}{5}C + 32$ for C.
26. Solve the formula $S = C + O + P$ for P.
27. Solve the formula $A = \frac{1}{2}bh$ for b.
28. Solve the formula $A = lw$ for w.

5-4 Word problems

OBJECTIVE: After completing this section, you should be able to translate stated problems into algebraic equations and solve them.

It is very important to learn how to translate verbal problems to equations, because most problems given to us in our everyday life are given verbally. Some helpful guidelines are given below.

1. Read the problem carefully to determine what you are asked to find as well as what information you are given.
2. Represent the quantity you are asked to find by a variable (such as x).
3. Translate the information of the problem into algebraic expressions.
4. Find two algebraic expressions that are equal and write them in an equation.
5. Solve the equation.
6. Check the solution in the original problem.

EXAMPLE 1

The sum of a number and 6 is equal to 3 times the number. Find the number.

Solution

Let x be the number.
The sum of the number and 6 is $x + 6$, and 3 times the number is $3x$.
Then the problem says $x + 6 = 3x$.
Solving, we find $6 = 2x$ or $x = 3$.

Check: $3 + 6$ is 3 times 3.

EXAMPLE 2

John Keller borrowed $12,000, part from the Port Royal Trust Co. and part from the Eastman Credit Union. If he borrowed 3 times as much from the trust company as he did from the credit union, how much did he borrow from each?

Solution

Let x represent the smaller amount.
Then $3x$ will represent the larger amount.
Then the total amount borrowed is $x + 3x = \$12,000$.
Solving, we find $4x = \$12,000$, or $x = \$3000$.
Thus, he borrowed $x = \$3000$ from the credit union and $3x = \$9000$ from the trust company.

EXAMPLE 3

Twenty years from now Jack will be 3 times his present age. What is his present age?

Solution

Let x represent Jack's present age.
His age in 20 years will be $x + 20$.
His age in 20 years will also be $3x$.
Thus $3x = x + 20$.
Solving, we find $2x = 20$ or $x = 10$.

Check: $3(10) = 10 + 20$.

EXAMPLE 4

A pipe which is 18 feet long is cut in two pieces so that one piece is twice the length of the other. What are the lengths of the pieces?

Solution

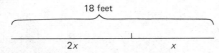

Let the shorter piece be represented by x and the longer by $2x$. The sum of the two pieces is 18 feet. Thus

$$x + 2x = 18$$

Solving the equation gives $3x = 18$ or $x = 6$. Thus the shorter piece is 6 feet and the longer piece is 12 feet.

Check: The sum of the two pieces is 18 feet.

EXAMPLE 5

Wallace Jones has $14,000 invested in two savings plans. One plan earns a 6-percent return and the other earns a 9-percent return. If his total return from the two investments is $1080 per year, how much does he have invested in each savings plan?

Solution

Let the amount invested at 6 percent be represented by x.
Then the amount invested at 9 percent is $\$14{,}000 - x$ since his total investment is $14,000.
The return from the 6 percent investment is $0.06x$.
The return from the 9 percent investment is $0.09(\$14{,}000 - x)$.
Thus the return from the two investments is

$$0.06x + 0.09(\$14{,}000 - x)$$

Then $0.06x + 0.09(\$14{,}000 - x) = \$1080$ (because his total return is $1080).
Solving the equation gives $0.06x + \$1260 - 0.09x = \$1080$ or $-0.03x = -\$180$.
Thus $x = \$6000$ and $\$14{,}000 - x = \$8000$.

Check: $0.06(\$6000) + 0.09(\$8000) = \$360 + \$720 = \$1080$

Exercise 5-4

Solve the following problems.

1. The sum of a number and 6 is 12. What is the number?
2. The sum of twice a number and 9 is 17. What is the number?
3. A number minus 16 is 37. What is the number?
4. Twice a number minus 7 is 53. What is the number?
5. Three times a number minus 14 is equal to the number. What is the number?
6. The sum of a number and 12 is equal to 3 times the number. What is the number?
7. When a number is divided by 3 and the quotient is decreased by 3, the result is 5. What is the number?
8. The length of a rectangle is 4 times the width. If the perimeter of the rectangle is 30 inches, what is the length and width?
9. The width of a rectangle is 15 feet and its area is 600 square feet. What is its length?
10. The sum of two numbers is 17. Find the numbers if twice the smaller is 4 more than the larger.
11. Jack has 5 more pens than John. If we double the number John has, he would then have 4 more than Jack. How many pens does John have?
12. Al has $3 more than Ed. If we double the amount Al has, he will then have $15 more than Ed. How many dollars does Ed have?
13. Bob is two years older than Al. Three years ago Al was half Bob's age. How old is each now?
14. The sum of two numbers is 19. The difference between twice the smaller and the larger is 8. What are the numbers?
15. A bank loaned $118,500 to two companies. It loaned one company $34,500 more than the other. How much did it loan each one?

16. A father's age is 8 years more than twice his son's age. The difference between their ages is 21 years. Find the age of each.
17. A board 16 feet long is cut into two pieces so that one piece is 3 times as long as the other. What is the length of each piece?
18. A used car dealer sold a car for $1550 and made a profit of $\frac{1}{4}$ of his cost. What was his cost?
19. A man has $23,500 invested in two rental properties. One earns 5 percent on the investment, and the other yields 6 percent. His total income from the two properties is $1275. How much is his income from each property?
20. Mr. Jackson borrowed money from his bank and on his life insurance to start a business. His interest rate on the bank loan is 5 percent, and his rate on the insurance loan is 6 percent. If the total amount borrowed is $10,000 and his total yearly interest payment is $545, how much did he borrow from the bank?

5-5 Mixture problems

OBJECTIVE: After completing this section, you should be able to solve stated problems involving mixtures.

In mixture problems it is important to remember that the **percent of the ingredient is equal to the amount of ingredient divided by the total amount of mixture.** The amount of ingredient in the final mixture is the sum of the ingredients from the two parts, and the total amount of substance is the sum of that present in both parts.

EXAMPLE 1

How much alcohol must be added to 100 cm³ (cubic centimeters) of a 40% alcohol solution to yield a 60% solution?

Solution

The 100 cm³ of 40% solution has 40 cm³ of alcohol. If x cm³ of alcohol is added, the new solution has a total of $(100 + x)$ cm³ and has $(40 + x)$ cm³ of alcohol. Then since the new mixture is a 60% mixture,

$$60\% = \frac{40 + x}{100 + x}$$

Solving the equation
$$0.60 = \frac{40 + x}{100 + x}$$
gives
$$60 + 0.60x = 40 + x$$
or
$$6000 + 60x = 4000 + 100x, \quad \text{so} \quad 2000 = 40x$$
Thus
$$50 = x$$

Check: Adding 50 cm³ of alcohol would make 90 cm³ of alcohol in 150 cm³ of solution which is a $\frac{90}{150} = 60\%$ solution.

EXAMPLE 2

How much cream that is 24% butterfat must be added to 500 pounds of skim milk which is 2% butterfat in order to obtain milk which is 4% butterfat?

Solution

The 500 pounds of 2% milk has 10 pounds of butterfat. The amount of cream to be added is x pounds, which has $0.24x$ pounds of butterfat. Then the new mixture has a total weight of $500 + x$ pounds with $10 + 0.24x$ pounds of butterfat. Thus

$$4\% = \frac{10 + 0.24x}{500 + x}$$

So
$$0.04(500 + x) = 10 + 0.24x$$
or
$$20 + 0.04x = 10 + 0.24x$$
So
$$2000 + 4x = 1000 + 24x \quad \text{or} \quad 1000 = 20x$$
Thus
$$x = 50 \text{ pounds}$$

Check: Adding 50 pounds of 24% mixture would increase the total to 550 pounds and the butterfat content to $10 + 0.24(50) = 22$ pounds; so the mixture is a $\frac{22}{550} = 4\%$ solution.

In dealing with mixture problems involving prices, the cost per pound (or other unit of measure) of the new mixture is the total cost divided by the total number of pounds purchased.

Linear Equations and Inequalities in One Variable

EXAMPLE 3

A store has some candy for 60 cents a pound and some for 90 cents a pound. How much of each must be purchased to have 10 pounds of a mixture worth 78 cents per pound?

Solution

Let us assume we purchased x pounds of the 60-cent candy and $10 - x$ pounds of the 90-cent candy. The x pounds of 60-cent candy costs $60x$. The $10 - x$ pounds of 90-cent candy costs $90(10 - x)$. Thus the cost per pound of the mixture is

$$\frac{60x + 90(10 - x)}{10}$$

Setting this equal to 78 cents and solving for x give the number of pounds of 60-cent candy purchased.

$$78 = \frac{60x + 90(10 - x)}{10}$$

So

$$780 = 60x + 900 - 90x, \quad \text{or} \quad -120 = -30x$$

Thus $x = 4$; so 4 pounds of 60-cent candy and $10 - 4 = 6$ pounds of 90-cent candy must be purchased.

Check: 4 pounds of 60-cent candy costs \$2.40 and 6 pounds of 90-cent candy costs \$5.40; so the total cost of the ten pounds is \$7.80, or 78 cents per pound.

Exercise 5-5

1. How many pints of water must be added to 5 pints of solution that is 30% alcohol to obtain a 20% solution?
2. How many gallons of water must be added to 6 gallons of an 80% alcohol solution to obtain a 60% solution?
3. How many quarts of pure alcohol must be added to 15 quarts of a 60% alcohol solution to obtain an 80% solution?
4. How many pints of pure acid must be added to 10 pints of a 30% acid solution to give a 50% solution?
5. How many quarts of an 80% alcohol solution must be added to 10 quarts of a 50% alcohol solution to give a 70% solution?
6. How many ounces of pure silver must be added to 20 ounces of an alloy that is 40% pure silver to make an alloy that is 60% pure silver?

7. How much copper must be added to 100 pounds of an alloy that is 40% copper to produce an alloy that is 60% copper?
8. How many tons of an alloy that is 50% zinc must be added to 40 tons of an alloy that is 30% zinc to obtain an alloy that is 40% zinc?
9. How many gallons of a mixture containing 20% acid should be added to 12 gallons of a 30% solution to give a 25% solution?
10. Two blocks of alloy containing 20% and 30% tin, respectively, were melted together to produce 50 pounds of an alloy which was 28% tin. How much did each of the original blocks weigh?
11. A store has one kind of candy worth 98 cents a pound and another kind worth $1.28 a pound. How much of each kind must he mix to get 4 pounds worth $1.18 a pound?
12. A wholesaler has 200 pounds of coffee worth $1.24 per pound. How many pounds of 90-cent coffee must he mix in to obtain a blend worth $1.00 per pound?
13. How much candy costing 76 cents per pound must be added to 5 pounds of candy costing 88 cents per pound to get a mix which costs 81 cents per pound?
14. A store has 100 pounds of mixed nuts priced to sell at $1.20 per pound. In order to reduce the price, peanuts priced at 75 cents per pound are added. How many pounds of peanuts should be added to bring the price down to 90 cents a pound?
15. An 18-quart radiator contains a solution that is 25% antifreeze. How much must be drained off and replaced by pure antifreeze to obtain a solution that is $33\frac{1}{3}$% antifreeze?
16. A 15-quart radiator contains a solution that is 10% antifreeze. How much must be drained off and replaced by pure antifreeze to obtain a solution that is 28% antifreeze?

5-6 Linear inequalities in one variable

OBJECTIVE: After completing this section, you should be able to graph and solve linear inequalities in one variable.

An **inequality** is a statement that one quantity is greater than (or less than) another quantity. $3x - 2 > 2x + 1$ is a first-degree (linear) inequality, which states that the left member is greater than the right member. Certain values of the variable will satisfy the inequality. The values form the solution set of the inequality. For example, we know that 4 is in the solution set of $3x - 2 > 2x + 1$ because $3 \cdot 4 - 2 > 2 \cdot 4 + 1$. But 2 is not in the solution set because $3 \cdot 2 - 2 \not> 2 \cdot 2 + 1$. **Solving** an inequality means finding its solution set, and two inequalities are **equivalent** if they have the same solution set. As with equations, we find the solutions to inequalities by finding equivalent inequalities from which the solutions can be easily seen. We use the following properties to reduce an inequality to a simple equivalent inequality.

Substitution Property of Inequalities. The inequality formed by substituting one expression for an equal expression in an equation is equivalent to the original inequality.

Example: $5x - 4x < 6$ is equivalent to $x < 6$ because $5x - 4x = x$. Thus $x < 6$ is the solution to $5x - 4x < 6$; that is, any number less than 6 is a solution to the inequality. We can write the solution set in set notation as $\{x: x < 6\}$.

Addition Property of Inequalities. The inequality formed by adding the same quantity to both sides of an inequality is equivalent to the original inequality.

Example: $x - 4 > 6$ is equivalent to $x > 10$ because 4 was added to both sides of $x - 4 > 6$. Thus $x > 10$ is the solution to $x - 4 > 6$; that is, any number greater than 10 is in the solution set of $x - 4 > 6$.

Note: This property permits us to use subtraction as well as addition to find equivalent inequalities. For example, adding -5 to both sides (or subtracting 5 from both sides), we see that $x + 5 < 12$ is equivalent to $x < 7$.

Although the properties of inequalities for substitution, addition, and subtraction are identical to those for equations, there is an important difference regarding the multiplication (and division) property for inequalities. The difference occurs when both sides of an inequality are multiplied (or divided) by a negative number. Thus we state two multiplication properties for inequalities.

Multiplication Property I for Inequalities. The inequality formed by multiplying (or dividing) both sides of an inequality by the same *positive* quantity is equivalent to the original inequality.

Example: $\frac{1}{2}x > 8$ is equivalent to $x > 16$ because both sides were multiplied by 2 (a positive number). $3x < 6$ is equivalent to $x < 2$ because both sides were divided by 3 (a positive number).

Multiplication Property II for Inequalities. The inequality formed by multiplying (or dividing) both sides of an inequality by a *negative* number and *reversing the sense* of the inequality is equivalent to the original inequality.

Example: $-3 < 2$ is equivalent to $3 > -2$ because multiplying both sides of $-3 < 2$ by -1 *and* reversing the inequality sign gives $3 > -2$. The inequality $-x < 6$ is equivalent to $x > -6$ (multiplying both sides by -1). Thus the solution set for $-x < 6$ contains $-5, -3, -1, 4$, and 10, to name a few. Checking shows $-(-5) < 6, -(10) < 6$, and so on.

We see also that $-3x > -27$ is equivalent to $x < 9$ because dividing both sides by -3 *and* reversing the inequality gives $x < 9$.

Note: We do not get an equivalent (or any) inequality if we multiply both sides of an inequality by 0. Clearly we cannot divide both sides by 0.

We may graph the solutions to inequalities in one unknown on the real-number line. The example, the graph of $x < 2$ consists of all points to the left of 2 on the number line. The open circle on the graph indicates that all points up to, but *not* including, 2 are in the solution set.

EXAMPLE 1

Solve the inequality $\frac{x}{-2} < -4$ and graph the solution set.

Solution

The solution set of $\frac{x}{-2} < -4$ is the same as that of $-2\left(\frac{x}{-2}\right) > -2(-4)$ and $x > 8$. Thus the solution set is $\{x: x > 8\}$. The graph of the solution set follows.

Some inequalities will require several operations to find their solution set. In this case, the order in which the operations are performed is the same as that used in solving linear equations.

EXAMPLE 2

Solve the inequality

$$2(x - 4) < \frac{x - 3}{3}$$

Solution

a. Remove fractions: $6(x - 4) < x - 3$
b. Remove parentheses: $6x - 24 < x - 3$
c. Perform additions and subtractions: $5x < 21$
d. Divide to find the solution: $x < \frac{21}{5}$

Check: $x = 4$ satisfies the inequality $2(4 - 4) < (4 - 1)/3$. $x = 5$ does not satisfy the inequality $2(5 - 4) \not< (5 - 3)/3$. Thus $x < \frac{21}{5}$ is a reasonable solution.

EXAMPLE 3

Solve the inequality

$$3(x + 2) > \frac{x}{4} + 1$$

and graph the solution set.

Solution

a. $12(x + 2) > x + 4$
b. $12x + 24 > x + 4$
c. $11x > -20$
d. $x > -\frac{20}{11}$

Check: $x = -1$ does satisfy the inequality, but $x = -2$ does not.

We may also solve inequalities of the form $a \leq b$. This means "a is less than b or $a = b$." The solution to $2x \leq 4$ is $x \leq 2$, since $x < 2$ is the solution of $2x < 4$ and $x = 2$ is the solution of $2x = 4$.

EXAMPLE 4

Solve the inequality $3x - 2 \leq 7$.

Solution

This inequality states that $3x - 2 = 7$ or that $3x - 2 < 7$. By solving in the usual manner, we get $3x \leq 9$ or $x \leq 3$.

Then $x = 3$ is the solution to $3x - 2 = 7$, and $x < 3$ is the solution to $3x - 2 < 7$; so the solution set for $3x - 2 \leq 7$ is $\{x: x \leq 3\}$.

The graph of the solution set includes the point $x = 3$ and all points $x < 3$.

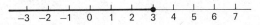

EXAMPLE 5

Solve the inequality

$$4x - 3 \geq \frac{x}{2}$$

Solution

$8x - 6 \geq x$ has the same solution set; $7x \geq 6$ has the same solution set; $x \geq \frac{6}{7}$ has the same solution set; so the solution set is $\{x: x \geq \frac{6}{7}\}$. The graph of the solution set follows.

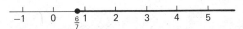

We use the notation $-2 < x < 4$ to state that x satisfies the two conditions $-2 < x$ and $x < 4$. This **compound inequality** represents an **interval** on the number line. The graph of this compound inequality, shown below, is called an **open interval** because neither **end point** (-2 or 4) is included in the graph.

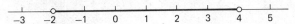

The graph of $-1 \leq x \leq 3$ is a **closed interval** because both end points are included.

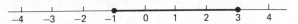

The graph of $-2 < x \leq 3$ is a **half-open interval** because one end point is included and one is not.

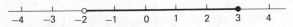

We can solve the compound inequality $5 < 2x - 1 < 9$ by solving the two inequalities $5 < 2x - 1$ and $2x - 1 < 9$ and writing the solution as the intersection of their solution sets. Solving $5 < 2x - 1$ gives $6 < 2x$ or $3 < x$. Solving $2x - 1 < 9$ gives $2x < 10$ or $x < 5$. The intersection of $\{x: x > 3\}$ and $\{x: x > 5\}$ is $\{x: x > 3 \text{ and } x < 5\}$.

This solution set can be written as $\{x: 3 < x < 5\}$. The solution set's graph is the open interval $3 < x < 5$.

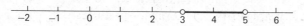

EXAMPLE 6

Solve and graph the solution set of $4x - 8 \leq 2x < 5x + 6$.

Solution

$4x - 8 \leq 2x < 5x + 6$ means $4x - 8 \leq 2x$ **and** $2x < 5x + 6$. Solving $4x - 8 \leq 2x$ gives $-8 \leq -2x$ or $4 \geq x$. Solving $2x < 5x + 6$ gives $-3x < 6$ or $x > -2$. Since x must satisfy both inequalities, the solution set is $\{x: x \leq 4 \text{ and } x > -2\}$. Thus the solution set is in the interval $-2 < x \leq 4$. The half-open interval is graphed below.

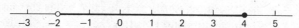

Exercise 5-6

Solve the following inequalities.

1. $x + 2 < 4$
2. $x - 3 > 5$
3. $x - 3 \geq 2x + 1$
4. $2x - 1 \leq 3x + 4$
5. $3(x - 1) < 2x - 1$
6. $2(x + 1) > x - 1$
7. $2(x - 1) - 3 > 4x + 1$
8. $7x + 4 \leq 2(x - 1)$
9. $\dfrac{x}{3} > x - 1$
10. $\dfrac{x - 3}{4} \geq 2x$
11. $\dfrac{3(x - 1)}{2} \leq x - 2$
12. $\dfrac{x - 1}{2} + 1 > x + 1$

Graph the solutions to the following inequalities.

13. $2x + 1 < x - 3$
14. $3x - 1 \geq 2x + 2$
15. $3(x - 1) < 2x$
16. $3(x + 2) \geq 4x + 1$
17. $2x + \tfrac{1}{2} \geq x - 3$
18. $5x - 2 > 3(x + 3)$

Determine if the graphs of the following inequalities are open intervals, closed intervals, or half-open intervals.

19. $-3 \leq x \leq 5$
20. $-5 < x < -3$
21. $-4 < x \leq 3$
22. $-6 \leq x < -4$
23. $4 \leq x \leq 6$
24. $-2 < x \leq -1$
25. $4 < x < 9$
26. $5 \leq x \leq 8$

Solve and graph the solution set of the following.

27. $5x + 4 < 6x < 5x + 7$
28. $x - 5 < 2x < x + 2$
29. $2x - 4 < 3x \leq 2x + 3$
30. $2x - 3 \leq 5x < 2x$
31. $3x < 4x - 1 < 5x$
32. $2x + 1 < x - 3 < 3x + 6$
33. $4x - 1 \leq 4 - x \leq x + 2$
34. $2x \leq 3 - x \leq 4$
35. $3x - 2 \leq 2x \leq 5x - 1$
36. $5x - 3 \leq 6x - 1 \leq x - 6$

5-7 Equations and inequalities involving absolute values

OBJECTIVES: After completing this section, you should be able to:
a. Solve equations involving absolute values.
b. Solve inequalities involving absolute values.

In Section 1-6, we stated that the absolute value of every number except 0 is positive and that the absolute value of 0 is 0. Thus $|5| = 5$, $|-4| = 4$, and $|0| = 0$. But if we are asked to evaluate the absolute value of an expression involving a variable, we cannot be sure if the expression is positive, negative, or 0. Thus we must consider two cases (letting a represent the expression).

1. The expression is positive or 0: We say

$$|a| = a \quad \text{if } a \geq 0$$

2. The expression is negative: If the expression is negative, we can make it positive by multiplying it by -1. We say

$$|a| = -a \quad \text{if } a < 0$$

Note: $-a$ represents a positive number if a represents a negative number.

We may combine these two cases to define the **absolute value**:

$$|a| = \begin{cases} a & \text{if } a \geq 0 \\ -a & \text{if } a < 0 \end{cases}$$

EXAMPLE 1

Evaluate the following absolute values.

a. $|4|$
b. $|0|$
c. $|-3|$
d. $|x - 1|$

Solution

a. $|4| = 4$
b. $|0| = 0$
c. $|-3| = -(-3) = 3$
d. $|x - 1| = x - 1$ if $x - 1 \geq 0$ and $|x - 1| = -(x - 1) = -x + 1$ if $x - 1 < 0$

We frequently have two solutions to equations involving absolute values.

EXAMPLE 2

Solve the equation $|x + 2| = 4$.

Solution

Because $|x + 2|$ is equal to either $x + 2$ or $-(x + 2)$, depending on whether $x + 2 \geq 0$ or $x + 2 < 0$, we consider two cases.

Case I: $x + 2 \geq 0$. Then $|x + 2| = 4$ is equivalent to $x + 2 = 4$; so $x = 2$ is a solution of $|x + 2| = 4$.

Case II: $x + 2 < 0$. Then $|x + 2| = 4$ is equivalent to $-(x + 2) = 4$; so $-x - 2 = 4$. Thus $-x = 6$ or $x = -6$ is a solution of $|x + 2| = 4$. Thus, the solution set for $|x + 2| = 4$ contains -6 and 2.

Check: $|(-6) + 2| = |-4| = 4$; $|(2) + 2| = |4| = 4$

EXAMPLE 3

Solve $|2x - 1| = 3x + 4$ for x.

Solution

Case I: $2x - 1 \geq 0$. Then $2x - 1 = 3x + 4$. Solving gives $-5 = x$.

Check: $|2(-5) - 1| = |-11| = 11$ and $3(-5) + 4 = -11$; so $x = -5$ does not check. Note that if $x = -5$, $2x - 1 = -11$, which is **not** ≥ 0. Thus -5 is **not** a solution to $|2x - 1| = 3x + 4$.

Case II: $2x - 1 < 0$. Then $|2x - 1| = 3x + 4$ is equivalent to $-(2x - 1) = 3x + 4$ or $-2x + 1 = 3x + 4$. Solving gives $-3 = 5x$ or $-\frac{3}{5} = x$.

Check: $|2(-\frac{3}{5}) - 1| = |-\frac{11}{5}| = \frac{11}{5}$ and $3(-\frac{3}{5}) + 4 = -\frac{9}{5} + 4 = \frac{11}{5}$. Thus the solution set contains only $-\frac{3}{5}$.

We may solve inequalities involving absolute values in a manner similar to that used for equations.

EXAMPLE 4

Solve the following inequality and graph the solution set: $|2x + 1| < 5$.

Solution

Case I: $2x + 1 \geq 0$ (so $x \geq -\frac{1}{2}$). Then $|2x + 1| < 5$ is equivalent to $2x + 1 < 5$. Solving gives $2x < 4$, or $x < 2$. Thus if $2x + 1 \geq 0$ ($x \geq -\frac{1}{2}$), the inequality is true when $x < 2$. Thus the solution set of Case I is $\{x: -\frac{1}{2} \leq x \text{ and } x < 2\}$ or $\{x: -\frac{1}{2} \leq x < 2\}$.

Case II: $2x + 1 < 0$ (so $x < -\frac{1}{2}$). Then $|2x + 1| < 5$ is equivalent to $-(2x + 1) < 5$ or $-2x - 1 < 5$. Solving gives $-2x < 6$ or $x > -3$. Thus if $2x + 1 < 0$ ($x < -\frac{1}{2}$), the inequality $x > -3$ is true. Thus the solution set for Case II is $\{x: -3 < x < -\frac{1}{2}\}$.

The complete solution of $|2x + 1| < 5$ is the union of the solutions in the two different cases. The complete solution is

$\{x: -\frac{1}{2} \leq x < 2\} \cup \{x: -3 < x < -\frac{1}{2}\}$

The union of these two sets is

$\{x: -3 < x < 2\}$

Thus the real numbers in the open interval $-3 < x < 2$ are in the solution set. The graphs of the solutions to Case I and Case II follow.

Their union gives the graph of the solution set:

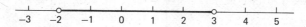

EXAMPLE 5

Solve $|3x + 1| \geq 4x + 6$.

Solution

Case I: $3x + 1 \geq 0$ $(x \geq \frac{1}{3})$. Then $|3x + 1| \geq 4x + 6$ is equivalent to $3x + 1 \geq 4x + 6$. Solving gives $-x \geq 5$ or $x \leq -5$. Thus if $3x + 1 \geq 0$ $(x \geq -\frac{1}{3})$, the inequality is true when $x < -5$. Thus the solution set for Case I is $\{x: -\frac{1}{3} \leq x$ and $x \leq -5\}$. But there is no number greater than $-\frac{1}{3}$ and less than -5. Thus the solution set for Case I is the empty set $\emptyset$.

Case II: $3x + 1 < 0$ $(x < -\frac{1}{3})$. Then $|3x + 1| \geq 4x + 6$ is equivalent to $-(3x + 1) \geq 4x + 6$ or $-3x - 1 \geq 4x + 6$. Solving gives $-7x \geq 7$ or $x \leq -1$. Thus if $3x + 1 < 0$ $(x < -\frac{1}{3})$, the inequality is true when $x \leq -1$. Thus the solution set for Case II is $\{x: x < -\frac{1}{3}$ and $x < -1\} = \{x: x < -1\}$.

The complete solution of $|3x + 1| \geq 4x + 6$ is the union of the solutions in the two different cases. The complete solution is

$$\emptyset \cup \{x: x < -1\} = \{x: x < -1\}$$

Thus the real numbers less than -1 are solutions to $|3x + 1| \geq 4x + 6$.

Exercise 5-7

Evaluate the following absolute values.

1. $|3|$
2. $|-8|$
3. $|-16|$
4. $|+4|$
5. $|0|$
6. $|-7|$
7. $|x|$ if $x > 0$
8. $|y|$ if $y < 0$
9. $|x - 3|$ if $x < 3$
10. $|y + 4|$ if $y \geq -4$

Solve the following equations.

11. $|x - 1| = 3$
12. $|x + 3| = 4$
13. $|y - 3| = 6$
14. $|z - 6| = 3$
15. $|x + 5| = 5$
16. $|x - 8| = 8$
17. $|x + 1| = 4x - 1$
18. $|2x + 2| = 3x - 1$
19. $|3y + 2| = y - 4$
20. $|z - 2| = 5z + 3$

Solve the following inequalities.

21. $|x| > 4$
22. $|x| < 9$
23. $|3x| < 6$
24. $|4x| > 2$
25. $|2x - 1| > 5$
26. $|3y + 2| < 8$
27. $|2x + 1| > 3x - 1$
28. $|5x - 1| < 4x + 3$
29. $|x - 3| \geq 2x + 1$
30. $|y + 2| \leq 3y - 2$

CHAPTER TEST

Are the following equations identities?

1. $2(x - 3) = 2x - 6$
2. $5x - 4 = 2(3x - 4)$

Write the following statements as equations.

3. The sum of twice a number and 6 is 14.
4. Twice a number minus 5 is 15.
5. The formula for the area of a triangle is $A = \frac{1}{2}bh$. Find the area of a triangle whose base is 4 feet and whose height is 8 feet.

Solve the following equations.

6. $x - 4 = 5$
7. $z + 4 = 8$
8. $y + 7 = 3$
9. $14x = 42$
10. $\frac{z}{4} = -16$
11. $5x - 3 = 12$
12. $3x + 4 = x - 6$
13. $5x = 3(x - 2)$

14. $\dfrac{x}{4} + 4 = \dfrac{5}{6}$

15. $\dfrac{3x}{2} = \dfrac{4(x-5)}{3}$

16. $0.5x - 0.76 = 0.3x + 1$

Solve for x.

17. $3x + 2b = 5x - 3b$
18. $5x - \dfrac{b}{2} = \dfrac{x-1}{3}$
19. Solve the formula $V = K + 32t$ for t.
20. The sum of twice a number and 9 is 35. What is the number?
21. One man weighs 25 pounds more than another. Their combined weights add up to 395 pounds. How much does each man weigh?
22. Jack has $5 more than Sam. If we double the amount Sam has, he will have $11 more than Jack. How much does each have?
23. How many liters of a 50-percent alcohol solution must be added to 10 liters of a solution that is 25 percent alcohol to obtain a 30-percent solution?

Solve the following inequalities.

24. $x - 3 < 2x + 4$
25. $\dfrac{x-2}{5} \leq 3x$
26. $3x - 1 \leq x - 1 < 2x + 1$
27. Solve $|3x - 4| = 2$ for x.
28. Solve $|4x - 1| < 2x + 3$ for x.

6

exponents and radicals

6-1 Exponents

OBJECTIVE: After completing this section, you should be able to use laws of exponents to find products and quotients of powers having the same base, and to raise products, quotients, and powers to higher powers.

As we learned in Chapter 2, an exponent indicates how many times a base should be taken as a factor. For example, a^5 indicates that a should be used as a factor five times. That is, $a^5 = a \cdot a \cdot a \cdot a \cdot a$. We derived rules for multiplication and division with powers of the same base. We state these rules again.

Rule 1. To multiply two powers having the same base, add the exponents.

$$a^m \cdot a^n = a^{m+n}$$

Rule 2a. To divide two powers having the same base when the exponent of the dividend is greater than the exponent of the divisor, subtract the exponent of the divisor from the exponent of the dividend and write the difference as the new exponent of the base.

$$a^m \div a^n = a^{m-n} \quad \text{(if } m > n \text{ and } a \neq 0\text{)}$$

Rule 2b. To divide two powers having the same base when the exponent of the dividend is less than the exponent of the divisor, subtract the exponent of the dividend from that of the divisor and write the quotient as a fraction that has 1 as its numerator and the base to the new exponent as its denominator.

$$a^m \div a^n = \frac{1}{a^{n-m}} \quad \text{(if } m < n \text{ and } a \neq 0)$$

Rule 2c. To divide two powers that have the same exponents in the dividend and divisor, write the quotient as 1:

$$a^m \div a^m = 1 \quad \text{(if } a \neq 0)$$

EXAMPLE 1

Perform the following operations.

a. $x^5 \cdot x^9$
b. $x^9 \div x^3$
c. $x^3 \div x^9$
d. $x^9 \div x^9$

Solution

a. $x^5 \cdot x^9 = x^{5+9} = x^{14}$
b. $x^9 \div x^3 = x^{9-3} = x^6$
c. $x^3 \div x^9 = \dfrac{1}{x^{9-3}} = \dfrac{1}{x^6}$
d. $x^9 \div x^9 = 1$

In Chapter 2 we also used the following rule.

Rule 3. To raise a product to a power, raise each factor to the power:

$$(ab)^m = a^m b^m$$

In addition, we see that

$$\left(\frac{a}{b}\right)^3 = \frac{a}{b} \cdot \frac{a}{b} \cdot \frac{a}{b} = \frac{a \cdot a \cdot a}{b \cdot b \cdot b} = \frac{a^3}{b^3}$$

which illustrates the following rule.

Rule 4. To raise an indicated quotient to a power, raise both the dividend and divisor to that power:

$$\left(\frac{a}{b}\right)^m = \frac{a^m}{b^m} \quad \text{if } b \neq 0$$

EXAMPLE 2

Compute

a. $(2xy)^3$

b. $\left(\dfrac{x}{y}\right)^4$

Solution

a. $(2xy)^3 = 2^3 x^3 y^3 = 8x^3 y^3$

b. $\left(\dfrac{x}{y}\right)^4 = \dfrac{x^4}{y^4}$

Now consider the problem $(x^2)^3$. Clearly $(x^2)^3 = x^2 \cdot x^2 \cdot x^2$. But $x^2 \cdot x^2 \cdot x^2 = x^6$. Thus $(x^2)^3 = x^6$, which illustrates Rule 5.

Rule 5. To raise to a power a number which has an exponent, multiply the exponents and write the base to the new exponent.

$$(a^m)^n = a^{mn}$$

EXAMPLE 3

Compute

a. $(x^2)^5$
b. $(2x^2 y^3)^4$

Solution

a. $(x^2)^5 = x^{10}$
b. $(2x^2 y^3)^4 = 2^4 x^8 y^{12} = 16 x^8 y^{12}$

EXAMPLE 4

Compute

a. $\left(\dfrac{a^2}{b^2}\right)^3$

b. $\left(\dfrac{2a^2}{3b^3}\right)^3$

Solution

a. $\left(\dfrac{a^2}{b^2}\right)^3 = \dfrac{a^6}{b^6}$

b. $\left(\dfrac{2a^2}{3b^3}\right)^3 = \dfrac{2^3 a^6}{3^3 b^9} = \dfrac{8a^6}{27b^9}$

Exercise 6-1

Compute the following.

1. $x^8 \cdot x^4$
2. $x^3 \cdot x^5$
3. $x^5 \cdot x^3$
4. $x^6 \cdot x^3$
5. $6x^3 \cdot 5xy$
6. $4x^2 \cdot 6x^3$
7. $4y^3 \cdot 3y$
8. $10x^3 \cdot 4x$
9. $x^{15} \div x^3$
10. $x^{16} \div x^4$
11. $x^6 \div x^6$
12. $y^3 \div y^3$
13. $x^6 \div x^8$
14. $z^2 \div z^3$
15. $y^4 \div y^6$
16. $m^3 \div m^6$
17. $(3m)^3$
18. $(-4x)^2$
19. $(mn)^4$
20. $(xy)^3$
21. $\left(\dfrac{x}{y}\right)^3$
22. $\left(\dfrac{w}{y}\right)^4$
23. $\left(\dfrac{-2}{w}\right)^3$
24. $\left(\dfrac{x}{4}\right)^2$
25. $(x^3)^3$
26. $(-y^2)^4$
27. $(x^5)^3$
28. $(w^3)^4$
29. $(3a^2)^3$
30. $(6b^3)^2$
31. $(6a^2 b)^3$
32. $(3x^3 y^4)^4$
33. $(-4ab^2)^2$
34. $(-2a^2 m^3)^3$
35. $\left(\dfrac{a^2}{b}\right)^3$
36. $\left(\dfrac{x^3}{y^2}\right)^4$
37. $\left(\dfrac{4}{a^2}\right)^2$
38. $\left(\dfrac{8}{a^3}\right)^3$
39. $\left(\dfrac{6a^2 b}{5x^2}\right)^2$
40. $\left(\dfrac{-3x^2 y^3}{7mn}\right)^2$

6-2 Zero and negative exponents

OBJECTIVES: After completing this section, you should be able to simplify algebraic expressions involving:

a. 0 as an exponent.
b. negative exponents.

Although it has not yet been discussed, it is possible to have negative exponents. If we divide two powers having the same base by subtracting the exponent of the divisor from the exponent of the dividend, we may sometimes get the base to a negative exponent. Until now we have not accepted negative exponents as meaningful. Now we want to assign them meaning by defining them in a way that will be consistent with the previous rules for exponents.

EXAMPLE 1

Compute $x^3 \div x^4$, $x \neq 0$.

Solution

Subtraction of the divisor's exponent from that of the dividend gives $x^{3-4} = x^{-1}$. But division using Rule 2b gives the quotient $1/x$. Thus $x^{-1} = 1/x$.

Consider the following series of powers of a ($\neq 0$) which result from division.

$$a^4 = \frac{a^5}{a} = a^{5-1}$$

$$a^3 = \frac{a^5}{a^2} = a^{5-2}$$

$$a^2 = \frac{a^5}{a^3} = a^{5-3}$$

$$a^1 = \frac{a^5}{a^4} = a^{5-4}$$

$$1 = \frac{a^5}{a^5} = a^{5-5} = a^0$$

$$\frac{1}{a} = \frac{a^5}{a^6} = a^{5-6} = a^{-1}$$

$$\frac{1}{a^2} = \frac{a^5}{a^7} = a^{5-7} = a^{-2}$$

$$\frac{1}{a^3} = \frac{a^5}{a^8} = a^{5-8} = a^{-3}$$

This example suggests two additional rules for exponents.

Rule 6. Any power with exponent 0 and a nonzero base is equal to 1:

$$a^0 = 1 \quad \text{if } a \neq 0$$

Rule 7. Any power with nonzero base a and negative exponent $-m$ is equal to $1/a^m$:

$$a^{-m} = \frac{1}{a^m}, \quad a \neq 0$$

EXAMPLE 2

Compute (for $x \neq 0$)

a. 4^0
b. $(4x)^0$
c. $4x^0$

Solution

a. $4^0 = 1$
b. $(4x)^0 = 1$
c. $4x^0 = 4 \cdot 1 = 4$

EXAMPLE 3

Write the following terms with positive exponents.

a. x^{-2}
b. $(2x)^{-1}$
c. $2(x^2)^{-2}$

Solution

a. $x^{-2} = \dfrac{1}{x^2}$

b. $(2x)^{-1} = \dfrac{1}{2x}$

c. $2(x^2)^{-2} = 2x^{-4} = \dfrac{2}{x^4}$

EXAMPLE 4

Compute

a. $4m^3 \cdot m^0 \quad (m \neq 0)$
b. $(2^0 m^2)^2$
c. $\dfrac{6p^3}{6^0}$

Solution

a. $4m^3 \cdot m^0 = 4m^3 \cdot 1 = 4m^3$
b. $(2^0 m^2)^2 = (1 \cdot m^2)^2 = m^4$
c. $\dfrac{6p^3}{6^0} = \dfrac{6p^3}{1} = 6p^3$

EXAMPLE 5

Compute

a. $x^5 \cdot x^{-3}$
b. $\dfrac{x^4}{x^{-2}}$
c. $(x^{-2}y^{-1})^{-3}$

Solution

a. $x^5 \cdot x^{-3} = x^{5+(-3)} = x^2$
b. $\dfrac{x^4}{x^{-2}} = x^{4-(-2)} = x^6$
c. $(x^{-2}y^{-1})^{-3} = x^{+6}y^{+3} = x^6 y^3$

EXAMPLE 6

Write the following terms with positive exponents.

a. $\left(\dfrac{x}{y}\right)^{-1}$

b. $\left(\dfrac{x}{y}\right)^{-2}$

c. $\left(\dfrac{x^{-2}}{y}\right)^{-1}$

Solution (two methods)

a. (i) $\left(\dfrac{x}{y}\right)^{-1} = \dfrac{x^{-1}}{y^{-1}} = \dfrac{\frac{1}{x}}{\frac{1}{y}} = \dfrac{1}{x} \cdot \dfrac{y}{1} = \dfrac{y}{x}$

or

(ii) $\left(\dfrac{x}{y}\right)^{-1} = \dfrac{1}{\frac{x}{y}} = 1 \cdot \dfrac{y}{x} = \dfrac{y}{x}$

Thus raising a fraction x/y to the -1 power gives its reciprocal y/x.

b. (i) $\left(\dfrac{x}{y}\right)^{-2} = \left[\left(\dfrac{x}{y}\right)^{-1}\right]^2 = \left[\dfrac{y}{x}\right]^2 = \dfrac{y^2}{x^2}$

or

(ii) $\left(\dfrac{x}{y}\right)^{-2} = \dfrac{x^{-2}}{y^{-2}} = \dfrac{\frac{1}{x^2}}{\frac{1}{y^2}} = \dfrac{1}{x^2} \cdot \dfrac{y^2}{1} = \dfrac{y^2}{x^2}$

c. (i) $\left(\dfrac{x^{-2}}{y}\right)^{-1} = \dfrac{y}{x^{-2}} = y \cdot x^2 = x^2 y$

or

(ii) $\left(\dfrac{x^{-2}}{y}\right)^{-1} = \dfrac{x^2}{y^{-1}} = x^2 \cdot y = x^2 y$

As Example 1 shows, we can use negative exponents to write quotients.

EXAMPLE 7

Find the quotients and write them two ways.

a. $14x^2y \div 7xy^3$
b. $36m^2n^2 \div 6m^3n$

Solution

a. $\dfrac{14x^2y}{7xy^3} = 2xy^{-2} \quad \left(\text{or } \dfrac{2x}{y^2}\right)$

b. $\dfrac{36m^2n^2}{6m^3n} = 6m^{-1}n \quad \left(\text{or } \dfrac{6n}{m}\right)$

Exercise 6-2

Compute and simplify the following.

1. 16^0
2. 47^0
3. $(-2)^0$
4. $(-16)^0$
5. $(2x)^0 \quad (x \ne 0)$
6. $(3xy)^0 \quad (x, y \ne 0)$
7. $(-2mn)^0 \quad (m, n \ne 0)$
8. $(5x^3y^2)^0 \quad (x, y \ne 0)$
9. $2xy^0 \quad (y \ne 0)$
10. $17xz^0 \quad (z \ne 0)$
11. $(4x^0y)^2 \quad (x \ne 0)$
12. $(3x^0z)^4 \quad (x \ne 0)$

Compute and simplify so that only positive exponents remain.

13. x^{-2}
14. x^{-4}
15. $6x^{-3}$
16. $5y^{-4}$

17. $(3x)^{-2}$
18. $(2y)^{-3}$
19. $3(xy)^{-1}$
20. $3(4x)^{-3}$
21. $4(x^2)^{-2}$
22. $5(y^3)^{-3}$
23. $6(x^3)^{-1}$
24. $3(x^{-1})^{-5}$
25. $(2x^{-1})^{-1}$
26. $(2x^2y)^{-3}$
27. $(3m^{-2}n)^{-2}$
28. $(4x^{-3}y)^{-2}$
29. $(4^0x^2)^{-1}$
30. $(3x^{-2})^0$
31. $x^4 \cdot x^{-5}$
32. $y^3 \cdot y^{-2}$
33. $(2xy^{-1})(3x^{-2}y)$
34. $(-3m^{-2}y^{-1}) \cdot (2m^{-3}y^{-2})$
35. $x^5 \div x^{-3}$
36. $x^{-3} \div x^3$
37. $(2x^{-2}y) \div (x^{-1}y^{-2})$
38. $(-4m^2n^{-3}) \div (2m^{-3}y^2)$
39. $\left(\dfrac{a}{b^2}\right)^{-4}$
40. $\left(\dfrac{b}{x^3}\right)^{-2}$
41. $\left(\dfrac{a^{-2}}{b^{-1}}\right)^{-2}$
42. $\left(\dfrac{x^{-2}}{y}\right)^{-3}$

Express the following quotients in two ways:

a. with positive exponents.
b. with no denominators.

43. $35m^3n \div 7m^4n^2$
44. $55x^2y \div 11xy^3$
45. $-49mx \div 7m^2y$
46. $-36xz^3 \div 9x^2z$
47. $16x^3y \div (-8xy^2)$
48. $32pq^2 \div 4pq^3$

6-3 Roots

OBJECTIVE: After completing this section, you should be able to find the indicated roots of numbers and expressions.

The inverse process of "raising to a power" is the process of "taking a root." We use roots to answer questions like "what number squared is 9?" We say that 3 is a square root of 9 because $3^2 = 9$ and that 2 is a cube root of 8 because $2^3 = 8$.

The general notation for the nth root of a is $\sqrt[n]{a}$. The n is called the index, $\sqrt{}$ is the radical, and a is called the radicand.

(If $n = 2$, the index is not usually written; that is, $\sqrt{9}$ is the square root of 9.)

Because there are two square roots of 9 [$3^2 = 9$ and $(-3)^2 = 9$], we introduce the convention of **principal root**.

1. **The principal nth root of a positive number is the positive root.**
   Thus the principal square root is $\sqrt{9} = 3$; $-\sqrt{9} = -3$ is also a square root, but not the principal root.
2. **The principal nth root of 0 is 0.**
   Thus $\sqrt{0} = 0$ and $\sqrt[3]{0} = 0$.
3. **The principal nth root of a negative number is the negative root when n is odd.**
   Thus the principal cube root of -27 is $\sqrt[3]{-27} = -3$.

EXAMPLE 1

Find the principal roots of the following.

a. $\sqrt{16}$
b. $\sqrt[3]{-8}$
c. $\sqrt[4]{0}$

Solution

a. $\sqrt{16} = 4$ because $4^2 = 16$
b. $\sqrt[3]{-8} = -2$ because $(-2)^3 = -8$
c. $\sqrt[4]{0} = 0$ because $0^4 = 0$

When we are asked for the root of a number, we give the principal root.

EXAMPLE 2

Find the following roots.

a. $\sqrt{49}$
b. $\sqrt[3]{-125}$
c. $\sqrt[3]{64}$

Solution

a. $\sqrt{49} = 7$ because $7^2 = 49$
b. $\sqrt[3]{-125} = -5$ because $(-5)^3 = -125$
c. $\sqrt[3]{64} = 4$ because $4^3 = 64$

We find the roots of algebraic expressions in the same manner as we find roots of numbers.

EXAMPLE 3

Find the following roots.

a. $\sqrt{x^2}$ $(x \geq 0)$
b. $\sqrt{x^4}$
c. $\sqrt[3]{y^6}$

Solution

a. $\sqrt{x^2} = x$ because $(x)^2 = x^2$.
b. $\sqrt{x^4} = x^2$ because $(x^2)^2 = x^4$.
c. $\sqrt[3]{y^6} = y^2$ because $(y^2)^3 = y^6$.

EXAMPLE 4

Find the following roots.

a. $\sqrt{4x^4}$
b. $\sqrt[3]{-27x^6y^3}$

Solution

a. $\sqrt{4x^4} = 2x^2$ because $(2x^2)^2 = 4x^4$.
b. $\sqrt[3]{-27x^6y^3} = -3x^2y$ because $(-3x^2y)^3 = -27x^6y^3$.

Exercise 6-3

Find the indicated roots.

1. $\sqrt{4}$
2. $\sqrt{16}$
3. $\sqrt{49}$
4. $\sqrt{64}$
5. $\sqrt[3]{8}$
6. $\sqrt[3]{27}$
7. $\sqrt[3]{-64}$
8. $\sqrt[3]{-8}$
9. $\sqrt[3]{0}$
10. $\sqrt[4]{0}$
11. $\sqrt{y^4}$
12. $\sqrt{z^6}$ $(z \geq 0)$
13. $\sqrt[3]{x^9}$
14. $\sqrt[3]{-y^{12}}$
15. $\sqrt{16x^6}$ $(x \geq 0)$
16. $\sqrt[3]{8x^9}$
17. $\sqrt[3]{27y^3}$
18. $\sqrt{9z^4}$
19. $\sqrt{4x^4y^2}$ $(y > 0)$
20. $\sqrt[3]{125x^3y^6}$
21. $\sqrt[4]{16}$
22. $\sqrt[4]{81}$
23. $\sqrt[4]{16x^8}$
24. $\sqrt[4]{x^4y^8}$ $(x \geq 0)$

6-4 Fractional exponents and radicals

OBJECTIVES: After completing this section, you should be able to:

a. Convert from radical notation to fractional exponent notation and vice versa.
b. Perform operations involving fractional exponents.

We have seen that by definition $(\sqrt{a})^2 = a$. Note that if we let $\sqrt{a} = a^{1/2}$, we have $(a^{1/2})^2 = a^1 = a$. In general, we represent $\sqrt[n]{a}$ as $a^{1/n}$. Thus $\sqrt[3]{x}$ can be written as $x^{1/3}$.

EXAMPLE 1

Write the following with fractional exponents.

a. $\sqrt{x}$
b. $\sqrt[3]{5}$
c. $\sqrt{2x}$

Solution

a. $\sqrt{x} = x^{1/2}$
b. $\sqrt[3]{5} = 5^{1/3}$
c. $\sqrt{2x} = (2x)^{1/2}$

We can also convert expressions with fractional exponents to radical expressions.

EXAMPLE 2

Write the following expressions in radical form.

a. $x^{1/4}$
b. $a^{1/2}$
c. $(4y)^{1/3}$

Solution

a. $x^{1/4} = \sqrt[4]{x}$
b. $a^{1/2} = \sqrt{a}$
c. $(4y)^{1/3} = \sqrt[3]{4y}$

In order to write a term such as $\sqrt[3]{x^2}$ with a fractional exponent, we define

$$a^{m/n} = \sqrt[n]{a^m}$$

6-4 Fractional exponents and radicals

We may state this as follows:

If a number has a fractional exponent, the numerator of the fractional exponent indicates the power to which the base is to be raised, and the denominator indicates the root to be taken.

EXAMPLE 3

Write the following terms in radical form.

a. $5^{3/4}$
b. $y^{3/2}$
c. $(6m)^{2/3}$

Solution

a. $5^{3/4} = \sqrt[4]{5^3}$
b. $y^{3/2} = \sqrt{y^3}$
c. $(6m)^{2/3} = \sqrt[3]{(6m)^2} = \sqrt[3]{36m^2}$

EXAMPLE 4

Write the following terms without radicals.

a. $\sqrt{x^3}$
b. $\sqrt[3]{b^2}$
c. $\sqrt{(ab)^3}$

Solution

a. $\sqrt{x^3} = x^{3/2}$
b. $\sqrt[3]{b^2} = b^{2/3}$
c. $\sqrt{(ab)^3} = (ab)^{3/2}$

Note that we may also write $\sqrt[n]{a^m}$ in the form $(\sqrt[n]{a})^m$ if m and n have no common factor greater than 1. For example, $\sqrt[3]{x^2} = (x^2)^{1/3} = x^{2/3} = (x^{1/3})^2 = (\sqrt[3]{x})^2$. In dealing with fractional exponents, it is sometimes easier to take a root before raising the base to a power. This is especially true if we are dealing with large numbers. For example, $\sqrt[3]{8^2} = \sqrt[3]{64} = 4$ and $(\sqrt[3]{8})^2 = 2^2 = 4$.

EXAMPLE 5

Simplify

a. $27^{2/3} = (\sqrt[3]{27})^2$
b. $27^{2/3} = \sqrt[3]{27^2}$

Solution

a. $27^{2/3} = (\sqrt[3]{27})^2 = 3^2 = 9$
b. $27^{2/3} = \sqrt[3]{27^2} = \sqrt[3]{729} = 9$

We can perform operations with fractional exponents just as we did with integer exponents. The rules for exponents, which we discussed for integer exponents, also apply to fractional exponents.

EXAMPLE 6

Compute the following.

a. $a^{1/2} \cdot a^{1/2}$
b. $a^{3/2} \div a^{1/2}$
c. $(ab)^{3/2}$
d. $\left(\dfrac{a}{b}\right)^{1/2}$
e. $(a^{3/2})^{1/2}$

Solution

a. By Rule 1, $a^{1/2} \cdot a^{1/2} = a^{1/2+1/2} = a^1 = a$
b. By Rule 2a, $a^{3/2} \div a^{1/2} = a^{3/2-1/2} = a^1 = a$
c. By Rule 3, $(ab)^{3/2} = a^{3/2}b^{3/2}$
d. By Rule 4, $\left(\dfrac{a}{b}\right)^{1/2} = \dfrac{a^{1/2}}{b^{1/2}}$
e. By Rule 5, $(a^{3/2})^{1/2} = a^{(3/2)\cdot(1/2)} = a^{3/4}$

EXAMPLE 7

Compute the following.

a. $a^{1/3} \cdot a^{1/4}$
b. $x^{1/2} \div x^{1/3}$
c. $(a^{2/3}b^{1/3})^{1/2}$

Solution

a. $a^{1/3} \cdot a^{1/4} = a^{1/3+1/4} = a^{7/12}$
b. $x^{1/2} \div x^{1/3} = x^{1/2-1/3} = x^{1/6}$
c. $(a^{2/3}b^{1/3})^{1/2} = a^{2/3 \cdot 1/2}b^{1/3 \cdot 1/2} = a^{1/3}b^{1/6}$

Exercise 6-4

Write each of the following without radicals by using fractional exponents.

1. $\sqrt{3}$
2. $\sqrt[3]{5}$
3. $\sqrt[3]{6}$
4. $\sqrt[4]{9}$
5. $\sqrt{p}$
6. $\sqrt[3]{m}$
7. $\sqrt{x}$
8. $\sqrt[5]{y}$
9. $\sqrt[3]{3x}$
10. $\sqrt{5y}$
11. $\sqrt{6m}$
12. $\sqrt[4]{8x}$
13. $\sqrt{m^3}$
14. $\sqrt[3]{y^2}$
15. $\sqrt{5^3}$
16. $\sqrt[3]{9^2}$
17. $\sqrt[3]{(ax)^2}$
18. $\sqrt{(6x)^3}$
19. $\sqrt[3]{(3x)^4}$
20. $\sqrt{(5m)^3}$

Write each of the following in radical form.

21. $6^{1/2}$
22. $4^{1/3}$
23. $9^{1/3}$
24. $12^{1/2}$
25. $x^{1/3}$
26. $y^{1/4}$
27. $p^{1/2}$
28. $m^{1/3}$
29. $(3x)^{1/2}$
30. $(4m)^{1/3}$
31. $(5y)^{1/4}$
32. $(6p)^{1/2}$
33. $6^{3/2}$
34. $9^{2/3}$
35. $m^{3/4}$
36. $y^{5/3}$
37. $(4x)^{2/3}$
38. $(5x)^{3/2}$
39. $(a^2x)^{2/3}$
40. $(6y^2)^{4/3}$

Simplify the following.

41. $16^{3/2}$
42. $25^{3/2}$
43. $64^{2/3}$
44. $125^{2/3}$

Compute the following.

45. $x^{1/3} \cdot x^{1/3}$
46. $y^{1/3} \cdot y^{1/2}$
47. $m^{1/4} \cdot m^{2/5}$
48. $z^3 \cdot z^{1/3}$
49. $(2a)^{1/3} \cdot (2a)^{2/3}$
50. $(ax)^{1/3} \cdot (ax)^{3/2}$
51. $x^{1/4} \div x^{1/8}$
52. $m^{1/2} \div m^{1/4}$
53. $(ab)^{4/5} \div (ab)^{1/5}$
54. $y^{1/4} \div y^{1/2}$
55. $(xy)^{1/3}$
56. $(mn)^{3/2}$
57. $\left(\dfrac{6}{x}\right)^{1/2}$
58. $\left(\dfrac{m}{n}\right)^{1/3}$
59. $(x^{2/3})^{1/2}$ $\quad (x \geq 0)$
60. $(m^{1/2})^{2/3}$
61. $(a^{1/3}b)^{1/2}$
62. $(x^{1/2}y^{2/3})^{1/3}$

6-5 Simplifying radicals; multiplication

OBJECTIVES: After completing this section, you should be able to:
a. **Simplify radicals by taking appropriate roots.**
b. **Simplify radicals by removing from the radicand factors whose roots are present.**
c. **Multiply radicals with the same index.**

158 Exponents and Radicals

We can perform operations with radicals by converting the expressions to fractional exponent notation, performing the operation with exponents, and then converting the answer back to radical form. But this is time-consuming and unnecessary. The rules given below are for operations with radicals. These rules make it easier to work with radicals and are consistent with the rules of exponents.

Given $a \geq 0$, $b \geq 0$.

Rule 1. $\sqrt[n]{a^n} = (\sqrt[n]{a})^n = a$

using exponents: $(a^n)^{1/n} = (a^{1/n})^n = a^1 = a$

Rule 2. $\sqrt[n]{a} \cdot \sqrt[n]{b} = \sqrt[n]{ab}$

using exponents: $a^{1/n} \cdot b^{1/n} = (ab)^{1/n}$

Rule 3. $\dfrac{\sqrt[n]{a}}{\sqrt[n]{b}} = \sqrt[n]{\dfrac{a}{b}}$

using exponents: $\dfrac{a^{1/n}}{b^{1/n}} = \left(\dfrac{a}{b}\right)^{1/n}$

Rule 1 [$\sqrt[n]{a^n} = (\sqrt[n]{a})^n = a$] simply states that the nth root of a nonnegative number to the nth power is the number.

EXAMPLE 1

Simplify

a. $\sqrt[3]{2^3}$
b. $(\sqrt[4]{16})^4$
c. $\sqrt{64}$

Solution

a. $\sqrt[3]{2^3} = 2$ by Rule 1
b. $(\sqrt[4]{16})^4 = 16$ by Rule 1
c. $\sqrt{64} = \sqrt{8^2} = 8$

Rule 2 ($\sqrt[n]{a} \cdot \sqrt[n]{b} = \sqrt[n]{ab}$) provides a procedure for simplifying radicals. If we write the rule in the form $\sqrt[n]{ab} = \sqrt[n]{a} \cdot \sqrt[n]{b}$, we see that it is possible to remove part of the radicand from under the radical. We can frequently simplify radicals by removing from the radicand all factors whose indicated roots are present.

EXAMPLE 2

Simplify $\sqrt{12}$.

Solution

$4 = 2^2$ is a factor of 12. Thus $\sqrt{12} = \sqrt{4 \cdot 3} = \sqrt{4} \cdot \sqrt{3} = 2\sqrt{3}$.

EXAMPLE 3

Simplify the following radicals.

a. $\sqrt[3]{81}$
b. $\sqrt{18}$
c. $\sqrt{x^5}$ $(x \geq 0)$

Solution

a. $\sqrt[3]{81} = \sqrt[3]{3 \cdot 3 \cdot 3 \cdot 3} = \sqrt[3]{3^3 \cdot 3} = \sqrt[3]{3^3} \sqrt[3]{3} = 3\sqrt[3]{3}$
b. $\sqrt{18} = \sqrt{3 \cdot 3 \cdot 2} = \sqrt{3^2 \cdot 2} = \sqrt{3^2} \cdot \sqrt{2} = 3\sqrt{2}$
c. $\sqrt{x^5} = \sqrt{x^4 \cdot x} = \sqrt{x^4} \cdot \sqrt{x} = \sqrt{(x^2)^2} \sqrt{x} = x^2 \sqrt{x}$

EXAMPLE 4

Simplify the following radicals.

a. $\sqrt{36a^3b^2}$ $(b \geq 0)$
b. $\sqrt[3]{72a^3b^4}$

Solution

a. $\sqrt{36a^3b^2} = \sqrt{6^2a^2 \cdot ab^2} = \sqrt{6^2} \sqrt{a^2} \sqrt{b^2} \sqrt{a} = 6ab\sqrt{a}$
b. $\sqrt[3]{72a^3b^4} = \sqrt[3]{8 \cdot 9a^3b^3b} = \sqrt[3]{8} \cdot \sqrt[3]{a^3} \cdot \sqrt[3]{b^3} \cdot \sqrt[3]{9b}$
$= 2ab\sqrt[3]{9b}$

Rule 2 ($\sqrt[n]{a} \cdot \sqrt[n]{b} = \sqrt[n]{ab}$) also provides a procedure for multiplying two roots with the same index.

EXAMPLE 5

Multiply the following and simplify the answers if possible.

a. $\sqrt{5} \cdot \sqrt{3}$
b. $\sqrt[3]{9} \cdot \sqrt[3]{6}$
c. $\sqrt{6x} \cdot \sqrt{7y}$

Solution

a. $\sqrt{5} \cdot \sqrt{3} = \sqrt{5 \cdot 3} = \sqrt{15}$
b. $\sqrt[3]{9} \cdot \sqrt[3]{6} = \sqrt[3]{54} = \sqrt[3]{27 \cdot 2} = \sqrt[3]{27}\,\sqrt[3]{2} = 3\sqrt[3]{2}$
c. $\sqrt{6x} \cdot \sqrt{7y} = \sqrt{42xy}$

EXAMPLE 6

Multiply the following and simplify the answers.

a. $\sqrt{8} \cdot \sqrt{18}$
b. $\sqrt{3x^3} \cdot \sqrt{15x}$
c. $\sqrt[3]{2xy} \cdot \sqrt[3]{4x^2y}$

Solution

a. $\sqrt{8} \cdot \sqrt{18} = \sqrt{8 \cdot 18} = \sqrt{144} = 12$
b. $\sqrt{3x^3} \cdot \sqrt{15x} = \sqrt{3x^3 \cdot 15x}$
$\qquad = \sqrt{45x^4} = \sqrt{9}\,\sqrt{x^4}\,\sqrt{5} = 3x^2\sqrt{5}$
c. $\sqrt[3]{2xy} \cdot \sqrt[3]{4x^2y} = \sqrt[3]{2xy \cdot 4x^2y} = \sqrt[3]{8x^3y^2} = \sqrt[3]{8}\,\sqrt[3]{x^3}\,\sqrt[3]{y^2}$
$\qquad = 2x\,\sqrt[3]{y^2}$

Note that it is possible to simplify the radicals before multiplying. This rarely will give the final answer in simplest form, but it may make the problem easier to simplify, especially if there are large numbers in the radicand.

EXAMPLE 7

Multiply $\sqrt{96} \cdot \sqrt{8}$

Solution

The two numbers in the radicand can be multiplied, giving $\sqrt{96 \cdot 8} = \sqrt{768}$, which now must be simplified.

$\sqrt{768} = \sqrt{4}\,\sqrt{192} = \sqrt{4}\,\sqrt{4}\,\sqrt{48} = \sqrt{4}\,\sqrt{4}\,\sqrt{4}\,\sqrt{12}$
$\qquad = \sqrt{4}\,\sqrt{4}\,\sqrt{4}\,\sqrt{4}\,\sqrt{3} = 2 \cdot 2 \cdot 2 \cdot 2\,\sqrt{3} = 16\sqrt{3}$

The answer could have also been found by simplifying the two radicals first:

$\sqrt{96} \cdot \sqrt{8} = \sqrt{4}\,\sqrt{24} \cdot \sqrt{4}\,\sqrt{2}$ or $\sqrt{4}\,\sqrt{4}\,\sqrt{6} \cdot \sqrt{4}\,\sqrt{2}$

Thus

$\sqrt{96} \cdot \sqrt{8} = 4\sqrt{6} \cdot 2\sqrt{2} = 8\sqrt{12} = 8 \cdot 2 \cdot \sqrt{3} = 16\sqrt{3}$

Exercise 6-5

Simplify the following.

1. $\sqrt[3]{64}$
2. $\sqrt{81}$
3. $\sqrt[4]{125^4}$
4. $(\sqrt{49})^2$
5. $(\sqrt{32})^2$
6. $\sqrt[3]{98^3}$
7. $\sqrt{75}$
8. $\sqrt{48}$
9. $\sqrt{40}$
10. $\sqrt{54}$
11. $\sqrt[3]{54}$
12. $\sqrt[3]{24}$
13. $\sqrt[3]{135}$
14. $\sqrt[3]{128}$
15. $\sqrt{x^7}$
16. $\sqrt{a^3}$
17. $\sqrt[3]{m^5}$
18. $\sqrt[3]{w^7}$
19. $\sqrt{49x^3y}$ $(x \geq 0)$
20. $\sqrt{63m^4n^3}$
21. $\sqrt[3]{54x^3y^5}$
22. $\sqrt[3]{32pq^6}$

Multiply the following and simplify when possible.

23. $\sqrt{15} \cdot \sqrt{2}$
24. $\sqrt{14} \cdot \sqrt{3}$
25. $\sqrt{8} \cdot \sqrt{6}$
26. $\sqrt{12} \cdot \sqrt{6}$
27. $\sqrt[3]{12} \cdot \sqrt[3]{6}$
28. $\sqrt[3]{16} \cdot \sqrt[3]{4}$
29. $\sqrt{6x^3y} \cdot \sqrt{3xy}$ $(y \geq 0)$
30. $\sqrt{12x^3y} \cdot \sqrt{3x^2y}$
31. $\sqrt{72x^3} \cdot \sqrt{36xy^3}$
32. $\sqrt{64x^4z} \cdot \sqrt{8x^3z^3}$
33. $\sqrt[3]{16x^2y^4} \cdot \sqrt[3]{4xy^2}$
34. $\sqrt[3]{9x^4y} \cdot \sqrt[3]{3x^2y}$

6-6 Simplifying radicals; division

OBJECTIVES: After completing this section, you should be able to:

a. Divide radicals with the same index and simplify the quotient.
b. Simplify radicals by rationalizing their denominators.
c. Find the decimal approximations of indicated quotients of radicals.

Rule 3 ($\sqrt[n]{a}/\sqrt[n]{b} = \sqrt[n]{a/b}$) indicates how to find the quotient of two roots with the same index.

EXAMPLE 1

Find the quotients and simplify the answers.

a. $\dfrac{\sqrt{8}}{\sqrt{2}}$

Exponents and Radicals

b. $\dfrac{\sqrt[3]{32}}{\sqrt[3]{4}}$

c. $\dfrac{\sqrt{16a^3x}}{\sqrt{2ax}}$

Solution

a. $\dfrac{\sqrt{8}}{\sqrt{2}} = \sqrt{\dfrac{8}{2}} = \sqrt{4} = 2$

b. $\dfrac{\sqrt[3]{32}}{\sqrt[3]{4}} = \sqrt[3]{\dfrac{32}{4}} = \sqrt[3]{8} = 2$

c. $\dfrac{\sqrt{16a^3x}}{\sqrt{2ax}} = \sqrt{\dfrac{16a^3x}{2ax}} = \sqrt{8a^2} = 2a\sqrt{2}$

As might be expected, the divisor will not always be a factor of the dividend when we are dividing radicals. For example, $\sqrt{9}$ cannot be divided into $\sqrt{7}$. We can simplify $\sqrt{7}/\sqrt{9}$, however, by removing the radical in the denominator; that is, $\sqrt{7}/\sqrt{9} = \sqrt{7}/3$, or $\sqrt{7}/\sqrt{9} = \tfrac{1}{3}\sqrt{7}$.

To write a radical expression in the simplest form, we can have no radicals in the denominator of a fraction and no fractions in the radicand of the expression.

Thus neither $\sqrt{7}/\sqrt{9}$ nor $\sqrt{7/9}$ is in the simplest form, while $\sqrt{7}/3$ and $\tfrac{1}{3}\sqrt{7}$ are in the simplest form.

EXAMPLE 2

State the following in simplest form.

a. $\dfrac{\sqrt{15}}{\sqrt{4}}$

b. $\sqrt{\dfrac{12}{25}}$

c. $\sqrt{\dfrac{9}{12}}$

Solution

a. $\dfrac{\sqrt{15}}{\sqrt{4}} = \dfrac{\sqrt{15}}{2}$ or $\tfrac{1}{2}\sqrt{15}$

b. $\sqrt{\dfrac{12}{25}} = \dfrac{\sqrt{12}}{\sqrt{25}} = \dfrac{\sqrt{12}}{5} = \tfrac{1}{5}\sqrt{12} = \tfrac{1}{5}\sqrt{4}\sqrt{3} = \tfrac{2}{5}\sqrt{3}$

6-6 Simplifying radicals; division

c. $\sqrt{\frac{9}{12}}$ could be written as $\frac{\sqrt{9}}{\sqrt{12}} = \frac{3}{\sqrt{12}}$, but $\frac{3}{\sqrt{12}}$ is not in simplest form.

$\sqrt{\frac{9}{12}}$ can also be written as $\sqrt{\frac{3}{4}} = \frac{\sqrt{3}}{2}$ or $\frac{1}{2}\sqrt{3}$

EXAMPLE 3

Simplify $\frac{\sqrt{5}}{\sqrt{2}}$.

Solution

To remove the radical from the denominator, we can multiply both the numerator and denominator by $\sqrt{2}$:

$$\frac{\sqrt{5}}{\sqrt{2}} \cdot \frac{\sqrt{2}}{\sqrt{2}} = \frac{\sqrt{10}}{\sqrt{4}} = \frac{\sqrt{10}}{2}$$

Note: We have changed the denominator into a rational number in Example 9. This process is called **rationalizing the denominator**. Rationalizing the denominator makes it much easier to find the decimal approximation of a radical expression.

EXAMPLE 4

Find the decimal approximation of $\frac{\sqrt{5}}{\sqrt{2}}$.

Solution

a. If we do not rationalize the denominator, we must find the decimal approximations of $\sqrt{5}$ and $\sqrt{2}$ from Table I in Appendix I. Then

$$\frac{\sqrt{5}}{\sqrt{2}} = \frac{2.2361}{1.4142} = 1.5811 \quad \text{(to four decimal places)}$$

(Note the difficulty in dividing these numbers.)

b. If we rationalize the denominator and look up the approximation of $\sqrt{10}$, we get

$$\frac{\sqrt{5}}{\sqrt{2}} = \frac{\sqrt{10}}{2} = \frac{3.1623}{2} = 1.5811 \quad \text{(to four decimal places)}$$

EXAMPLE 5

Simplify each of the following, rationalizing the denominator if necessary.

a. $\dfrac{\sqrt{20}}{\sqrt{3}}$

b. $\dfrac{\sqrt{20}}{\sqrt{24}}$

c. $\dfrac{\sqrt{4x}}{\sqrt{3xy}}$

Solution

a. $\dfrac{\sqrt{20}}{\sqrt{3}} = \dfrac{\sqrt{20}}{\sqrt{3}} \cdot \dfrac{\sqrt{3}}{\sqrt{3}} = \dfrac{\sqrt{60}}{3} = \dfrac{2\sqrt{15}}{3}$ (or $\tfrac{2}{3}\sqrt{15}$)

b. $\dfrac{\sqrt{20}}{\sqrt{24}} = \sqrt{\dfrac{20}{24}} = \sqrt{\dfrac{5}{6}} = \dfrac{\sqrt{5}}{\sqrt{6}} = \dfrac{\sqrt{5}}{\sqrt{6}} \cdot \dfrac{\sqrt{6}}{\sqrt{6}} = \dfrac{\sqrt{30}}{6}$ (or $\tfrac{1}{6}\sqrt{30}$)

c. $\dfrac{\sqrt{4x}}{\sqrt{3xy}} = \sqrt{\dfrac{4x}{3xy}} = \sqrt{\dfrac{4}{3y}} = \dfrac{\sqrt{4}}{\sqrt{3y}} = \dfrac{\sqrt{4}}{\sqrt{3y}} \cdot \dfrac{\sqrt{3y}}{\sqrt{3y}} = \dfrac{\sqrt{12y}}{3y} = \dfrac{2\sqrt{3y}}{3y}$

Exercise 6-6

Compute the following quotients and simplify when possible.

1. $\dfrac{\sqrt{32}}{\sqrt{2}}$

2. $\dfrac{\sqrt{192}}{\sqrt{3}}$

3. $\dfrac{\sqrt{18}}{\sqrt{2}}$

4. $\dfrac{\sqrt[3]{81}}{\sqrt[3]{3}}$

5. $\dfrac{\sqrt{84}}{\sqrt{3}}$

6. $\dfrac{\sqrt{96}}{\sqrt{3}}$

7. $\dfrac{\sqrt[3]{48}}{\sqrt[3]{3}}$

8. $\dfrac{\sqrt[3]{80}}{\sqrt[3]{5}}$

9. $\dfrac{\sqrt{8x^3y}}{\sqrt{2xy}}$ $(x \geq 0)$

10. $\dfrac{\sqrt{32m^3n^4}}{\sqrt{2mn}}$

11. $\dfrac{\sqrt{45pq^3}}{\sqrt{9pq}}$ $(q > 0)$

12. $\dfrac{\sqrt{18m^3n^3}}{\sqrt{2m^2n^2}}$

13. $\dfrac{\sqrt[3]{36x^3y^2}}{\sqrt[3]{4xy^2}}$

14. $\dfrac{\sqrt[3]{54m^4n}}{\sqrt[3]{2m}}$

Simplify the following, rationalizing the denominator when necessary.

15. $\dfrac{\sqrt{8}}{\sqrt{9}}$ 16. $\dfrac{\sqrt{6}}{\sqrt{16}}$

17. $\dfrac{\sqrt{10}}{\sqrt{4}}$ 18. $\dfrac{\sqrt{12}}{\sqrt{9}}$

19. $\dfrac{\sqrt{6}}{\sqrt{5}}$ 20. $\dfrac{\sqrt{9}}{\sqrt{2}}$

21. $\sqrt{\dfrac{5}{7}}$ 22. $\sqrt{\dfrac{4}{5}}$

23. $\dfrac{\sqrt{10}}{\sqrt{15}}$ 24. $\dfrac{\sqrt{40}}{\sqrt{12}}$

25. $\dfrac{\sqrt{24}}{\sqrt{10}}$ 26. $\dfrac{\sqrt{6}}{\sqrt{18}}$

27. $\dfrac{\sqrt{5x}}{\sqrt{6x^2}}$ 28. $\dfrac{\sqrt{3xy}}{\sqrt{2x}}$

29. $\dfrac{\sqrt{5x^3w}}{\sqrt{4xw^2}}$ 30. $\dfrac{\sqrt{m^2x}}{\sqrt{mx^2}}$

6-7 Simplifying radicals; addition and subtraction

OBJECTIVES: After completing this section, you should be able to:

a. **Add terms containing like radicals.**
b. **Subtract terms containing like radicals.**
c. **Simplify fractions containing radicals.**
d. **Find decimal approximations to fractions containing radicals.**

We learned in Chapter 2 that we can add like terms in algebra by adding their coefficients. We can also combine radical terms if they are **like radicals**. Like radicals are those that have the same index and the same radicand. Thus $4\sqrt{3}$ and $6\sqrt{3}$ are like radicals, while $\sqrt{3}$ and $\sqrt{6}$ are not. We add like radicals by adding their coefficients just as we add like terms in algebra.

EXAMPLE 1

Add the following.

a. $3x + 5x + 2y$
b. $3\sqrt{2} + 5\sqrt{2} + 2\sqrt{6}$

Solution

a. We can add the coefficients of the terms $3x$ and $5x$ since they are like terms. The sum is $3x + 5x + 2y = 8x + 2y$.
b. Because two terms have the same order and radicand, we add their coefficients. The sum is $3\sqrt{2} + 5\sqrt{2} + 2\sqrt{6} = 8\sqrt{2} + 2\sqrt{6}$.

EXAMPLE 2

Add the following.

a. $6\sqrt{3} + 2\sqrt{3} + 5\sqrt{3} + 2\sqrt{5}$
b. $\sqrt[3]{5} + 2\sqrt{5} + 6\sqrt{5}$

Solution

a. Combining the like radicals gives $13\sqrt{3} + 2\sqrt{5}$.
b. Combining like radicals gives $\sqrt[3]{5} + 8\sqrt{5}$.

Note that $\sqrt[3]{5}$ and $\sqrt{5}$ are not like radicals.

We can perform subtractions of like radicals in the same manner.

EXAMPLE 3

Compute the following.

a. $5\sqrt{15} - 3\sqrt{15}$
b. $6\sqrt{5} - 8\sqrt{5}$

Solution

a. Combining the coefficients of the like radicals gives $2\sqrt{15}$.
b. Combining the coefficients of the like radicals gives $-2\sqrt{5}$.

We can sometimes transform unlike radicals to like radicals by simplifying them.

EXAMPLE 4

Compute the following.

a. $3\sqrt{8} + 2\sqrt{2}$
b. $3\sqrt{27} + 6\sqrt{3} - \sqrt{12}$

Solution

a. $3\sqrt{8} + 2\sqrt{2} = 3(2\sqrt{2}) + 2\sqrt{2} = 6\sqrt{2} + 2\sqrt{2} = 8\sqrt{2}$
b. $3\sqrt{27} + 6\sqrt{3} - \sqrt{12} = 3(3\sqrt{3}) + 6\sqrt{3} - 2\sqrt{3}$
$= 9\sqrt{3} + 6\sqrt{3} - 2\sqrt{3} = 15\sqrt{3} - 2\sqrt{3}$
$= 13\sqrt{3}$

We shall have occasion to simplify fractions in which the numerator contains a radical. Consider the following example.

EXAMPLE 5

Simplify the following fractions.

a. $\dfrac{2 + \sqrt{8}}{2}$

b. $\dfrac{6 - \sqrt{18}}{3}$

Solution

a. We first simplify the radical:

$$\dfrac{2 + \sqrt{8}}{2} = \dfrac{2 + 2\sqrt{2}}{2}$$

Now it is apparent that 2 can be divided into both the numerator (*both* terms of it!) and the denominator. Thus

$$\dfrac{2 + \sqrt{8}}{2} = \dfrac{2 + 2\sqrt{2}}{2} = \dfrac{2(1 + 1\sqrt{2})}{2} = \dfrac{1 + \sqrt{2}}{1} = 1 + \sqrt{2}$$

b. Using the same technique as in a, we get

$$\dfrac{6 - \sqrt{18}}{3} = \dfrac{6 - 3\sqrt{2}}{3} = 2 - \sqrt{2}$$

EXAMPLE 6

Simplify the following fraction.

$$\dfrac{18 - \sqrt{180}}{9}$$

Solution

$$\dfrac{18 - \sqrt{180}}{9} = \dfrac{18 - 6\sqrt{5}}{9} = \dfrac{3(6 - 2\sqrt{5})}{3(3)}$$

168 Exponents and Radicals

The largest factor that will divide into **both** terms of the numerator and the denominator is 3. The simplified fraction is

$$\frac{6 - 2\sqrt{5}}{3}$$

Exercise 6-7

Compute the following.

1. $3\sqrt{2} + 5\sqrt{2}$
2. $6\sqrt{7} + 2\sqrt{7}$
3. $4\sqrt{6} - 3\sqrt{6}$
4. $6\sqrt{13} - 8\sqrt{13}$
5. $6\sqrt{3} + 4\sqrt{3} + 6\sqrt{5}$
6. $2\sqrt{6} + 4\sqrt{3} + 4\sqrt{6}$
7. $2\sqrt{13} + 2\sqrt[3]{13} + 4\sqrt{13}$
8. $5\sqrt[3]{6} + \sqrt[3]{6} + \sqrt[3]{6}$
9. $2\sqrt{3} + 3\sqrt{3} - 6\sqrt{3}$
10. $5\sqrt{10} - 2\sqrt{10} + 6\sqrt{10}$
11. $3\sqrt{12} + 4\sqrt{3}$
12. $6\sqrt{32} - 3\sqrt{2}$
13. $5\sqrt{128} + 6\sqrt{8}$
14. $5\sqrt{63} + 5\sqrt{28}$
15. $2\sqrt{27} + 8\sqrt{3} - \sqrt{12}$
16. $8\sqrt{5} - 6\sqrt{125} + 3\sqrt{45}$

Simplify the following fractions.

17. $\dfrac{6 + \sqrt{32}}{2}$
18. $\dfrac{9 - \sqrt{27}}{3}$
19. $\dfrac{5 + \sqrt{75}}{5}$
20. $\dfrac{4 + \sqrt{72}}{2}$
21. $\dfrac{6 - \sqrt{56}}{6}$
22. $\dfrac{16 - \sqrt{128}}{16}$
23. $\dfrac{14 - \sqrt{98}}{21}$
24. $\dfrac{15 - \sqrt{162}}{6}$

6-8 Operations with expressions containing radicals

OBJECTIVES: After completing this section, you should be able to:

a. **Multiply expressions containing radicals.**
b. **Rationalize denominators containing two terms.**
c. **Divide expressions containing radicals.**

In Section 6-5, we multiplied two radicals. We now concern ourselves with the multiplication of irrational expressions containing more than one term. To perform the multiplications, we apply the usual rules of multiplication.

EXAMPLE 1

Multiply $\sqrt{2}(\sqrt{18} + \sqrt{6})$.

Solution

Multiply each term of $\sqrt{18} + \sqrt{6}$ by $\sqrt{2}$ and simplify.

$$\sqrt{2}(\sqrt{18} + \sqrt{6}) = \sqrt{2}\sqrt{18} + \sqrt{2}\sqrt{6} = \sqrt{36} + \sqrt{12}$$
$$= 6 + 2\sqrt{3}$$

EXAMPLE 2

Multiply $(2 + \sqrt{3})(4 - 3\sqrt{3})$.

Solution

Using the FOIL method of simplifying gives

$$(2 + \sqrt{3})(4 - 3\sqrt{3}) = \overset{F}{2 \cdot 4} + \overset{O}{2 \cdot (-3\sqrt{3})} + \overset{I}{\sqrt{3} \cdot 4}$$
$$+ \overset{L}{\sqrt{3}(-3\sqrt{3})}$$
$$= 8 - 6\sqrt{3} + 4\sqrt{3} - 3\sqrt{9}$$
$$= 8 - 2\sqrt{3} - 9$$
$$= -1 - 2\sqrt{3}$$

EXAMPLE 3

Multiply $(\sqrt{5} - \sqrt{2})(2\sqrt{5} - 3\sqrt{2})$.

Solution

$$(\sqrt{5} - \sqrt{2})(2\sqrt{5} - 3\sqrt{2}) = \sqrt{5} \cdot 2\sqrt{5} + \sqrt{5}(-3\sqrt{2})$$
$$+ (-\sqrt{2})(2\sqrt{5})$$
$$+ (-\sqrt{2})(-3\sqrt{2})$$
$$= 10 - 3\sqrt{10} - 2\sqrt{10} + 6$$
$$= 16 - 5\sqrt{10}$$

EXAMPLE 4

Find the value of $x^2 + 8x + 13$ when $x = -4 + \sqrt{3}$.

Solution

Substituting $-4 + \sqrt{3}$ for x gives

$$(-4 + \sqrt{3})^2 + 8(-4 + \sqrt{3}) + 13 = 16 - 8\sqrt{3} + 3 - 32 \\ + 8\sqrt{3} + 13 = 0$$

EXAMPLE 5

Multiply $(4 - \sqrt{6})(4 + \sqrt{6})$.

Solution

$$(4 - \sqrt{6})(4 + \sqrt{6}) = 16 + 4\sqrt{6} - 4\sqrt{6} - 6 \\ = 16 - 6 \\ = 10$$

Note: The factors in Example 5 are conjugates [just as $(a - b)$ and $(a + b)$ are], and their product is the difference of two squares, $4^2 - (\sqrt{6})^2$.

EXAMPLE 6

Multiply $(\sqrt{a} - \sqrt{b})(\sqrt{a} + \sqrt{b})$.

Solution

Because the factors are conjugates, their product is the difference of two squares, $(\sqrt{a})^2 - (\sqrt{b})^2 = a - b$.

Conjugate factors are quite useful in rationalizing (removing square-root radicals from) denominators that contain two terms.

EXAMPLE 7

Simplify $\dfrac{2\sqrt{3} - \sqrt{2}}{\sqrt{3} + \sqrt{2}}$.

Solution

Multiplying the numerator and denominator by the conjugate of $\sqrt{3} + \sqrt{2}$ will give the difference of two squares (and remove the radicals) in the denominator.

$$\frac{2\sqrt{3} - \sqrt{2}}{\sqrt{3} + \sqrt{2}} \cdot \frac{\sqrt{3} - \sqrt{2}}{\sqrt{3} - \sqrt{2}} = \frac{6 - 2\sqrt{6} - \sqrt{6} + 2}{3 - 2} \\ = \frac{8 - 3\sqrt{6}}{1} = 8 - 3\sqrt{6}$$

We perform division with radicals by rationalizing the denominator. Thus the quotient of $(2\sqrt{3} - \sqrt{2}) \div (\sqrt{3} + \sqrt{2})$ is $8 - 3\sqrt{6}$, in simplest terms.

EXAMPLE 8
Divide $(4 + 2\sqrt{5})$ by $(5 - \sqrt{5})$.

Solution

$$\frac{4 + 2\sqrt{5}}{5 - \sqrt{5}} \cdot \frac{5 + \sqrt{5}}{5 + \sqrt{5}} = \frac{20 + 4\sqrt{5} + 10\sqrt{5} + 10}{25 - 5} = \frac{30 + 14\sqrt{5}}{20}$$

$$= \frac{2(15 + 7\sqrt{5})}{2 \cdot 10}$$

$$= \frac{15 + 7\sqrt{5}}{10}$$

We can divide a radical into an expression having several terms by dividing each term of the expression by the radical.

EXAMPLE 9
Divide $\sqrt{32} + \sqrt{24} - 3\sqrt{6}$ by $\sqrt{2}$.

Solution

$$\frac{\sqrt{32} + \sqrt{24} - 3\sqrt{6}}{\sqrt{2}} = \frac{\sqrt{32}}{\sqrt{2}} + \frac{\sqrt{24}}{\sqrt{2}} - \frac{3\sqrt{6}}{\sqrt{2}}$$
$$= \sqrt{16} + \sqrt{12} - 3\sqrt{3}$$
$$= 4 + 2\sqrt{3} - 3\sqrt{3}$$
$$= 4 - \sqrt{3}$$

Exercise 6-8

Multiply the following.

1. $\sqrt{2}(3 - \sqrt{2})$
2. $\sqrt{5}(5 + \sqrt{5})$
3. $\sqrt{6}(4 + 2\sqrt{2})$
4. $\sqrt{3}(2 - 3\sqrt{12})$
5. $\sqrt{3}(2\sqrt{2} - \sqrt{3})$
6. $\sqrt{5}(4\sqrt{5} - \sqrt{3})$
7. $\sqrt{6}(\sqrt{27} + 3\sqrt{3})$
8. $\sqrt{10}(2\sqrt{5} - 5\sqrt{2})$
9. $(\sqrt{3} + 1)(\sqrt{3} + 2)$
10. $(\sqrt{2} - 3)(\sqrt{2} + 1)$
11. $(2\sqrt{3} + 2)(\sqrt{3} - 3)$
12. $(3\sqrt{5} - 1)(\sqrt{5} - 3)$
13. $(\sqrt{3} + \sqrt{2})(\sqrt{3} + 1)$
14. $(\sqrt{2} + \sqrt{5})(\sqrt{2} + \sqrt{5})$
15. $(2\sqrt{3} + \sqrt{2})(\sqrt{3} + 2\sqrt{2})$
16. $(3\sqrt{5} - \sqrt{2})(\sqrt{5} - \sqrt{2})$

Exponents and Radicals

17. $(\sqrt{3} - \sqrt{2})(\sqrt{3} + \sqrt{2})$
18. $(\sqrt{5} + \sqrt{2})(\sqrt{5} - \sqrt{2})$
19. Evaluate $x^2 - 3x$ if $x = 2 - \sqrt{3}$.
20. Evaluate $x^2 - 2x$ if $x = 3 - 2\sqrt{2}$.
21. Evaluate $x^2 - x + 8$ if $x = 2 - \sqrt{2}$.
22. Evaluate $x^2 + 2x + 4$ if $x = 1 - \sqrt{3}$.

Simplify

23. $\dfrac{2 - \sqrt{3}}{2 + \sqrt{3}}$
24. $\dfrac{\sqrt{5} - 1}{\sqrt{5} + 1}$
25. $\dfrac{\sqrt{3} + 2\sqrt{5}}{\sqrt{5} - \sqrt{3}}$
26. $\dfrac{3 - \sqrt{2}}{\sqrt{6} - \sqrt{2}}$

Divide the following.

27. $(\sqrt{6} - \sqrt{2}) \div (\sqrt{6} + \sqrt{2})$
28. $(5 - \sqrt{3}) \div (3 - \sqrt{3})$
29. $\dfrac{\sqrt{12} - 2\sqrt{6} + 3\sqrt{27}}{\sqrt{3}}$
30. $\dfrac{3\sqrt{5} - 2\sqrt{10} - \sqrt{15}}{\sqrt{5}}$
31. $\dfrac{\sqrt{8} + \sqrt{24} - \sqrt{6}}{\sqrt{2}}$
32. $\dfrac{4\sqrt{14} - \sqrt{28} + 3\sqrt{7}}{\sqrt{7}}$

CHAPTER TEST

Compute the following.

1. $x^5 \cdot x^4$
2. $y^6 \div y^3$
3. $(-2x)^4$
4. $(3a^2b)^3$
5. $\left(\dfrac{4a}{b^2}\right)^2$
6. $(-2x^2)^0$ $(x \neq 0)$

Compute and simplify so that only positive exponents remain.

7. $6x^{-3}$
8. $(3m^{-1}n)^{-2}$
9. $(2xy^{-1})(5x^{-2}y^2)$
10. $(-6x^3y) \div (3x^{-2}y^{-2})$

Compute and simplify the following.

11. $y^{2/3} \cdot y^{3/4}$
12. $16^{3/2} + 4^{5/2}$
13. $(x^{2/5})^{7/2}$
14. $\sqrt{128}$
15. $\sqrt[3]{16p^4q^6}$
16. $\sqrt{8x^3} \cdot \sqrt{2x^2}$
17. $\dfrac{\sqrt{72}}{\sqrt{8}}$
18. $\dfrac{\sqrt{54x^3y}}{\sqrt{9xy}}$ $(x > 0)$
19. Simplify $\dfrac{\sqrt{2x}}{\sqrt{3y}}$ by rationalizing the denominator.

Compute the following.

20. $2\sqrt{5} - \sqrt{5} + 3\sqrt{5}$
21. $3\sqrt{54} + 2\sqrt{24}$
22. Multiply $(\sqrt{3} + 3)(2\sqrt{3} - 2)$
23. Multiply $(\sqrt{7} + \sqrt{3})(\sqrt{7} - \sqrt{3})$
24. Simplify $\dfrac{1 - \sqrt{2}}{1 + \sqrt{2}}$
25. Divide $\dfrac{3\sqrt{8} - 3\sqrt{6} - \sqrt{10}}{\sqrt{2}}$

7

quadratic equations in one variable

7-1 Solution of quadratic equations by factoring

OBJECTIVE: After completing this section, you should be able to solve quadratic equations in a single variable by factoring.

A **second-degree polynomial equation** is a polynomial equation that has at least one term with degree two and no term with degree higher than two. A second-degree polynomial equation is also called a **quadratic equation**. In this chapter we shall solve quadratic equations in a single variable. The standard form for such equations is $ax^2 + bx + c = 0$, where a, b, and c are constants and $a \neq 0$. Solution by factoring is based on the fact that if $a \cdot b = 0$, then a or b must equal 0.

EXAMPLE 1

Solve the equation $(x - 2)(x + 4) = 0$.

Solution

Because the product of $x - 2$ and $x + 4$ is 0, the equation can be true only if $x - 2 = 0$ or $x + 4 = 0$. But if $x - 2 = 0$ makes the equation true, then $x = 2$ is a solution to the equation. Also if $x + 4 = 0$ makes the equation true, then $x = -4$ is a solution to the equation.

Thus we have two solutions: $x = 2$, $x = -4$.

Check: If $x = 2$, $(2 - 2)(2 + 4) = 0(6) = 0$. If $x = -4$, $(-4 - 2)(-4 + 4) = -6(0) = 0$.

7-1 Solution of quadratic equations by factoring

We can use the following procedure to solve equations in one variable by factoring.

PROCEDURE	EXAMPLE
To solve a quadratic equation by factoring:	Solve $x^2 - 2x = 3$
1. Write the equation with 0 on the right side.	1. $x^2 - 2x - 3 = 0$
2. Factor the left side of the equation.	2. $(x+1)(x-3) = 0$
3. Set each factor equal to 0 and solve for x.	3. $x + 1 = 0$; so $x = -1$ $x - 3 = 0$; so $x = 3$
4. Check the solutions in the original equations.	4. Check for $x = -1$: $(-1)^2 - 2(-1) = 3$ Check for $x = 3$: $(3)^2 - 2(3) = 3$

EXAMPLE 2

Solve $y^2 - 5y = -6$ for y.

Solution

1. $y^2 - 5y + 6 = 0$
2. Factor: $(y - 3)(y - 2) = 0$
3. $y - 3 = 0$; so $y = 3$ is a solution.
   $y - 2 = 0$; so $y = 2$ is a solution.

Check: For $y = 3$: $(3)^2 - 5(3) = -6$. For $y = 2$: $(2)^2 - 5(2) = -6$.

EXAMPLE 3

Solve $6x^2 + 3x = 4x + 2$.

Solution

1. The proper form for factoring is $6x^2 - x - 2 = 0$.
2. Factoring gives $(3x - 2)(2x + 1) = 0$.

176 Quadratic Equations in One Variable

3. $3x - 2 = 0$; so $x = \frac{2}{3}$ is a solution.
$2x + 1 = 0$; so $x = -\frac{1}{2}$ is a solution.

Check: For $x = \frac{2}{3}$: Does $6(\frac{2}{3})^2 + 3(\frac{2}{3}) = 4(\frac{2}{3}) + 2$? Yes, because $6(\frac{4}{9}) + 2 = \frac{8}{3} + 2$. For $x = -\frac{1}{2}$: Does $6(-\frac{1}{2})^2 + 3(-\frac{1}{2}) = 4(-\frac{1}{2}) + 2$? Yes, because $6(\frac{1}{4}) - \frac{3}{2} = -2 + 2$.

EXAMPLE 4

Solve $(y - 3)(y + 2) = -4$ for y.

Solution

The left side of the equation is factored, but the right member is not 0. Therefore, we multiply the factors and put the equation in standard form.

$$y^2 - y - 6 = -4 \quad \text{or} \quad y^2 - y - 2 = 0$$

Factoring gives $(y - 2)(y + 1) = 0$.
Thus $y - 2 = 0$ or $y = 2$ and $y + 1 = 0$ or $y = -1$.
The solutions are $y = 2$, $y = -1$.

Check: If $y = 2$, $(2 - 3)(2 + 2) = -4$. If $y = -1$, $(-1 - 3)(-1 + 2) = -4$.

Exercise 7-1

Solve the following equations for x, y, or z.

1. $x(x - 1) = 0$
2. $x(x - 2) = 0$
3. $x(x + 3) = 0$
4. $x(x + 2) = 0$
5. $(x + 1)(x + 3) = 0$
6. $(x + 2)(x + 3) = 0$
7. $(x - 2)(x - 3) = 0$
8. $(x - 1)(x - 6) = 0$
9. $(x + 3)(x - 2) = 0$
10. $(x + 3)(x - 5) = 0$
11. $x^2 - 9 = 0$
12. $x^2 - 25 = 0$
13. $y^2 - 16 = 0$
14. $z^2 - 64 = 0$
15. $z^2 - 4z = 0$
16. $y^2 + 5y = 0$
17. $x^2 - 7x = 0$
18. $y^2 - 4y = 0$
19. $x^2 + 3x + 2 = 0$
20. $x^2 - 4x + 3 = 0$
21. $y^2 - 17y + 70 = 0$
22. $z^2 + 12z + 35 = 0$
23. $x^2 - 4x - 21 = 0$
24. $z^2 - 3z - 10 = 0$
25. $x^2 + 2x - 15 = 0$
26. $y^2 + 5y - 6 = 0$
27. $2x^2 + 5x - 12 = 0$
28. $9y^2 + 6y - 24 = 0$

29. $x^2 + 6x + 9 = 0$
30. $3y^2 + 2y - 1 = 0$
31. $3x^2 - 11x + 6 = 0$
32. $x^2 - 8x + 16 = 0$
33. $x^2 - x = x + 3$
34. $x^2 + 10x = 18x - 15$
35. $x^2 + 5x = 21 + x$
36. $x^2 + 17x = 8x - 14$
37. $1 - 2x + x^2 = 0$
38. $2 - x^2 + x = 0$
39. $6x - 4 - 2x^2 = 0$
40. $3x + 10 - x^2 = 0$
41. $2x^2 + 12x + 18 = 0$
42. $3x^2 - 12x + 9 = 0$
43. $4x^2 + 12x + 8 = 0$
44. $5x^2 - 20x + 15 = 0$
45. $(x - 3)(x - 1) = -1$
46. $(x + 4)(x - 2) = -9$
47. $(x - 1)(x + 5) = 7$
48. $(x + 2)(x + 3) = 30$
49. $x^2 = 4$
50. $x^2 = 16$

7-2 Solution of quadratic equations by completing the square

OBJECTIVES: After completing this section, you should be able to:

a. **Solve pure quadratic equations.**
b. **Use completing the square to solve quadratic equations.**

Quadratic equations that contain only the second-degree terms and constants (with no first-degree terms) are called **pure quadratic equations**. Examples of pure quadratic equations are $x^2 - 9 = 0$ and $x^2 = 3$. We can solve some of these equations easily by factoring. For example, $x^2 - 9 = 0$ is equivalent to $(x - 3)(x + 3) = 0$; so its solutions are $x = 3$ and $x = -3$. We can also use the following procedure to solve pure quadratic equations.

PROCEDURE	EXAMPLE
To solve a pure quadratic equation:	Solve $5x^2 - 30 = x^2 - 5$.
1. Write the equation so that all terms containing the variable are on one side and all other terms are on the other side, and simplify.	1. $5x^2 - x^2 = 30 - 5$ $4x^2 = 25$
2. Divide both sides by the numerical coefficient of the variable.	2. $x^2 = \frac{25}{4}$

3. Take the square root of both sides of the equation (we want **both** roots in this case, not just the principal root).

3. $x = \pm\sqrt{\frac{25}{4}}$ or $x = \pm\frac{5}{2}$. That is, $x = \frac{5}{2}$ and $x = -\frac{5}{2}$ are the solutions.

4. Check the solutions.

4. $5(\frac{5}{2})^2 - 30 = (\frac{5}{2})^2 - 5$ and $5(-\frac{5}{2})^2 - 30 = (-\frac{5}{2})^2 - 5$

EXAMPLE 1

Solve $x^2 - 16 = 0$ for x by two methods.

Solution

a. By factoring: $(x - 4)(x + 4) = 0$; so $x = 4$ and $x = -4$ are solutions of $x^2 - 16 = 0$.
b. By taking the square root: $x^2 = 16$; so $x = \pm\sqrt{16}$ or $x = \pm 4$. That is, $x = 4$ and $x = -4$ are solutions of $x^2 - 16 = 0$.

EXAMPLE 2

Solve $3x^2 - 15 = 0$.

Solution

$3x^2 = 15$; so $x^2 = 5$. Thus $x = \pm\sqrt{5}$; that is, $x = \sqrt{5}$ or $x = -\sqrt{5}$.

Check: $3(\sqrt{5})^2 - 15 = 0$ and $3(-\sqrt{5})^2 - 15 = 0$

Note that the solutions in Example 2 are irrational numbers. We did not arrive at irrational solutions in the problems we solved by factoring because they were "special" problems that could be factored easily. One method, which makes it possible to solve **all** quadratic equations in one variable, is the method called **completing the square.**

The method of completing the square depends upon the fact that

$$(a + b)^2 = a^2 + 2ab + b^2$$

Recall that in factoring perfect trinomial squares, we observed that the "middle" term was twice the product of the square roots of the other two terms.

EXAMPLE 3

What term must be added to make $x^2 + 6x$ a perfect trinomial square?

Solution

Let the b^2 represent the missing square term. Then 2 times the product of the roots is the middle term. Thus $2 \cdot x \cdot b = 6x$, or $b = 3$. Thus the missing term is $b^2 = 9$.

Check: $x^2 + 6x + 9 = (x+3)^2$

We may use completing the square to solve quadratic equations as follows.

PROCEDURE	EXAMPLE
To solve a quadratic equation by completing the square:	Solve $2x^2 - 7x = x + 14$.
1. Write the equation with like terms collected, with the constant term on the right side of the equation, and with 1 as the coefficient of the second-degree term.	1. $2x^2 - 8x = 14$ $x^2 - 4x = 7$
2. Find half the coefficient of the first-degree term. Squaring it gives the term that completes the square.	2. $\frac{1}{2}(-4) = -2$ $(-2)^2 = 4$ completes the square.
3. Add the term that will complete the square to **both** sides of the equation.	3. $x^2 - 4x + 4 = 7 + 4$
4. Write the left side of the equation as a square.	4. $(x-2)^2 = 11$
5. Take the square root of both sides, writing both square roots of the right member.	5. $x - 2 = \pm\sqrt{11}$. That is, $x - 2 = +\sqrt{11}$ or $x - 2 = -\sqrt{11}$
6. Solve the two resulting linear equations.	6. $x = 2 + \sqrt{11}$ or $x = 2 - \sqrt{11}$

Quadratic Equations in One Variable

EXAMPLE 4

Solve $x^2 - 4x + 3 = 0$ for x by completing the square.

Solution

1. $x^2 - 4x = -3$
2. $\frac{1}{2}(-4) = -2$. $(-2)^2 = 4$ will complete the square.
3. Completing the square gives $x^2 - 4x + 4 = -3 + 4$ or $x^2 - 4x + 4 = 1$.
4. So $(x - 2)^2 = 1$.
5. Then $(x - 2) = \pm 1$; That is, $x - 2 = 1$ or $x - 2 = -1$.
6. Thus $x = 3$ or $x = 1$.

Check: $3^2 - 4(3) + 3 = 0$ and $1^2 - 4(1) + 3 = 0$

EXAMPLE 5

Solve $2y^2 - 18y + 3 = 0$ for y by completing the square.

Solution

Dividing both sides by 2 gives $y^2 - 9y + \frac{3}{2} = 0$; so $y^2 - 9y = -\frac{3}{2}$.
Completing the square gives $y^2 - 9y + (-\frac{9}{2})^2 = -\frac{3}{2} + (-\frac{9}{2})^2$.
So $(y - \frac{9}{2})^2 = -\frac{3}{2} + \frac{81}{4}$, or $(y - \frac{9}{2})^2 = \frac{75}{4}$.
Then $y - \frac{9}{2} = \pm \frac{\sqrt{75}}{2}$; so $y = \frac{9}{2} \pm \frac{\sqrt{75}}{2}$.
Thus $y = \frac{9 \pm 5\sqrt{3}}{2}$; so $y = \frac{9 + 5\sqrt{3}}{2}$ or $y = \frac{9 - 5\sqrt{3}}{2}$.

Exercise 7-2

Solve the following pure quadratic equations.

1. $x^2 = 9$
2. $x^2 = 16$
3. $x^2 = 12$
4. $x^2 = 20$
5. $6x^2 - 36 = 0$
6. $5y^2 - 40 = 0$
7. $7y^2 - 28 = 0$
8. $8x^2 - 40 = 0$
9. $3x^2 - 12 = 0$
10. $5x^2 - 20 = 0$
11. $3x^2 - 4 = 0$
12. $2x^2 - 5 = 0$
13. $5y^2 - 8 = 0$
14. $8z^2 - 12 = 0$
15. $5x^2 - 3 = 3x^2 + 2$
16. $4y^2 + 2 = y^2 + 3$
17. $6x^2 - 12 = 0$
18. $3y^2 - 36 = 0$
19. $5x^2 - 3 = 3x^2 + 6$
20. $6x^2 + 4 = 4z^2 + 7$

Solve the following by completing the square.

21. $x^2 - 6x + 4 = 0$
22. $x^2 - 4x - 3 = 0$
23. $y^2 - 12y + 3 = 0$
24. $z^2 - 10z + 4 = 0$
25. $x^2 - 5x + 2 = 0$
26. $y^2 + 9z - 3 = 0$
27. $2x^2 - 8x + 1 = 0$
28. $3x^2 - 6x - 5 = 0$
29. $2y^2 - 3y + 1 = 0$
30. $3y^2 - 2y - 3 = 0$

7-3 Solution of quadratic equations by formula

OBJECTIVES: After completing this section, you should be able to:

a. Write a quadratic equation in the general form $ax^2 + bx + c = 0$.
b. Substitute the values for *a, b,* and *c,* from the general form of a quadratic equation into the quadratic formula to solve the equation.

The general form of the quadratic equation is $ax^2 + bx + c = 0$, where x is the variable (or unknown), a is the coefficient of the variable squared, b is the coefficient of the variable to the first power, and c is the term that does not contain the variable. Note that a cannot equal 0 if the equation is quadratic.

EXAMPLE 1

Write the equation $3x^2 = 2x + 1$ in general form.

Solution

Writing the equation in general form gives $3x^2 - 2x - 1 = 0$.

EXAMPLE 2

State the values that represent $a, b,$ and c of the general formula in the following equation.

a. $3x^2 + 2x + 5 = 0$
b. $4x^2 - 3x + 2 = 0$
c. $6y^2 + 2y - 3 = 0$

Solution

a. a is 3; b is +2; and c is +5.
b. a is 4; b is −3; and c is +2.
c. a is 6; b is +2; and c is −3.

We have seen in Section 7-2 how to complete the square to solve quadratic equations. We shall now complete the square with the general quadratic equation. This will give us a formula which we can use to solve quadratic equations.

The general quadratic equation is $ax^2 + bx + c = 0$. Completing the square gives

$$ax^2 + bx = -c \qquad \text{(placing constant on right)}$$

$$x^2 + \frac{b}{a}x = -\frac{c}{a} \qquad \text{(dividing by } a\text{)}$$

$$x^2 + \frac{b}{a}x + \left(\frac{b}{2a}\right)^2 = -\frac{c}{a} + \left(\frac{b}{2a}\right)^2 \qquad \text{(completing square)}$$

$$\left(x + \frac{b}{2a}\right)^2 = -\frac{c}{a} \cdot \frac{4a}{4a} + \frac{b^2}{4a^2} \qquad \text{(simplifying)}$$

$$\left(x + \frac{b}{2a}\right)^2 = \frac{b^2 - 4ac}{4a^2} \qquad \text{(simplifying)}$$

$$x + \frac{b}{2a} = \pm\sqrt{\frac{b^2 - 4ac}{4a^2}} \qquad \text{(taking square root of both sides)}$$

$$x + \frac{b}{2a} = \pm\frac{\sqrt{b^2 - 4ac}}{\sqrt{4a^2}} \qquad \text{(simplifying)}$$

$$x + \frac{b}{2a} = \pm\frac{\sqrt{b^2 - 4ac}}{2a} \qquad \text{(simplifying)}$$

$$x = -\frac{b}{2a} \pm \frac{\sqrt{b^2 - 4ac}}{2a} \qquad \text{(solving for } x\text{)}$$

$$x = \frac{-b + \sqrt{b^2 - 4ac}}{2a}$$

Thus the quadratic formula for the solution of a quadratic equation is

$$x = \frac{-b \pm \sqrt{b^2 - 4ac}}{2a}$$

We can solve a quadratic equation in general form by substituting the values of a, b, and c into the quadratic formula.

EXAMPLE 3

Use the quadratic formula to solve $x^2 - 5x + 4 = 0$.

Solution

The equation is in general form, with $a = 1$, $b = -5$, and $c = 4$.

$$x = \frac{-(-5) \pm \sqrt{(-5)^2 - 4(1)(4)}}{2(1)}$$

$$x = \frac{+5 \pm \sqrt{25 - 16}}{2}$$

so

$$x = \frac{5 \pm \sqrt{9}}{2} \quad \text{or} \quad x = \frac{5 \pm 3}{2}$$

There are two values for x:

$$x = \frac{5 + 3}{2} = 4 \quad \text{or} \quad x = \frac{5 - 3}{2} = 1$$

Note that we could have solved this quadratic equation by factoring:

$$(x - 1)(x - 4) = 0$$

so

$$x = 1 \quad \text{or} \quad x = 4$$

Use of the quadratic formula makes it unnecessary to complete the square to solve a quadratic equation that will not factor. However, we shall find that completing the square will be very useful in graphing quadratic equations in Chapter 12.

EXAMPLE 4

Solve $2y^2 + 3y - 6 = 0$ for y.

Solution

The equation is in general form, with $a = 2$, $b = +3$, and $c = -6$.

$$y = \frac{-3 \pm \sqrt{3^2 - 4(2)(-6)}}{2(2)} = \frac{-3 \pm \sqrt{9 + 48}}{4}$$

Thus

$$y = \frac{-3 \pm \sqrt{57}}{4}$$

so

$$y = \frac{-3 + \sqrt{57}}{4} \quad \text{or} \quad y = \frac{-3 - \sqrt{57}}{4}$$

EXAMPLE 5

Use the quadratic formula to solve $2x^2 = 5 - 3x$.

Solution

We must first write the equation in general form:

$$2x^2 + 3x - 5 = 0$$

Thus $a = 2$, $b = 3$, and $c = -5$; so

$$x = \frac{-3 \pm \sqrt{3^2 - 4(2)(-5)}}{2(2)} = \frac{-3 \pm \sqrt{49}}{4}$$

Thus

$$x = \frac{-3 + 7}{4} = 1 \quad \text{or} \quad x = \frac{-3 - 7}{4} = -\frac{5}{2}$$

The solutions, $x = 1$ and $x = -\frac{5}{2}$, could also be found by factoring.

Exercise 7-3

Write the following equations in general form.

1. $x^2 + 5x = 3$
2. $y^2 + 4y = 2$
3. $x^2 = 6x - 2$
4. $z^2 = 4 - 3z$
5. $3x^2 = 4x$
6. $w^2 + 4 = 0$
7. $3y = y^2 - 2$
8. $2 = 4z - z^2$
9. $2x = 1 - x^2$
10. $3x^2 = 2x - 3$

Use the quadratic formula to solve the following equations.

11. $x^2 - 5x + 6 = 0$
12. $x^2 - 3x + 2 = 0$
13. $y^2 + 7y + 10 = 0$
14. $y^2 - 8y + 16 = 0$
15. $x^2 + 2x - 15 = 0$
16. $x^2 - x - 2 = 0$
17. $3y^2 + 5y - 2 = 0$
18. $4y^2 - 4x - 3 = 0$
19. $6x^2 - x - 1 = 0$
20. $4w^2 - 8w - 5 = 0$
21. $x^2 - 9x - 3 = 0$
22. $w^2 - 4w - 2 = 0$
23. $3x^2 - 4x + 1 = 0$
24. $3y^2 - 4y - 1 = 0$
25. $y^2 - 2y - 1 = 0$
26. $3w^2 - 7w + 4 = 0$
27. $2x^2 - 6x + 3 = 0$
28. $3x^2 + 5x + 1 = 0$
29. $3x^2 + 5x = 2$
30. $4y^2 = 12y - 1$
31. $4w^2 = 72w - 1$
32. $3w^2 + 1 = 5w$
33. $4x^2 - 12x = 5$
34. $x^2 - 1 = 4x$
35. $5x = 2x^2 - 3$
36. $4x = 3 - 2x^2$
37. $3x^2 - 4 = 0$
38. $4x^2 - 3 = 0$
39. $2x^2 - 3x = 0$
40. $5x^2 - 3x = 0$
41. $2x^2 - 3x = 2x - 1$
42. $3x^2 + 2x = 1 - 3x$

43. $\dfrac{x^2}{3} - \dfrac{x}{2} + \dfrac{1}{6} = 0$ 44. $\dfrac{x^2}{4} + \dfrac{x}{3} = \dfrac{1}{6}$

45. $x - \dfrac{3}{x} = 4$ 46. $2 + \dfrac{1}{x} = 4x$

7-4 Imaginary and complex numbers

OBJECTIVES: After completing this section, you should be able to:

a. Identify imaginary numbers.
b. Solve pure quadratic equations with imaginary solutions.

The numbers we have used up to this point are real numbers (either rational numbers such as 2, −3, $\frac{5}{8}$, and $-\frac{2}{3}$ or irrational numbers such as $\sqrt{3}$, $\sqrt[3]{6}$, and π).

The following example will show that some quadratic equations cannot be solved with real numbers.

If
$$x^2 + 1 = 0$$
then
$$x^2 = -1$$

But there is no real number which, when squared, will equal −1. Thus $x^2 + 1$ has no solution in the real numbers. There is a larger set of numbers, however, in which this equation does have a solution. This set of numbers is called the set of **complex numbers**. The set of real numbers is a part of the system of complex numbers. Another part of the complex number system is the set of **imaginary numbers**. The imaginary unit is the number i having the property

$$i^2 = -1$$

Thus
$$\sqrt{-1} = \sqrt{i^2} = i$$
Then
$$\sqrt{-4} = \sqrt{-1} \cdot \sqrt{4} = 2i$$
and
$$\sqrt{-5} = \sqrt{-1}\,\sqrt{5} = i\sqrt{5}$$

A **number** of the form bi, where b is a real number, is called an **imaginary number**. We see that the solution to the equation $x^2 = -1$ is $x = \pm\sqrt{-1} = \pm i$.

EXAMPLE 1

Solve $x^2 + 9 = 0$.

Solution

$$x^2 + 9 = 0$$
$$x^2 = -9$$
$$x = \pm\sqrt{-9} = \pm\sqrt{-1}\ \sqrt{9} = \pm 3i$$

Check: $(3i)^2 + 9 = 9i^2 + 9 = 9(-1) + 9 = 0;\ (-3i)^2 + 9 = 9i^2 + 9 = 9(-1) + 9 = 0$

A **number** of the form $a + bi$, in which a and b are real numbers, is called a **complex number**. The a is the **real part** and the bi is the **imaginary part**. The number $4 + 2i$ is a complex number with real part 4 and imaginary part $2i$. It is easily seen that a real number a may be written as $a + 0i$, so that a real number is also a complex number. Similarly, the imaginary number bi may be written as $0 + bi$; so an imaginary number is also a complex number.

EXAMPLE 2

Tell whether each of the following numbers is real, imaginary, and/or complex.

a. -4
b. $3i$
c. $2 - 5i$
d. $2 - i^2$

Solution

a. -4 is real and complex.
b. $3i$ is imaginary and complex.
c. $2 - 5i$ is complex.
d. $2 - i^2 = 2 - (-1) = 3$. Thus, $2 - i^2$ is real and complex.

In Chapter 1 we indicated that the set of natural numbers is a subset of the set of integers, the set of integers is a subset of the set of rational numbers, and the set of rational numbers is a subset of the real numbers. We have now seen that the set of real numbers is a subset of the complex numbers. Another subset of the set of complex numbers is the set of imaginary numbers. We can illustrate this with the following set diagram (Figure 7-1).

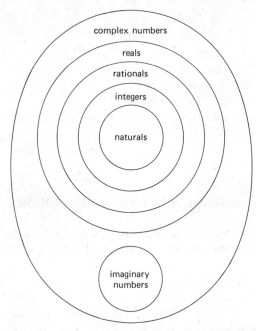

Figure 7-1 Set diagram of complex-number system.

EXAMPLE 3

a. Is a natural number also a rational number?
b. Is an integer also a complex number?
c. Is a complex number an imaginary number?

Solution

a. Yes. The natural number 4 can also be written as $\frac{4}{1}$.
b. Yes. The integer -3 can be written as $-3 + 0i$.
c. Not necessarily. The complex number $-2 + 3i$ is not imaginary because it has a real part.

188 Quadratic Equations in One Variable

Exercise 7-4

Tell whether each of the following numbers is imaginary.

1. $\sqrt{-4}$
2. $\sqrt{-16}$
3. $\sqrt[3]{-8}$
4. $\sqrt[3]{-27}$
5. $\sqrt{-8}$
6. $\sqrt{-12}$

Solve the following quadratic equations.

7. $x^2 + 16 = 0$
8. $x^2 + 4 = 0$
9. $x^2 + 12 = 0$
10. $y^2 + 27 = 0$
11. $y^2 + 64 = 0$
12. $z^2 + 81 = 0$
13. $x^2 = -24$
14. $z^2 = -36$
15. Is a rational number an imaginary number?
16. Is a natural number a real number?
17. Is an imaginary number a complex number?
18. Is an integer a rational number?

7-5 Quadratic equations with complex solutions

OBJECTIVE: After completing this section, you should be able to find and simplify complex solutions of quadratic equations.

In section 7-4 we saw that some pure quadratic equations have solutions that are imaginary numbers. It is also possible to have quadratic equations whose solutions are complex numbers, as the following example shows.

EXAMPLE 1

Solve $x^2 - 3x + 4 = 0$ for x.

Solution

By the quadratic equation

$$x = \frac{3 \pm \sqrt{9 - 16}}{2} = \frac{3 \pm \sqrt{-7}}{2}$$

That is,

$$x = \frac{3 + i\sqrt{7}}{2} \quad \text{or} \quad x = \frac{3 - i\sqrt{7}}{2}$$

There are many applications of complex numbers in the physical and applied sciences, but early mathematicians felt there were no applications if the solutions were not real. Thus the solutions were given the name imaginary by these mathematicians.

EXAMPLE 2

Solve $2x^2 + 3x + 4 = 0$ for x.

Solution

By the quadratic formula,

$$x = \frac{-3 \pm \sqrt{9 - 32}}{4} = \frac{-3 \pm \sqrt{-23}}{4}$$

That is,

$$x = \frac{-3 + i\sqrt{23}}{4} \quad \text{or} \quad x = \frac{-3 - i\sqrt{23}}{4}$$

EXAMPLE 3

Solve $y^2 - 2y + 4 = 0$ for y.

Solution

By the quadratic formula,

$$y = \frac{2 \pm \sqrt{4 - 16}}{2} = \frac{2 \pm \sqrt{-12}}{2} = \frac{2 \pm \sqrt{-1}\sqrt{4}\sqrt{3}}{2}$$
$$= \frac{2 \pm i \cdot 2\sqrt{3}}{2}$$

That is,

$$y = \frac{2 \pm 2i\sqrt{3}}{2} = \frac{2(1 \pm i\sqrt{3})}{2} = 1 \pm i\sqrt{3}$$

Thus

$$y = 1 + i\sqrt{3} \quad \text{or} \quad y = 1 - i\sqrt{3}$$

Exercise 7-5

Solve the following quadratic equations.

1. $x^2 - 3x + 4 = 0$
2. $x^2 + 5x + 7 = 0$

3. $x^2 - 2x + 2 = 0$
4. $x^2 + 3x + 5 = 0$
5. $y^2 + 4y + 5 = 0$
6. $y^2 - 4y + 6 = 0$
7. $z^2 - 2z + 3 = 0$
8. $z^2 + 2z + 4 = 0$
9. $2x^2 - 4x + 3 = 0$
10. $3x^2 + 5x + 3 = 0$
11. $3x^2 + 5x + 2 = 0$
12. $4x^2 + 6x + 1 = 0$
13. $5y^2 + 4y + 1 = 0$
14. $4y^2 + 5y + 4 = 0$
15. $3x^2 - 5x + 4 = 0$
16. $2x^2 - 6x + 5 = 0$

7-6 Word problems

OBJECTIVE: After completing this section, you should be able to solve word problems involving quadratic equations.

Many types of problems lead to quadratic equations. When quadratic equations are solved, they yield two possible solutions for a problem. The solutions must be checked carefully in the original problem as only one solution will apply in some cases.

EXAMPLE 1

Find a number such that its square is 6 less than 7 times the number.

Solution

Let x represent the number. Then x^2 is 6 less than $7x$, or $x^2 = 7x - 6$.

Solving the equation gives

$$x^2 - 7x + 6 = 0$$
$$(x - 6)(x - 1) = 0$$
$$x = 6 \quad \text{or} \quad x = 1$$

Check: $(1)^2 = 7(1) - 6$; 1 checks in the problem. $(6)^2 = 7(6) - 6$; 6 checks in the problem.

EXAMPLE 2

The sum of two numbers is 16. Their product is 48. Find the numbers.

Solution

Let $x =$ one of the numbers.
Then $16 - x =$ the other number.
The product of the two numbers is $x(16 - x) = 48$.
  Solving the equation gives

$$16x - x^2 = 48$$

$$16x - x^2 - 48 = 0 \quad \text{or} \quad x^2 - 16x + 48 = 0$$

$$(x - 4)(x - 12) = 0$$

Thus $x = 4$ or $x = 12$.
If $x = 4$, $16 - x = 12$.
If $x = 12$, $16 - x = 4$.
Thus the numbers are 4 and 12.

Check: The sum of 4 and 12 is 16. The product of 4 and 12 is 48.

Note: In factoring this quadratic equation we are answering the question, "What two numbers have 16 as their sum and 48 as their product?" The example gives practice setting of a quadratic equation, but the problem could have been solved without the use of algebra.

EXAMPLE 3

Find the dimensions of a rectangular room that is 5 feet longer than it is wide and that has an area of 300 square feet. (The formula for area is $A =$ length $\times$ width.)

Solution

Let w represent the width of the room.
Then $w + 5$ represents the length.
The area is $w(w + 5) = 300$.
  Solving this equation gives

$$w^2 + 5w = 300$$

$$w^2 + 5w - 300 = 0$$

$$(w - 15)(w + 20) = 0$$

$$w = 15 \quad \text{or} \quad w = -20$$

But the negative value has no meaning as a width of a room. If the width is $w = 15$, then the length is $w + 5 = 20$.

Check: The area of a room 15 feet by 20 feet is 300 square feet, and 20 is 5 more than 15.

EXAMPLE 4

A formula for the height h (above ground level) of a projectile launched from the top of a 300-foot building is $h = 300 + 52t - 16t^2$ (where t represents time in seconds). How many seconds after the projectile is launched will it strike the ground?

Solution

The value of h will be 0 when the object strikes the ground; so solving the equation $0 = 300 + 52t - 16t^2$ will give the length of time which passes before the projectile strikes the ground.

Solving this equation, we obtain

$16t^2 - 52t - 300 = 0$

$4(4t^2 - 13t - 75) = 0$

$4(4t - 25)(t + 3) = 0$

Since 4 cannot equal 0, either $4t - 25 = 0$ or $t + 3 = 0$.
If $4t - 25 = 0$, $t = \frac{25}{4} = 6\frac{1}{4}$. If $t + 3 = 0$, $t = -3$.

$t = -3$ does not fit the problem. Therefore, $t = 6\frac{1}{4}$ seconds is the length of time which passes before the projectile strikes the ground.

Check: If $t = \frac{25}{4}$, $300 + 52(\frac{25}{4}) - 16(\frac{25}{4})^2 = 0$.

Exercise 7-6

1. A number squared is equal to 9 times the number. What is the number?
2. A number squared is 10 more than 3 times the number. What is the number?
3. A number squared is 14 less than 9 times the number. What is the number?
4. A number squared is 4 less than 5 times the number. What is the number?
5. The sum of two numbers is 10, and their product is 24. What are the numbers?
6. The sum of two numbers is 15, and their product is 56. What are the numbers?

7. The difference between two numbers is 2, and their product is 15. What are the numbers?
8. The difference between two numbers is 3, and their product is 28. What are the numbers?
9. Find two consecutive positive integers whose product is 30.
10. Find two consecutive positive integers whose product is 72.
11. Find two consecutive negative integers whose product is 12.
12. Find two consecutive negative integers whose product is 132.
13. Find the dimensions of a rectangular room that is 5 feet longer than it is wide, if its area is 176 square feet.
14. A rectangular bulletin board is 6 inches longer than it is high. If the area of the board is 432 square inches, find its dimensions.
15. A rectangular piece of plywood is 24 inches longer than it is wide. Its area ia 1152 square inches. What is the width?
16. The length of a rectangle is 4 inches more than its width. If its area is 320 square inches, what are the dimensions?
17. A flower bed 32 feet by 42 feet is bordered by a walk of uniform width. The area of the walk is 656 feet. Find the width of the walk.

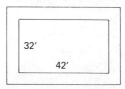

18. The flower border around a rectangular plot of grass is of uniform width, and its area is $\frac{2}{3}$ the area of the grass. If the plot of grass is 18 by 30 feet, what is the width of the border?
19. A square piece of tin is formed into a box by cutting 4-inch squares from the corners and folding the sides up. If the volume of the box is 256 cubic inches, what were the dimensions of the original piece of tin?

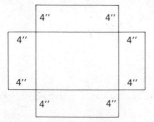

194 Quadratic Equations in One Variable

20. A rectangular piece of tin is 4 inches longer than it is wide. We can form a box from this tin by cutting 5-inch squares from each corner and folding up each side. If the volume of the box is 700 cubic inches, what were the original dimensions of the tin?

7-7 Fractional equations

OBJECTIVE: After completing this section, you should be able to solve fractional equations which can be converted to quadratic equations.

In Section 5-9 we solved fractional equations which converted into linear equations when the fractions were removed. Recall that we removed the fractions by multiplying both sides of the equation by the L.C.D. of all fractions in the equation.

EXAMPLE 1

Solve $1 + \dfrac{1}{x} = \dfrac{1}{2}$.

Solution

The L.C.D. of the fractions is $2x$. Then

$$2x\left(1 + \dfrac{1}{x}\right) = 2x\left(\dfrac{1}{2}\right) \quad \text{(Multiply both sides by } 2x.)$$

so $2x + 2 = x$ or $x + 2 = 0$. Thus $x = -2$.

Check: Does $1 + (-\tfrac{1}{2}) = \tfrac{1}{2}$? Yes, because $1 - \tfrac{1}{2} = \tfrac{1}{2}$.

Many fractional equations result in quadratic equations when the fractions are eliminated.

EXAMPLE 2

Solve the equation $x + 5 + \dfrac{4}{x} = 0$.

Solution

The L.C.D. is x.
Multiplying both sides of the equation by x gives

$$x \cdot x + x \cdot 5 + x \cdot \frac{4}{x} = x \cdot 0 \quad \text{or} \quad x^2 + 5x + 4 = 0$$

Factoring gives $(x + 1)(x + 4) = 0$.
So $x + 1 = 0$ and $x + 4 = 0$.
Thus the solutions are $x = -1$, $x = -4$.

Check: If $x = -1$, $-1 + 5 + \frac{4}{-1} = 0$. If $x = -4$, $-4 + 5 + \frac{4}{-4} = 0$.

It is especially important to check the solutions of fractional equations because multiplying both sides of the equation by factors involving the variable may introduce new solutions which are not really solutions of the original equation.

EXAMPLE 3

Solve $\dfrac{x^2 - 4x + 3}{x - 3} = -2$.

Solution

The L.C.D. is $x - 3$.
Multiplying both sides by $x - 3$ gives $x^2 - 4x + 3 = -2x + 6$.
So $x^2 - 2x - 3 = 0$.
Factoring gives $(x + 1)(x - 3) = 0$.
Thus $x = -1$ and $x = 3$ are possible solutions.

Check: Is -1 a solution? Yes, because

$$\frac{(-1)^2 - 4(-1) + 3}{-1 - 3} = -2$$

Is 3 a solution? No, because

$$\frac{3^2 - 4(3) + 3}{3 - 3} = \frac{0}{0} \neq -2$$

(Recall division by 0 is not possible.)

A solution to a converted equation that does not check because it is not a solution to the original equation (as $x = 3$ above) is called an **extraneous solution**.

EXAMPLE 4

Solve $\dfrac{6}{y - 8} - 1 = \dfrac{2}{y - 3}$.

Solution

The L.C.D. is $(y-8)(y-3)$.
Multiplying both sides of the equation by $(y-8)(y-3)$ gives

$$(y-8)(y-3) \cdot \frac{6}{y-8} - (y-8)(y-3) \cdot 1$$
$$= (y-8)(y-3) \cdot \frac{2}{y-3}$$

or

$$(6y - 18) - (y^2 - 11y + 24) = 2y - 16$$

Putting this equation in standard form gives

$$-y^2 + 15y - 26 = 0 \quad \text{or} \quad y^2 - 15y + 26 = 0$$

Factoring gives $(y-2)(y-13) = 0$.
Thus $y - 2 = 0$ or $y = 2$ and $y - 13 = 0$ or $y = 13$.

Check: Does $\frac{6}{2-8} - 1 = \frac{2}{2-3}$? Yes, because $-1 - 1 = -2$.

Does $\frac{6}{13-8} - 1 = \frac{2}{13-3}$? Yes, because $\frac{6}{5} - 1 = \frac{2}{10}$.

EXAMPLE 5

The sum of a number and its reciprocal is $\frac{5}{2}$. What is the number?

Solution

Let x be the number and $1/x$ the reciprocal of the number.
Then $x + (1/x)$ is the sum.
So $x + (1/x) = \frac{5}{2}$.
Multiplying both sides of the equation by the L.C.D., $2x$ gives

$$2x^2 + 2 = 5x \quad \text{or} \quad 2x^2 - 5x + 2 = 0$$

Thus $(2x - 1)(x - 2) = 0$; so $2x - 1 = 0$ or $x - 2 = 0$.
The numbers are $\frac{1}{2}$ and 2.

Check: $\frac{1}{2} + 2 = \frac{5}{2}$; $2 + \frac{1}{2} = \frac{5}{2}$

EXAMPLE 6

A driver increases his original speed by 8 miles an hour and saves 1 hour on a 240-mile trip. What was his original rate of speed? (distance = rate · time)

Solution

original rate = x
faster rate = $x + 8$
distance for trip = 240 miles

We can solve the distance formula ($d = rt$) for t getting $t = d/r$.
The time for the original trip is $t_1 = 240/x$.
The time for the faster trip is $t_2 = 240/(x + 8)$.
Since the original trip takes 1 hour longer,

$$t_1 - 1 = t_2 \quad \text{or} \quad \frac{240}{x} - 1 = \frac{240}{x+8}$$

Multiplying both sides of the equation by the L.C.D., $x(x+8)$, gives

$$x(x+8)\left(\frac{240}{x} - 1\right) = x(x+8)\left(\frac{240}{x+8}\right)$$

or
$$240x + 1920 - x^2 - 8x = 240x$$
so
$$0 = x^2 + 8x - 1920 = 0$$

Then $0 = (x - 40)(x + 48)$; so the original rate is $x = 40$.

Check: $x = -48$ is not a foward rate of speed. If $x = 40$ miles/hr, $40t = 240$; so $t = 6$ hours. Then 8 miles an hour faster (48 miles/hr) for $t - 1$ hours (5) is $48 \cdot 5 = 240$.

Exercise 7-7

Solve for x, y, or z.

1. $\dfrac{1}{x} = 4$

2. $\dfrac{3}{y} = 6$

3. $\dfrac{16}{z} = 4z$

4. $\dfrac{100}{x} = 25x$

5. $\dfrac{1}{x} = \dfrac{x}{4}$

6. $\dfrac{2}{y} = \dfrac{y}{8}$

7. $1 + \dfrac{2}{x} = 3$

8. $3 + \dfrac{1}{y} = 4$

9. $x - 9 = \dfrac{-8}{x}$

10. $x - 2 + \dfrac{1}{x} = 0$

11. $\dfrac{x}{2} + \dfrac{1}{2x} = 1$

12. $\dfrac{x}{2} + \dfrac{4}{x} = 3$

13. $\dfrac{x}{4} + \dfrac{3}{x} = 2$

14. $\dfrac{x}{4} + \dfrac{8}{x} = -3$

15. $\dfrac{y-2}{y-4} = \dfrac{y}{y-3}$

16. $\dfrac{2}{z-3} = \dfrac{3}{z+2}$

17. $\dfrac{5}{z+4} - \dfrac{3}{z-2} = 4$

18. $\dfrac{3}{x-1} + \dfrac{2}{x-2} = 2$

19. $\dfrac{1}{x} + \dfrac{1}{x+2} = \dfrac{5}{12}$

20. $\dfrac{x+2}{x^2-9} + \dfrac{2}{x+3} = 2$

21. $\dfrac{1}{y} + \dfrac{2}{y-1} - \dfrac{8}{y+3} = 0$

22. $\dfrac{3}{z-3} + \dfrac{8}{z-8} = \dfrac{7}{z+3}$

23. $\dfrac{2x+6}{x+3} = \dfrac{x}{x-2}$

24. $\dfrac{10x+5}{2x+1} = \dfrac{6}{x+4}$

25. $\dfrac{x^2+5x}{x+5} = 3$

26. $\dfrac{2x^2-4x}{x-2} = 8$

27. $\dfrac{x^2-3x}{x-3} = \dfrac{x^2}{x+2}$

28. $\dfrac{x^2-4x-5}{x-5} = \dfrac{x^2+5x+4}{x+4}$

29. The reciprocal of a number is 4 times the number.
30. The reciprocal squared of a number is 16. What is the number?
31. The sum of a number and its reciprocal is $\tfrac{13}{6}$. What is the number?
32. The reciprocal of a number is $\tfrac{1}{4}$ the number. What is the number?
33. Twice the reciprocal of a number is $\tfrac{1}{8}$ of the number. What is the number?
34. A man averaged 54 miles/hr on a trip to Scottsville and 60 miles/hr on the return trip. If the return trip took one hour less time, how many miles was the trip one way?
35. One plane travels 60 miles/hr faster than another and saves one hour on a 1200-mile trip. How fast does each plane travel?
36. When two capacitors c_1 and c_2 are connected in series in an electrical circuit, their equivalent capacitance c is given by

$$\dfrac{1}{c} = \dfrac{1}{c_1} + \dfrac{1}{c_2}$$

If a capacitor of 300 microfarads (μF) is connected in series with a capacitor of 500 microfarads, what is the equivalent capacitance?

CHAPTER TEST

Solve the following quadratic equations.

1. $x^2 = 36$
2. $3x^2 - 15 = 0$
3. $x^2 - 5x - 6 = 0$
4. $y^2 - 16y + 48 = 0$
5. $x^2 + 7x = 10x + 10$
6. $2x^2 + 5x - 12 = 0$
7. $(y - 2)(y - 5) = -2$
8. $x^2 - 5x + 3 = 0$
9. $y^2 = 4y + 3$
10. $2 = 4z - z^2$
11. $4x^2 = 5x - 1$
12. $x^2 + 4 = 0$
13. $x^2 + 25 = 0$
14. $2x^2 - 5x + 4 = 0$
15. $3y^2 + 6y + 5 = 0$

Identify each of the following numbers as one or more of natural number, rational number, real number, imaginary number, or complex number.

16. $\frac{5}{7}$
17. $\sqrt{-4}$
18. $\frac{15}{3}$
19. $5 + \sqrt{5}$
20. $6 - 3i$
21. The square of a number is 6 more than 5 times the number. What is the number?
22. The width of a rectangular picture is 3 inches less than its length. Find the dimensions of the picture if its area is 180 square inches.
23. Solve $\frac{x}{2} - 1 = \frac{12}{x}$ for x.
24. Solve $\frac{1}{y-2} = \frac{y+3}{y+10}$ for y.
25. The sum of a number and its reciprocal is $\frac{37}{6}$. What is the number?
26. One person drives his auto 10 miles/hr faster than another and saves one hour on a 300-mile trip. How fast was each auto traveling?

linear equations in two unknowns

8-1 Solution of linear equations in two variables

OBJECTIVES: After completing this section, you should be able to:

a. Give a partial listing of the solutions of a linear equation in two unknowns.
b. Plot points and graph linear equations on a rectangular coordinate system.

We studied linear equations in one unknown in Chapter 3. We shall now consider linear equations (or first-degree equations) in **two** unknowns. The equation $3x + 2y = 12$ is a linear equation in two variables. **Any pair of values for x and y that satisfies the equation is called a solution to the equation.** We write the solution as an **ordered pair** of numbers (a, b), where the first number (a) is the replacement for x and the second number is the replacement for y. Thus the ordered pair $(2, 3)$ is **a solution** to the equation $3x + 2y = 12$. But there are other solutions to the equation; for example $(0, 6)$, $(4, 0)$ and $(\frac{2}{3}, 5)$ are also solutions to $3x + 2y = 12$. It is impossible to list all the ordered pairs in the solution set of a linear equation in two variables since there are an infinite number of them. A partial listing of the solutions can be given. This listing is frequently given in the form of a table.

EXAMPLE 1

Use a table to give a partial listing of the solutions to the equation $y = (6 - x)/2$.

Solution

Assigning arbitrary values for x and solving the equation for y gives a partial list of solutions. Letting x equal each integer between -1 and 7 gives a partial list of solutions to $x + 2y = 6$.

$(-1, \frac{7}{2})$, $(0, 3)$, $(1, \frac{5}{2})$, $(2, 2)$, $(3, \frac{3}{2})$, $(4, 1)$, $(5, \frac{1}{2})$, $(6, 0)$, $(7, -\frac{1}{2})$

x	y
-1	$\frac{7}{2}$
0	3
1	$\frac{5}{2}$
2	2
3	$\frac{3}{2}$
4	1
5	$\frac{1}{2}$
6	0
7	$-\frac{1}{2}$

The x-value is written as the first **component**, and the y-value is written as the second component of the ordered pairs which are solutions. In Example 1, $(4, 1)$ is a solution but $(1, 4)$ is not. We found the partial listing of solutions by substituting values in for x and determining the corresponding values for y.

Another method of describing the solutions of a linear equation is to draw a graph containing points that satisfy the equation. We use a rectangular coordinate system to make graphs of linear equations. We construct a rectangular coordinate system by drawing a horizontal axis (usually called the x-axis) and a vertical axis (usually called the y-axis). The point where the axes intersect is called the **origin**. The axes divide the graph into four parts, called **quadrants**. On the x-axis, values to the right of the origin are positive, and values to the left are negative. Values above the origin on the y-axis are positive, and values below are negative. A scale is established on each of the lines, starting at the origin. (See Figure 8-1.)

A point in the plane is designated by a pair of numbers (a, b), where a is the signed perpendicular distance from the y-axis (called the **abscissa**) and b is the signed perpendicular distance from the x-axis (called the **ordinate**). We say that a and b are the **coordinates** of the point. To **plot** a point $P(a, b)$ means to **plot** the point with the x-coordinate a and y-coordinate b.

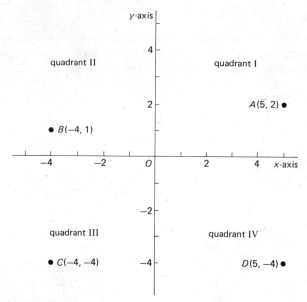

Figure 8-1

EXAMPLE 2

Plot the points $A(5, 2)$, $B(-4, 1)$, $C(-4, -4)$, and $D(5, -4)$.

Solution

The points are plotted in Figure 8-1.

The **graph** of an equation is the picture that is drawn by plotting the points whose coordinates (x, y) satisfy the equation. Since there are an infinite number of points which satisfy an equation, we cannot hope to plot them all. But we can plot enough points to determine the general outline of the graph, and we can then connect the points with a smooth curve.

EXAMPLE 3

Graph the equation $y = 3x + 2$.

Solution

To graph this equation, we choose some sample values of x and find the corresponding values of y. Placing these values in a table, we have the coordinates of sample points to plot. When we have enough points to determine that the graph is a straight line, we connect the points to complete the graph. The table and graph are shown in Figure 8-2.

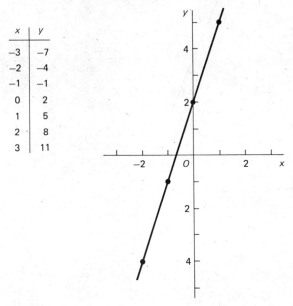

x	y
-3	-7
-2	-4
-1	-1
0	2
1	5
2	8
3	11

Figure 8-2

The graph of the equation of Example 3 is a straight line. All first-degree equations have straight lines as their graphs (thus the name **linear**). Only two points are required to determine the graph of a linear equation (two points determine a straight line). A third point should be plotted as a check on the graph.

EXAMPLE 4

Graph the equation $3x + y = 9$.

Solution

a. Plot any two points that satisfy the equation:
   i) If $x = 0$, $y = 9$; so $(0, 9)$ is one point.
   ii) If $x = 5$, $y = -6$; so $(5, -6)$ is another point.
b. Draw a straight line through the points (see Figure 8-3).
c. Plot a third point as a check. If $x = 2$, $y = 3$; so $(2, 3)$ should be on the line.

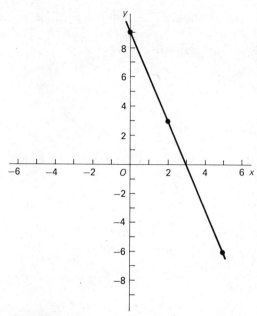

Figure 8-3

EXAMPLE 5

Graph $x = 2$.

Solution

We may think of this equation as $x + 0y = 2$, which indicates that $x = 2$ for all values of y. That is, y can have any value and x will still be 2. (See Figure 8-4.)

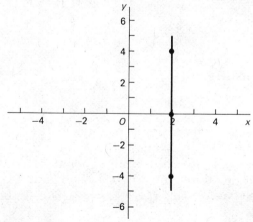

Figure 8-4

EXAMPLE 6

Graph $y = -1$.

Solution

We may write the equation $y = -1$ as $0x + y = -1$, which indicates $y = -1$ regardless of what value is substituted for x. That is, x can have any value and y will still equal -1. [Note that $(0, -1)$, $(2, -1)$, and $(-2, -1)$ satisfy the equation $y = -1$.] The graph is shown in Figure 8-5.

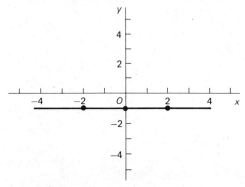

Figure 8-5

The two points where the graph of a linear equation crosses the x- and y-axis are called the **x-intercept** and the **y-intercept**, respectively. These two points are usually easy to find; so they are frequently used to determine the graph of a linear equation.

EXAMPLE 7

Find the x-intercept and y-intercept of the graph of $3x + 2y = 6$.

Solution

Because the x-intercept is on the x-axis, the y-coordinate of the point is 0. Substituting 0 for y in $3x + 2y = 6$ and solving give $x = 2$. Thus the graph meets the x-axis at $(2, 0)$. We say the x-intercept is 2.

Because the y-intercept is on the y-axis, the x-coordinate is 0. Substituting $x = 0$ into the equation gives $y = 3$. The line crosses the y-axis at $(0, 3)$. We say the y-intercept is 3.

EXAMPLE 8

Use the intercept method to graph $5x - 3y = 15$.

Solution

Letting $x = 0$, we find a y-intercept of -5.
Letting $y = 0$, we find an x-intercept of 3. (See Figure 8-6.)
A third point [such as $(6, 5)$] can be used as a check.

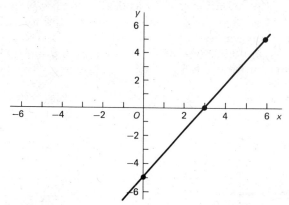

Figure 8-6

The intercept method may not always be the most convenient method to use when graphing a linear equation, especially if the intercepts are fractional values (for example, a y-intercept of $\frac{13}{25}$ would be difficult to plot). It should also be pointed out that the intercept method will not work if the line passes through the origin, because there would be intersection at only one point $(0, 0)$.

Exercise 8-1

Graph the following linear equations.

1. $y = 6x - 2$
2. $y = 2x + 8$
3. $y = 4x + 3$
4. $y = 3x + 2$
5. $y = 3x - 8$
6. $y = 2x - 1$
7. $y = \dfrac{5 - x}{2}$
8. $y = \dfrac{7 - x}{3}$
9. $y = \dfrac{10 - x}{2}$
10. $y = \dfrac{8 + 3x}{2}$
11. $3x - y = 6$
12. $5x - y = 10$
13. $2x - 3y = 12$
14. $3x + 2y = 6$
15. $6x + 5y = 9$
16. $4x + 5y = 8$
17. $y = 3$
18. $x = -5$

Use the intercept method to graph the following.

19. $3x + 2y = 12$
20. $3x - 2y = 12$
21. $4x + 2y = 8$
22. $5x - 3y = 15$
23. $6x - 3y = 12$
24. $x - 3y = 9$
25. $2x - y = 4$
26. $5x + 2y = 10$

Graph the following.

27. $5x + 2y = 6$
28. $4x + 5y = 10$
29. $5x - 3y = 9$
30. $3x - 4y = 6$
31. $5x - 3y = 8$
32. $3x - 4y = 2$

8-2 Slope of a line

OBJECTIVE: After completing this section, you should be able to find the slope of a line from its graph and from its equation.

The **slope** of a straight line is a measure of the steepness of the line. We find the slope of a line by determining how much the y-value changes on the line when x is changed by one unit. We can use the following formula to find the slope of a line:

$$m = \frac{\text{change in } y}{\text{change in } x}$$

If a line passes through the points $P_1(x_1, y_1)$ and $P_2(x_2, y_2)$ (see Figure 8-7), its slope is

$$m = \frac{y_2 - y_1}{x_2 - x_1}$$

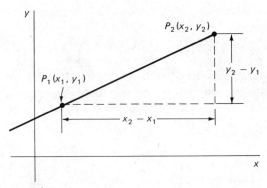

Figure 8-7

We use this formula to find the slope of any line (which is not vertical) if we know the coordinates of two points on the line.

We shall see that when we use the formula it does not matter which point we choose as $P_2(x_2, y_2)$.

EXAMPLE 1

Find the slope of the line that passes through $(-2, 1)$ and $(4, 3)$.

Solution

The graph of this line is shown in Figure 8-8. The slope is

$$m = \frac{3-1}{4-(-2)} = \frac{2}{6} = \frac{1}{3}$$

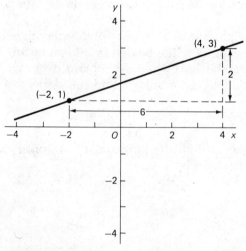

Figure 8-8

This means that a point 3 units to the right and 1 unit up from any point on the line is also on the line. Thus if $(-2, 1)$ is on the line, $(-2+3, 1+1) = (1, 2)$ is also on the line.

EXAMPLE 2

Find the slope of the line passing through $(-2, 3)$ and $(2, -1)$.

Solution

The graph of this line is shown in Figure 8-9. The slope is

$$m = \frac{-1-3}{2-(-2)} = \frac{-4}{4} = -1 \quad \text{or} \quad \frac{-1}{1}$$

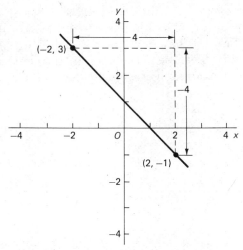

Figure 8-9

This slope tells us that a point which is **down** 1 unit and 1 unit to the **right** of any point on the line is also on the line.

Note that if we interchange the points in the formula, we shall get the same slope, **as long as** we find the change in y in the numerator and the **corresponding** change in x in the denominator. That is, we can also find the slope in the above example by

$$m = \frac{3-(-1)}{-2-2} = \frac{4}{-4} = -1$$

From the previous examples we see that the slope describes the direction of a line as follows:

1. The slope is positive if the line slopes upward toward the right.
2. The slope is negative if the line slopes downward toward the right.

Two additional facts should be stated:

3. The slope of a horizontal line is 0 because the y-values

are the same for any two points on the line, making $y_2 - y_1 = 0$.
4. The slope of a vertical line is undefined, because $x_2 - x_1$ would equal 0, and division by 0 is undefined.

We can find the slope of a line from its equation if it is written in the proper form. **The equation $y = mx + b$ is called the slope-intercept form of the equation of a line.** The coefficient of x, namely, m, is the slope of the line, and the constant b is the value of y where the graph crosses the y-axis (b is called the y-intercept).

EXAMPLE 3

Find the slope and y-intercept of the line whose equation is $y = 2x + 1$.

Solution

The slope is 2, or $\frac{2}{1}$, and the y-intercept is 1. Figure 8-10 shows the slope and the y-intercept.

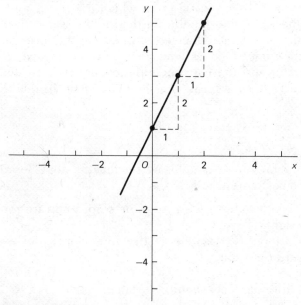

Figure 8-10

EXAMPLE 4

Find the slope of the line whose equation is $y = \frac{1}{2}x - 3$.

Solution

The slope is $\frac{1}{2}$.

EXAMPLE 5

Find the slope of the line whose equation is $3x + 2y = 4$.

Solution

We must solve the equation for y to put the equation in the slope-intercept form:

$$2y = -3x + 4 \quad \text{or} \quad y = \frac{-3}{2}x + 2$$

Thus the slope is $\frac{-3}{2}$. The y-intercept is 2.

Exercise 8-2

Find the slopes of the lines passing through the following points.

1. (1, 2) and (2, 4)
2. (2, 3) and (3, 4)
3. (−1, 3) and (0, 6)
4. (−1, 2) and (2, −3)
5. (2, 1) and (3, −4)
6. (−1, −2) and (−2, −3)
7. (3, 2) and (−1, −6)
8. (−4, 2) and (2, 4)
9. (7, 3) and (−6, 2)
10. (10, 2) and (5, 7)
11. (10, 2) and (5, 2)
12. (3, 6) and (3, 8)

Find the slopes and the y-intercepts of lines whose equations are

13. $y = 4x - 6$
14. $y = 3x + 1$
15. $y = 3x + 2$
16. $y = 2x - 1$
17. $y = \frac{1}{2}x - \frac{2}{3}$
18. $y = \frac{2}{3}x + \frac{1}{2}$
19. $y = -8x + 2$
20. $y = -\frac{1}{2}x + 2$
21. $2x - y = 3$
22. $2x + y = 2$
23. $2x + 3y = 6$
24. $4x - y = 3$
25. $3x + 6y = 1$
26. $2x - 3y = 2$
27. $4x - 2y = 3$
28. $y = 3$
29. $x = 4$
30. $y = -8$

8-3 Using slopes to graph lines

OBJECTIVE: After completing this section, you should be able to graph a line given the slope and the y-intercept or the slope and a point on the line.

If we are given the slope and the y-intercept of a line, we can use the following procedure to sketch the graph of the line (see Figure 8-11).

PROCEDURE	EXAMPLE
To graph a line given the slope and y-intercept:	Sketch the graph of the line which has slope 3 and y-intercept -2.
1. Plot the intercept on the y-axis.	1. Plot the point $(0, -2)$ (See Figure 8-11).
2. Write the slope as a fraction b/a. If the fraction is negative, place the minus sign in the numerator.	2. $m = 3 = \frac{3}{1}$
3. Plot a point which is a units to the right and b units above the y-intercept if b is positive, or b units below the y-intercept if b is negative.	3. Plot the point which is 1 unit to the right and 3 units above the point $(0, -2)$. The point is $(1, 1)$. (See Figure 8-11.)
4. Plot a second point which is a units to the right and b units above (or below) the point found in Step 3.	4. Plot the point which is 1 unit to the right and 3 units above the point $(1, 1)$. The point is $(2, 4)$. (See Figure 8-11.)
5. Draw the line through the points.	5. Figure 8-11 gives the graph.

EXAMPLE 1

Graph the line which has y-intercept 4 and slope $\frac{-1}{2}$.

Solution

The intercept is $(0, 4)$ and the slope is $\frac{-1}{2}$.

1. Plot the point $(0, 4)$.

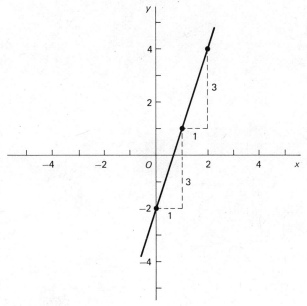

Figure 8-11

2. The slope indicates that y changes by -1 when x changes by 2.
3. Plot the point which is 2 units to the right and 1 unit **below** the point (0, 4). (We move **down** as we move to the right because the slope is negative.) This point is (2, 3).
4. Plot a second point which is 2 units to the right and 1 unit below the point (2, 3). This point is (4, 2).
5. The graph of the equation is shown in Figure 8-12.

It is also possible to graph a straight line if we have its slope and a point that it passes through. We simply plot the point which is given, and then use the slope to plot other points.

EXAMPLE 2

Graph the line which passes through $(-1, 2)$ with slope $\frac{2}{3}$.

Solution

We plot the point $(-1, 2)$ and then plot a second point, $(2, 4)$, by moving 3 units to the right and 2 units above $(-1, 2)$. We can plot a third point, $(5, 6)$, by moving in the same manner from

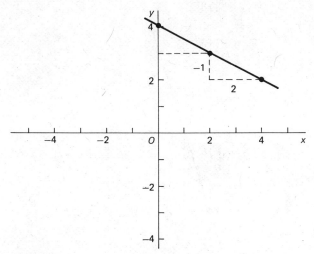

Figure 8-12

the second point. But we can also write the slope $\frac{2}{3}$ as $\frac{-2}{-3}$, which implies we can move 3 units to the left and 2 units down to plot the point $(-4, 0)$. Note that it is also a point on the graph. The graph is shown in Figure 8-13.

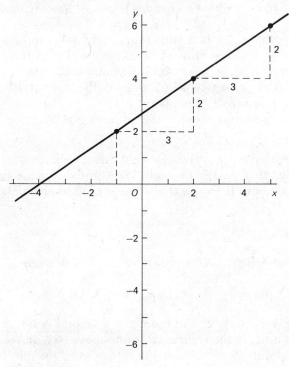

Figure 8-13

EXAMPLE 3

Graph a line with 0 slope that passes through (2, 1).

Solution

If the slope is 0, the y-values at every point must be the same; so the line is horizontal, passing through (2, 1). (See Figure 8-14.)

We can graph the equation of a straight line easily if the equation is written in the slope-intercept form, for we can read the slope and the y-intercept from the equation and plot points immediately.

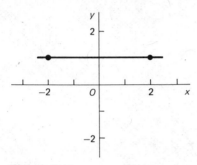

Figure 8-14

EXAMPLE 4

Graph the equation $y = 3x - \frac{1}{2}$.

Solution

The y-intercept is $-\frac{1}{2}$, and the slope is 3. The graph is as shown in Figure 8-15.

EXAMPLE 5

Graph the equation $6x - 3y = 4$.

Solution

We first solve the equation for y, which puts the equation in the slope-intercept form.

$$-3y = -6x + 4 \quad \text{or} \quad y = 2x - \tfrac{4}{3}$$

Thus the slope is 2 and the y-intercept is $\tfrac{-4}{3}$. The graph is shown in Figure 8-16.

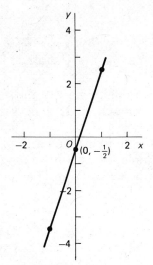

Figure 8-15

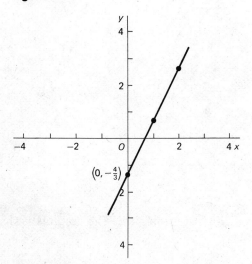

Figure 8-16

Exercise 8-3

Graph the line which has

1. slope 3 and y-intercept -2
2. slope 2 and y-intercept 3
3. slope $\frac{1}{2}$ and y-intercept 3
4. slope $\frac{2}{3}$ and y-intercept -1

5. slope -2 and y-intercept $\frac{1}{2}$
6. slope -1 and y-intercept $\frac{3}{2}$
7. slope 0 and y-intercept -4
8. slope 0 and y-intercept 0

Graph the line which

9. passes through (2, 0) with slope $\frac{1}{2}$
10. passes through (0, 0) with slope 2
11. passes through $(-1, 3)$ with slope -2
12. passes through (1, 1) with slope $-\frac{1}{3}$
13. passes through (2, 2) with slope -2
14. passes through $(3, -1)$ with slope 1
15. passes through (1, 1) with 0 slope
16. passes through $(-1, 1)$ with undefined slope
17. Graph the equation $y = 4x + 3$.
18. Graph the equation $y = \frac{-1}{2}x - 2$.
19. Graph the equation $y = 3x + \frac{1}{2}$.
20. Graph the equation $y = 4$.
21. Graph the equation $4x - y = -3$.
22. Graph the equation $3x - 2y = 6$.

8-4 Solution of linear equations in two unknowns

OBJECTIVE: After completing this section, you should be able to solve systems of linear equations in two unknowns.

In the previous sections, we graphed linear equations in two unknowns and observed that the graphs are straight lines. Each point on the graph represents a value of x and y that satisfies the equation. Clearly, there is an infinite number of values for x and y that will satisfy the equation. Now suppose we have two equations in two unknowns, and we wish to find the solution(s) that satisfy both equations. We can find the solution by graphing both equations and noting their point(s) of intersection. If the graphs intersect at a point, both equations are satisfied by the same pair of values, and they thus have a common solution. The equations are referred to as a **system of equations**, and values that satisfy both equations are the solutions to the system.

EXAMPLE 1

Use graphing to find the solution of $4x + 3y = 11$ and $2x - 5y = -1$.

Solution

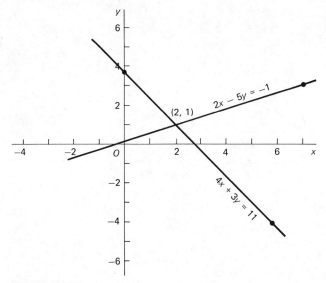

Figure 8-17

The graphs of the two equations intersect (meet) at the point (2, 1). (See Figure 8-17.) The solution of the system is $x = 2$, $y = 1$.

Note that these values satisfy both equations.

$$4(2) + 3(1) = 11 \quad \text{and} \quad 2(2) - 5(1) = -1$$

If the graphs of two equations are parallel, they have no point in common and thus the equations have no common solution. Such a system of equations is called **inconsistent.**

EXAMPLE 2

Find the solution, if it exists, for $4x + 3y = 4$ and $8x + 6y = 9$.

Solution

The graphs are parallel; so the equations are inconsistent. As Figure 8-18 shows, there is no solution to the system.

It is also possible to graph two equations and obtain only one line. When this happens, the equations are equivalent, and the system is called **dependent.**

8 - 4 Solution of linear equations in two unknowns   219

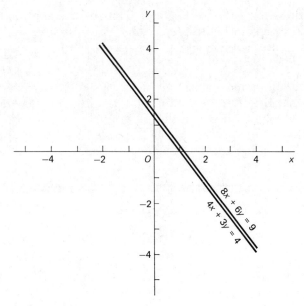

Figure 8-18

EXAMPLE 3

By means of a graph, find the solution of the equations $2x+3y=6$ and $4x+6y=12$.

Solution

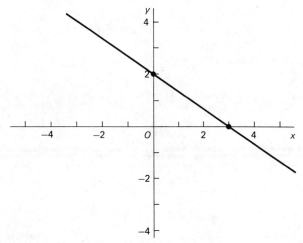

Figure 8-19

Both equations have the same graph. Thus every point on the line represents a solution to the system. The system is dependent. (See Figure 8-19.)

While graphic means of solving a system of equations illustrate what the solution represents, it is not very accurate. There are algebraic methods for solving the equations, including elimination of one of the variables by **substitution**. The procedure follows.

PROCEDURE	EXAMPLE
To solve a system of two equations in two unknowns by substitution:	Solve the system containing $2x + 3y = 4$ and $x - 2y = 3$.
1. Solve one of the equations for one of the variables in terms of the other.	1. $x = 2y + 3$
2. Substitute the value determined in Step 1 in the other equation to give one equation in one unknown.	2. $2(2y + 3) + 3y = 4$
3. Solve this linear equation for the unknown.	3. $4y + 6 + 3y = 4$ $7y = -2$ $y = -\frac{2}{7}$
4. Substitute this solution in the equation in Step 1 or in one of the original equations to solve for the other unknown.	4. $x = 2(\frac{-2}{7}) + 3$; so $x = \frac{17}{7}$ or $x - 2(\frac{-2}{7}) = 3$; so $x = \frac{17}{7}$
5. Check the solution by substituting for x and y in both original equations.	5. $2(\frac{17}{7}) + 3(\frac{-2}{7}) = 4$ or $\frac{34}{7} - \frac{6}{7} = 4$ and $\frac{17}{7} - 2(\frac{-2}{7}) = 3$

EXAMPLE 4

Solve the system:

$$4x + 5y = 18$$
$$x - 3y = -4$$

Solution

1. $x = 3y - 4$
2. $4(3y - 4) + 5y = 18$
3. $12y - 16 + 5y = 18$
   $17y = 34$
   $y = 2$
4. $x = 3(2) - 4$
   $x = 2$
5. $4(2) + 5(2) = 18$ and $(2) - 3(2) = -4$

Note that if we graphed the two equations, they would intersect at the point (2, 2).

We can also eliminate one of the variables in a system by **addition** or **subtraction**. If one of the variables has identical coefficients in both equations, we can eliminate that variable by subtracting one equation from the other.

EXAMPLE 5

Solve the system:

$3x + 2y = 7$ (1)

$4x + 2y = 5$ (2)

Solution

$-x + 0y = 2$ subtracting (2) from (1)

$\therefore x = -2$

$-6 + 2y = 7$ substituting in (1) to solve for y

$\therefore y = \frac{13}{2}$ the solution is $(-2, \frac{13}{2})$

$4(-2) + 2(\frac{13}{2}) = 5$ check in both equations

and

$3(-2) + 2(\frac{13}{2}) = 7$

Note that if we graphed Equations (1) and (2), the graphs would intersect at the point $(-2, \frac{13}{2})$.

If one of the variables has identical coefficients, except for signs, in both equations, we can eliminate that variable by adding the equations.

EXAMPLE 6

Solve the system:

$4x - 3y = 7$ \quad (1)

$2x + 3y = 5$ \quad (2)

Solution

$6x + 0y = 12$ \qquad adding (1) and (2)

$x = 2$

$4(2) - 3y = 7$ \qquad substituting in (1)

$y = \frac{1}{3}$

$4(2) - 3(\frac{1}{3}) = 7$ \qquad check in both equations

$2(2) + 3(\frac{1}{3}) = 5$

If neither of the variables has identical coefficients, we can multiply one or both equations by numbers that will make the coefficients of one of the variables identical.

EXAMPLE 7

Solve the system:

$2x + 3y = 3$ \quad (1)

$3x - 2y = 4$ \quad (2)

Solution

$4x + 6y = 6$ \qquad multiplying (1) by 2

$9x - 6y = 12$ \qquad multiplying (2) by 3

$13x = 18$ \qquad adding (1) and (2)

$x = \frac{18}{13}$

$2(\frac{18}{13}) + 3y = 3$ \qquad substituting in (1)

$3y = 3 - \frac{36}{13}$

$3y = \frac{3}{13}$

$y = \frac{1}{13}$

$3(\frac{18}{13}) - 2(\frac{1}{13}) = 4$ \qquad check in both equations

$2(\frac{18}{13}) + 3(\frac{1}{13}) = 3$

8-4 Solution of linear equations in two unknowns

We can now summarize the procedure for **solving a system of two equations in two unknowns by addition or subtraction.**

PROCEDURE	EXAMPLE
To solve a system of two equations in two unknowns by addition or subtraction:	Solve the system: $2x - 5y = 4$ $x + 2y = 3$
1. If necessary, multiply one or both equations by a number that will make the coefficients of one of the variables identical, except perhaps for signs.	1. $2x - 5y = 4$ $2x + 4y = 6$
2. Add or subtract the equations to eliminate one of the variables.	2. Subtracting gives $0x - 9y = -2$
3. Solve for the variable in the resulting equations.	3. $y = \frac{2}{9}$
4. Substitute the solution in one of the original equations and solve for the remaining variable.	4. $2x - 5(\frac{2}{9}) = 4$ $2x = 4 + \frac{10}{9} = \frac{36}{9} + \frac{10}{9}$ $x = \frac{23}{9}$
5. Check the solutions in both original equations.	5. $\frac{23}{9} + 2(\frac{2}{9}) = 3$ $2(\frac{23}{9}) - 5(\frac{2}{9}) = 4$

EXAMPLE 8

Solve the system:

$2x - 7y = 4$

$3x + 2y = 3$

Solution

1. $6x - 21y = 12$
   $6x + 4y = 6$
2. $ -25y = 6$
3. $y = -\frac{6}{25}$ (or -0.24)

4. $2x - 7(\frac{-6}{25}) = 4$
$$2x = 4 - \frac{42}{25}$$
$$2x = \frac{58}{25}$$
$$x = \frac{29}{25} \quad \text{(or 1.16)}$$
5. $3(\frac{29}{25}) + 2(\frac{-6}{25}) = 3 \quad$ since $\frac{87}{25} - \frac{12}{25} = \frac{75}{25}$

and

$2(\frac{29}{25}) - 7(\frac{-6}{25}) = 4 \quad$ since $\frac{58}{25} + \frac{42}{25} = \frac{100}{25}$

EXAMPLE 9

Solve the system:

$$4x - 2y = 3$$
$$2x - y = 5$$

Solution

a. $4x - 2y = 3$
$\underline{4x - 2y = 10}$

b. $0x - 0y = -7$; that is, $0 = 7$. The system is solved when $0 = 7$.

This is impossible; so there are no solutions to the system. The equations are inconsistent. (Their graphs are parallel lines.)

EXAMPLE 10

Solve the system:

$$6x - 2y = 4$$
$$3x - y = 2$$

Solution

a. $6x - 2y = 4$
$\underline{6x - 2y = 4}$

b. $0x - 0y = 0$; that is, $0 = 0$. The equations have the same solutions when $0 = 0$.

This is an identity; so the two equations share infinitely many solutions. The equations are dependent (their graphs coincide, or are the same).

EXAMPLE 11

Solve the system:

$$\frac{x}{3} + \frac{y}{4} = 3$$

$$\frac{x}{4} - y = -\frac{5}{2}$$

Solution

We first multiply both equations by their L.C.D.'s to remove the fractions:

$$12\left(\frac{x}{3} + \frac{y}{4}\right) = 12(3) \quad \text{or} \quad 4x + 3y = 36 \quad (1)$$

$$4\left(\frac{x}{4} - y\right) = 4\left(-\frac{5}{2}\right) \quad \text{or} \quad x - 4y = -10 \quad (2)$$

Solving Equation (2) for x and substituting the result in Equation (1) gives

$$4(4y - 10) + 3y = 36$$

Solving gives $16y - 40 + 3y = 36$, or $19y = 76$; so $y = 4$. Substituting in Equation (2) gives $x - 16 = -10$, or $x = 6$.

Check: $\frac{6}{3} + \frac{4}{4} = 3$ and $\frac{6}{4} - 4 = -\frac{5}{2}$.

Exercise 8-4

Solve the following systems of equations using graphical methods.

1. $4x + y = 5$
   $3x - 2y = 1$

2. $x - 2y = -1$
   $2x + y = 8$

3. $4x - 2y = 4$
   $x - 2y = -2$

4. $x - y = -2$
   $2x + y = -1$

5. $3x - y = 10$
   $6x - 2y = 5$

6. $2x - y = 3$
   $4x - 2y = 6$

Solve the following systems of equations using substitution, addition, or subtraction.

7. $4x - y = 3$
   $2x + 3y = 19$

8. $5x - 3y = 9$
   $x + 2y = 7$

9. $2x - y = 2$
$3x + 4y = 6$

10. $x - y = 4$
$3x - 2y = 5$

11. $3x - 2y = 6$
$4y = 8$

12. $5x - 2y = 4$
$2x - 3y = 5$

13. $x - 2y = 4$
$3x + 2y = 6$

14. $2x - y = 4$
$3x - y = 5$

15. $3m + 2n = 2$
$3m - 3n = 1$

16. $3u - 2v = 5$
$u + 3v = 6$

17. $7x - 3y = 6$
$2x - y = 3$

18. $3x - 2y = 6$
$x + 4y = 3$

19. $7x + 3y = 5$
$2x - 2y = 5$

20. $3x - 6y = 4$
$5x + 5y = 9$

21. $2w - 3z = 4$
$3w - 2z = 6$

22. $2m - 4n = 3$
$5m - 7n = 4$

23. $0.2x - 0.3y = 4$
$2.3x - y = 1.2$

24. $0.5x + y = 3$
$0.3x + 0.2y = 6$

25. $4x + 6y = 4$
$2x + 3y = 2$

26. $3x + y = 2$
$6x + 2y = 5$

27. $\dfrac{x}{4} + \dfrac{3y}{4} = 12$
$\dfrac{y}{2} - \dfrac{x}{3} = -4$

28. $\dfrac{x + y}{4} = 2$
$\dfrac{y - 1}{x} = 6$

8-5 Word problems

OBJECTIVE: After completing this section, you should be able to use two equations in two unknowns to solve word problems.

We can frequently use two equations in two unknowns to solve word problems. The major difficulty is in expressing the given relationships in algebraic equations.

EXAMPLE 1

Separate 60 into two parts such that one part is twice the other.

Solution

Let x be the smaller part.
Let y be the larger part.
Then $x + y = 60$ (their sum is 60) and $y = 2x$ (larger part is twice the smaller).
Substituting $2x$ for y in $x + y = 60$ gives $x + 2x = 60$ or $3x = 60$.
Thus $x = 20$ and $y = 2(20) = 40$.

Check: $20 + 40$ is 60, and 40 is twice 20.

EXAMPLE 2

The difference between two numbers is 6. Their sum is 20. Find the numbers.

Solution

Let x be the larger number.
Let y be the smaller number.
Then $x - y = 6$ (their difference is 6) and $x + y = 20$ (their sum is 20).
Adding the two equations gives $2x = 26$; so $x = 13$.
Then substituting 13 in for x in $x - y = 6$ gives $13 - y = 6$; so $y = 7$.

Check: The difference of 13 and 7 is 6, and the sum of 13 and 7 is 20.

EXAMPLE 3

The sum of the ages of Jack and his younger brother Jake is 48 years. The difference between their ages is 3 less than Jake's age. How old is each man?

Solution

Let x be Jack's age.
Let y be Jake's age (y is less than x).
Then $x + y = 48$ and $x - y = y - 3$.
Thus we have two equations in two unknowns:

$$x + y = 48$$
$$x - 2y = -3$$

Subtracting gives $3y = 51$ or $y = 17$. Jake is 17.
Substitution gives $x + 17 = 48$ or $x = 31$. Jack is 31.

Check: The sum of 31 and 17 is 48, and $31 - 17 = 17 - 3$.

EXAMPLE 4

Mr. Walters invested $4200, part at 6% and part at 8%. His yearly income from the two investments was $284. How much did he invest at each rate?

Solution

Let x be the amount invested at 6%.
Then $0.06x$ is the yearly income from this investment.
Let y be the amount invested at 8%.
Then $0.08y$ is the yearly income from this investment.
The sum of the investments is $4200: $x + y = \$4200$.
The yearly income is the sum of the incomes:

$$0.06x + 0.08y = \$284$$

Solving the two equations in two unknowns gives

$$x + y = 4200$$
$$\underline{6x + 8y = 28{,}400} \quad \text{(multiplying both sides of}$$
$$.06x + .08y = 284 \text{ by 100)}$$

or

$$6x + 6y = 25{,}200$$
$$\underline{6x + 8y = 28{,}400}$$
$$-2y = -3200$$

So $y = \$1600$ and $x = \$2600$

Check: Sum of investments is $\$1600 + \$2600 = \$4200$, and income is $0.06(\$2600) + 0.08(\$1600) = \$156 + \$128 = \$284$.

EXAMPLE 5

Three times the width of a rectangle is twice the length. The perimeter is 40 inches. What are the dimensions of the rectangle?

Solution

Let x be the width of the rectangle.
Let y be the length of the rectangle.

Then $3x = 2y$ and $2x + 2y = 40$.
Substituting $3x$ in for $2y$ in $2x + 2y = 40$ gives $2x + 3x = 40$ or $5x = 40$.
Thus $x = 8$.
Substituting 8 in for x in $3x = 2y$ gives $24 = 2y$ or $y = 12$.
Thus 8 is the width, and 12 is the length.

Check: Three times 8 is two times 12, and $2(8) + 2(12) = 40$ is the perimeter.

Note: Many verbal problems can be solved with one equation (see Section 5-4), but it is often easier to formulate the problem with two equations.

Exercise 8-5

1. Separate 51 into two parts such that one part is twice the other part.
2. Separate 61 into two parts such that one part is 21 more than the other part.
3. The sum of two numbers is 63, and one number is twice the other. What are the numbers?
4. The difference of two numbers is 38, and one number is $\frac{1}{2}$ of the other. What are the numbers?
5. The sum of two numbers is 30. Their difference is 10. Find the numbers.
6. The sum of two numbers is 81. Their difference is 47. Find the numbers.
7. Twice one number is 3 times another. The sum of the numbers is 20. What are the numbers?
8. Five times one number is 7 times another. The sum of the numbers is 24. What are the numbers?
9. Bill is 6 years older than his brother. The sum of their ages is 30. What is Bill's age?
10. Bob is 8 years older than his brother. In 5 years he will be twice as old as his brother. How old is each now?
11. A 21-foot pipe is cut into two parts. One part is twice the other. What are the lengths of the parts?
12. A 25-foot board is cut into two parts. The difference of the lengths of the pieces is 9. What is the length of each piece?
13. The length of a rectangle is 5 feet more than the width,

and the perimeter of the rectangle is 38 feet. What are the dimensions of the rectangle?

14. The width of a rectangle is 8 inches less than the length. If the perimeter of the rectangle is 48 square inches, what are the dimensions of the rectangle?
15. Jack has 10 coins, part nickels and part dimes. If the value of his coins is $0.80, how many dimes does he have?
16. Rob has 12 coins, part dimes and part quarters. If the value of his coins is $1.95, how many coins of each type does he have?
17. Mr. Stong invested $6000, part at 6 percent, part at 8 percent. Find the amount of each investment if the annual income from both investments is $430.
18. Mr. Fritz invested $8500, part at 7 percent, part at 8 percent. If the annual income from the two investments was $660, what was the amount of each investment?
19. The sum of the reciprocals of two numbers is $\frac{5}{6}$. The difference of their reciprocals is $\frac{1}{6}$. What are the numbers?
20. The difference of the reciprocals of two numbers is $\frac{1}{12}$, and the sum of their reciprocals is $\frac{5}{12}$. What are the numbers?

8-6 Motion problems

OBJECTIVE: After completing this section, you should be able to solve motion problems involving linear equations.

Motion problems can be solved by using the relationship distance = rate · time or $d = rt$. The information given in motion problems can usually be translated into two equations in two unknowns.

EXAMPLE 1

A man can row upstream 6 miles in 2 hours and return downstream in 40 minutes. What is his rate of rowing in still water and what is the rate of the stream's current?

Solution

The current is working against the man when he is rowing upstream; so the rate he is traveling (r) is his still-water rate (x) minus the rate of the current (y).

$$r_1 = x - y$$

Then
$$6 = (x - y) \cdot 2 \qquad (d = rt)$$

The current is working with him when he is rowing downstream; so the rate downstream is $r_2 = x + y$. Then

$$6 = (x + y) \cdot \tfrac{2}{3} \qquad (40 \text{ minutes} = \tfrac{2}{3} \text{ hours})$$

Thus we have two equations in two unknowns:

$$6 = 2x - 2y$$
$$6 = \tfrac{2}{3}x + \tfrac{2}{3}y$$

We may solve the system as follows:

$$6 = 2x - 2y$$
$$\underline{18 = 2x + 2y}$$
$$24 = 4x$$
$$x = 6$$

Then
$$y = 3$$

The man's still-water rate is 6 miles/hr. The current's rate is 3 miles/hr.

EXAMPLE 2

A plane flies into a 60 miles/hr wind and reaches Atown in 3 hours. The return trip, flying with the 60 miles/hr wind, took $1\tfrac{1}{2}$ hours. What was the still-air speed of the plane, if it was constant in both directions?

Solution

Let the still-air rate be x. Then the rate into the wind is $x - 60 = r_1$. If d_1 represents the distance to Atown,

$$d_1 = (x - 60)3 \qquad (d_1 = r_1 t_1)$$

The rate traveling with the wind is $x + 60 = r_2$. If d_2 represents the distance back, then

$$d_2 = (x + 60)\tfrac{3}{2} \qquad (d_2 = r_2 t_2)$$

But the distance to Atown equals the distance back, so $d_1 = d_2$. Thus

$$(x - 60)3 = (x + 60)\tfrac{3}{2}$$

Solving gives

$$3x - 180 = \tfrac{3}{2}x + 90$$
$$6x - 360 = 3x + 180$$
$$3x = 540$$
$$x = 180$$

The still-air speed of the plane is 180 miles/hr.

EXAMPLE 3

One train leaves at 6 A.M., and a second train leaves 2 hours later. If the second train travels 10 miles/hr faster and catches the first after the first had traveled for 10 hours, how fast was each train traveling?

Solution

Let x represent the rate of the first train. Let d_1 represent the distance the first train traveled. The distance traveled by the first train is $d_1 = x \cdot 10$. The second train travels 10 miles/hr faster; so let $x + 10$ represent the rate of the second train. The time the second train traveled was 2 hours less than that of the first, or 8 hours. The distance traveled by the second train is $d_2 = (x + 10)8$. Because both trains traveled the same distances, $d_1 = d_2$. Thus

$$x \cdot 10 = (x + 10)8$$

Solving gives

$$10x = 8x + 80$$
$$2x = 80$$
$$x = 40$$

Thus the first train traveled at 40 miles/hr, and the second traveled at 50 miles/hr.

Exercise 8-6

1. An airplane maintaining the same still-air speed can go 300 miles in 2 hours, but it takes 3 hours to return. If the rate of wind is constant, what is the still-air speed of the plane and what is the rate of the wind?

2. An airplane makes a 4200-mile trip in 6 hours but takes 7 hours to return. If the rate of the plane and the wind was constant throughout the trip, what was the still-air rate of the plane and the rate of the wind?
3. A boat goes 6 miles downstream to Portsmouth in 1 hour and returns in $1\frac{1}{2}$ hours. What is the rate of the boat in still water and what is the rate of the current?
4. In a boat that can travel 10 miles/hr in still water, a man goes upstream for 1 hour and then returns in 40 minutes. What is the rate of the current in the stream?
5. In an airplane with a still-air speed of 200 miles/hr, a pilot travels into the wind for 3 hours and returns with the wind in 2 hours. What is the speed of the wind?
6. Ship A leaves port and travels north. Four hours later, Ship B leaves port traveling north 10 knots faster than Ship A. If Ship B catches Ship A 8 hours after Ship B leaves port, how fast is each ship traveling?
7. Jack takes 2 hours longer than Bill to walk 12 miles. If Jack doubles his speed, he will walk the 12 miles in $\frac{1}{2}$ hour less than Bill. What was Jack's original speed?
8. A boat is rowed upstream against a current of 3 miles/hr and reaches its destination in 3 hours. The return trip, rowing with the 3 miles/hr current, takes $1\frac{1}{2}$ hours. What is the still-water rate of the boat?

CHAPTER TEST

1. Make a partial listing of the solutions to the equation $4x + 3y = 12$ for values of x from -4 to 3.
2. Make a partial listing of the solutions to $y = 3x + 4$ for $-3 \leq x \leq 1$.
3. Graph $y = 4x - 2$.
4. Graph $5x + 2y = 4$.
5. Find the x-intercept and the y-intercept of the graph of $2x - 3y = 12$.
6. Find the slope of the line passing through $(1, 2)$ and $(3, -4)$.
7. Find the slope and the y-intercept of the graph of $2x - y = 3$.
8. Graph the line that has slope $\frac{1}{2}$ and y-intercept -2.

9. Solve the system:

$$3x + 4y = 5$$
$$2x - y = 7$$

10. Solve the system:

$$x - 2y = 3$$
$$3x - 6y = 4$$

11. The sum of two numbers is 7, and twice the smaller is 1 less than the larger. What are the numbers?
12. Jack is 7 years older than his brother. The sum of their ages is 35. How old is each?
13. Jack can run 5 miles/hr faster than John. If John starts 4 minutes before Jack and Jack catches John in 8 minutes, how fast were John and Jack each running?

9

relations and functions

9-1 Relations

OBJECTIVES: After completing this section, you should be able to:

a. Graph relations.
b. State the domain and range of certain relations.

We say that any correspondence between two sets is a relation. In Chapter 1 we discussed the one-to-one correspondence between the real numbers and the points on a line. This one-to-one correspondence is a relation. Other relations we have discussed are "is less than," "is equal to," and other (nonmathematical) relations are "is a brother of," "is married to," and so on.

The relation that exists between two sets may be defined by

1. a set of ordered pairs: The **domain** of the relation is the set of first components of the ordered pairs, and the range is the set of second components;
2. a graph of the ordered pairs: The first components are the **abscissas**, and the second components are the ordinates of the points on the graph; or
3. a formula (or word statement).

For example, if $U = \{1, 2, 3\}$ and $V = \{2, 4, 6\}$, then the relation between U and V defined by $y = 2x$ (with $x \in U$, $y \in V$) is

$R_1 = \{(1, 2), (2, 4), (3, 6)\}$

Note that we could also write

$$R_1 = \{(x, y): y = 2x, x \in U, y \in V\}$$

The domain of relation R_1 is $U = \{1, 2, 3\}$, and the range is $V = \{2, 4, 6\}$.

The graph of R_1 is composed of the three points in Figure 9-1.

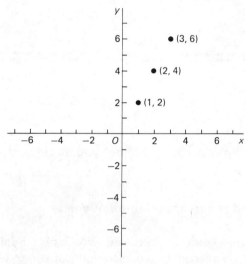

Figure 9-1

EXAMPLE 1

Graph the relation $R_2 = \{(x, y): x < y, x \in U, y \in U\}$, where $U = \{2, 4, 6, 8\}$.

Solution

$$R_2 = \{(2, 4), (2, 6), (2, 8), (4, 6), (4, 8), (6, 8)\}$$

The graph is given in Figure 9-2.

It is not necessary to use set notation to describe a relation. A stated formula or rule of correspondence, along with the domain and range, is sufficient to define a relation.

A graph of a relation may be regarded as a "picture" which shows the relationship between the variables. If the domain of the relation is the real numbers, or any large set, we cannot plot all the ordered pairs that satisfy the relation. In this case we make a table which contains a number of values from

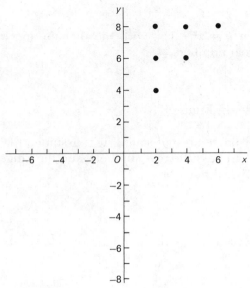

Figure 9-2

the domain, with the corresponding values from the range. These ordered pairs are plotted. If the domain is the set of real numbers, we draw a smooth curve passing through the plotted points to represent the graph of the relation (see Figure 9-3).

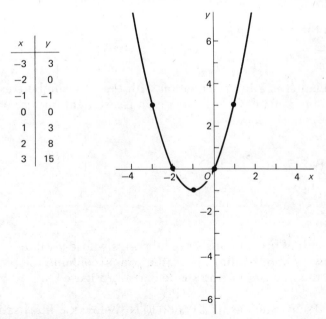

x	y
−3	3
−2	0
−1	−1
0	0
1	3
2	8
3	15

Figure 9-3

EXAMPLE 2

Graph the relation $y = x^2 + 2x$, with the domain and range chosen from the real numbers.

Solution

The graph is shown in Figure 9-3.

If the domain and range are not stated, it is assumed that the domain consists of all possible real numbers, and the range is a subset of the set of real numbers.

EXAMPLE 3

What is the domain and range of the relation $y = \sqrt{x - 4}$?

Solution

In order that all y-values are real numbers, no value of x may be used that will make $x - 4$ negative. Thus the domain of the relation is $\{x: x \geq 4\}$. The range will contain only nonnegative numbers; that is, the range is $\{y: y \geq 0\}$.

EXAMPLE 4

Graph the relation $y = 2x$.

Solution

The relation is a linear equation with the real numbers as its range and domain. Clearly, its graph is a straight line, given in Figure 9-4.

EXAMPLE 5

Graph the relation $y < 2x$.

Solution

The graph of $y < 2x$ consists of all points whose y-value is less than twice the x-value. Thus the graph contains all points below the line $y = 2x$ (the shaded area in Figure 9-5).

Note: Points such as (3, 5) and (1, 1) [below the line] satisfy the relation, while (2, 4) [on the line] and (3, 7) [above the

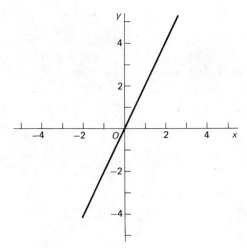

Figure 9-4

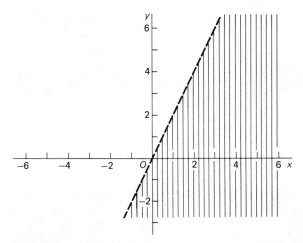

Figure 9-5

line] do not. The dotted line on the graph indicates that the points on the line do not satisfy the relation.

The relation graphed in Figure 9-5 is a **linear inequality in two unknowns**. The graph of a linear inequality is a half-plane. The graph in Figure 9-5 is an **open half-plane** because the line $y = 2x$ is not included in the graph.

EXAMPLE 6

Graph the inequality $y \geq 3x - 1$.

Solution

The graph of $y \geq 3x - 1$ consists of all points on or above the graph of $y = 3x - 1$. The graph of these points is in Figure 9-6. The graph is a **closed half-plane** because it contains the line.

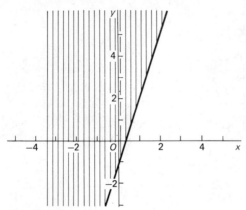

Figure 9-6

Exercise 9-1

Graph the following relations.

1. $\{(3, 5), (2, 4), (4, 6), (\frac{1}{2}, \frac{5}{2})\}$
2. $\{(1, 1), (-2, -3), (-2, -5), (3, 1)\}$
3. $\{(3, 1), (2, 0), (1, -1), (2, -2)\}$
4. $\{(5, 3), (4, 6), (2, 1), (6, -1)\}$
5. $\{(x, y): y \geq x, x \in U, y \in U\}$, where $U = \{0, 1, 2, 3\}$
6. $\{(x, y): y < 2x, x \in U, y \in U\}$, where $U = \{1, 2, 3, 6, 9\}$
7. $\{(x, y): x = 3y, x \in U, y \in U\}$, where $U = \{1, 2, 3, 6, 9\}$
8. $\{(x, y): x = \frac{1}{2}y, x \in U, y \in U\}$, where $U = \{1, 2, 3, 4, 5, 6\}$
9. $y = x - 1$
10. $y = 4 - x$
11. $y = \sqrt{x - 2}, x \geq 2$
12. $y = x^2 - 1$
13. $y > 2x$
14. $y \leq 3x$
15. $y < 3x + 4$
16. $y > 2x + 1$
17. $y \geq 4x - 3$
18. $y \leq 3x - 4$

State the domain and range of the following relations.

19. $y = x^2 + 4$
20. $y = x^2 - 1$
21. $y = \sqrt{x - 1}$
22. $y = \sqrt{x + 2}$
23. $y = \sqrt{x^2 + 1}$
24. $y = \sqrt{x^2 - 4}$

9-2 Functions

OBJECTIVES: After completing this section, you should be able to:

a. Determine if a relation is a function.
b. State the domain of a function.

We say that a relation *maps* its domain onto its range. Figure 9-7 shows two relations, R_1 and R_2, mapping their domains (A and C) onto their ranges (sets B and D). The arrows indicate the direction of the mapping.

One special type of relation is called a **function. A function is a set of ordered pairs (a relation) such that no two distinct ordered pairs have the same first component.** That is, for each element in the domain of a function, there is exactly one element in the range.

The relation R_1 in Figure 9-7 maps each element of the set A (the domain) to exactly one element of set B (the range). Thus R_1 is a function. The function R_1 may be written

$$R_1 = \{(a, 3), (b, 1), (c, 5), (d, 9), (e, 7), (f, 9)\}$$

Note that no two ordered pairs in R_1 have the same first components.

The relation R_2 in Figure 9-7 maps the element p of set c (the domain) to both f and g; so R_2 is **not** a function. The relation R_2 may be written

$$R_2 = \{(m, c), (n, a), (o, f), (p, f), (p, g), (q, g)\}$$

Note that two ordered pairs of R_2, namely, (p, f) and (p, g), have the same first component.

We may determine that a relation is a function by

1. inspecting its set of ordered pairs;
2. studying its formula;
3. inspecting its graph.

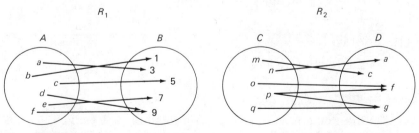

Figure 9-7

Relations and Functions

The following example illustrates again how to identify a function by inspecting its set of ordered pairs.

EXAMPLE 1

Is the relation $G = \{(1, 3), (2, 4), (4, 4), (6, 1)\}$ a function?

Solution

Yes, because no two ordered pairs have the same first component.

EXAMPLE 2

Is the relation $R = \{(2, 3), (4, 5), (3, 2), (2, 6)\}$ a function?

Solution

No, because two ordered pairs have the same first component [(2, 3) and (2, 6)]

The relation $A = \pi r^2$ is the formula for a relation between the radius of a circle and its area, with r representing the radius and A representing the area. By inspecting the formula, we see that there is exactly one value of A (the area) for each value of r (the radius). Thus the relation mapping the radius onto the area is a function. Because the function maps the radius, we say that A is a function of r. Because the value of A **depends** on the value of r, we say that r is the **independent variable** and A is the **dependent variable**. The set of possible values for r (the domain) is the set of positive real numbers (because a radius must be positive), and the range is the set of positive real numbers (all values of r will result in a positive value for A).

Examples 2 and 3 will further illustrate how to identify functions from equations (formulas).

EXAMPLE 3

Is the relation defined by $y = 2x^2$ a function?

Solution

If the domain and/or range of a relation is not given, it is assumed to contain all allowable real numbers. There is no real number which, when substituted for x in $y = 2x^2$, will give two values for y. Thus y is a function of x. The domain of the function is the set of real numbers, and the range is the set of nonnegative real numbers (the positive real numbers and 0).

EXAMPLE 4

Does $y^2 = 2x$ express y as a function of x?

Solution

No, because for some values of x, there is more than one value for y. For example, if $x = 8$, $y = \pm 4$. $y^2 = 2x$ expresses a relation between x and y, but y is not a function of x.

EXAMPLE 5

What is the domain of the function $y = 1 + 1/x$?

Solution

The domain is the set of all real numbers except 0, because $1 + 1/x$ is undefined if $x = 0$.

We can also determine if a relation is a function by inspecting its graph. If the relation is a function, no two points will have the same first coordinate (component); so there can be no two points on a vertical line.

EXAMPLE 6

Graph $y = 2x^2$ and test to see if y is a function of x.

Solution

Sample points of the relation from x to y are given in the table, and the graph is in Figure 9-8.

We can use the "vertical-line test" to determine if y is a function of x. The "test" consists of determining if any vertical line would intersect the graph of the relation in more than one point. If no such line can be drawn, the relation is a function. Clearly, y is a function of x if $y = 2x^2$. Note that we had reached the same conclusion concerning $y = 2x^2$ in Example 2.

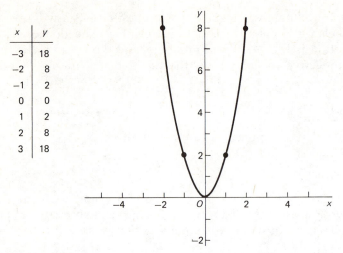

Figure 9-8

EXAMPLE 7

Graph $y^2 = 2x$ and determine if y is a function of x.

Solution

Sample points of the relation are given in the table. Note that no negative values are in the domain of the relation. Choosing points that can be easily graphed and completing the curve give the graph in Figure 9-9.

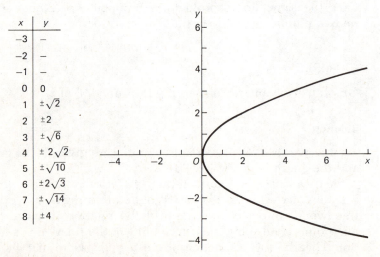

Figure 9-9

The vertical-line test indicates that y is not a function of x. For example, a vertical line at $x = 8$ would intersect the curve at $(8, 4)$ and $(8, -4)$. Note that we concluded in Example 3 that y was not a function of x if $y^2 = 2x$.

Exercise 9-2

Which of the following relations are functions?

1. $\{(1, 1), (2, 2), (3, 4)\}$
2. $\{(a, 1), (b, 5), (c, 6)\}$
3. $\{(1, 3), (2, 3), (4, 3)\}$
4. $\{(1, 5), (2, 5), (3, 5), (5, 5)\}$
5. $\{(1, 2), (2, 3), (1, 4)\}$
6. $\{(a, b), (c, d), (a, e), (b, d)\}$
7. $\{(1, 2), (2, 3), (3, 3)\}$
8. $\{(a, 5), (b, 6), (c, 5)\}$
9. $\{(1, 3), (2, 1), (1, 2), (3, 1)\}$
10. $\{(a, b), (b, a), (c, d), (d, c)\}$
11. If $y = 3x^3$, is y a function of x?
12. If $y = 6x^2$, is y a function of x?
13. If $y^2 = 3x$, is y a function of x?
14. If $y^2 = 10x^2$, is y a function of x?
15. If $y = \sqrt{2x}$, is y a function of x?
16. If $y = \sqrt{4x}$, is y a function of x?
17. If $y = \pm \sqrt{2x}$, is y a function of x?
18. If $y = \sqrt[3]{x}$, is y a function of x?
19. Is the relation of Figure 9-1 a function?
20. Is the relation of Figure 9-2 a function?
21. Is the relation of Figure 9-3 a function?
22. Is the relation of Figure 9-4 a function?
23. Is the relation of Figure 9-5 a function?
24. Is the relation of Figure 9-6 a function?

State the domain of the following functions.

25. $y = x^2$
26. $y = x^3 - 1$
27. $y = 3 - \dfrac{1}{x}$
28. $y = \dfrac{4x - 1}{x}$

29. $y = \dfrac{1}{4-x}$

30. $y = \dfrac{3}{x+3}$

9-3 Functional notation

OBJECTIVE: After completing this section, you should be able to use functional notation.

If we wish to state that area A is a function of radius r without stating the equation (or formula), we may write $A = f(r)$. This is read "A is a function of r," or "A equals f of r." To say that y is a function of x, we write $y = f(x)$. We may also use other letters to indicate functions, such as $y = g(x)$ or $y = h(x)$. If $y = 3x^2 + x + 1$, y is a function of x; so we may write $y = f(x) = 3x^2 + x + 1$ or $f(x) = 3x^2 + x + 1$. To express the value of y when $x = 2$, we may write $f(2)$; in the same manner, $f(-3)$ represents the value of y when $x = -3$ [**not** f times (-3)]. If $f(x) = 3x^2 + 2x + 1$, then $f(2) = 3(2)^2 + 2(2) + 1 = 15$; so $y = 15$ when $x = 2$.

EXAMPLE 1

If $y = f(x) = 2x^3 - 3x^2 + 1$, find

a. $f(0)$
b. $f(3)$
c. $f(-1)$

Solution

a. $f(0) = 2(0)^3 - 3(0)^2 + 1 = 1$; so $y = 1$ when $x = 0$.
b. $f(3) = 2(3)^3 - 3(3)^2 + 1 = 2(27) - 3(9) + 1 = 28$; so $y = 28$ when $x = 3$.
c. $f(-1) = 2(-1)^3 - 3(-1)^2 + 1 = 2(-1) - 3(1) + 1 = -4$; so $y = -4$ when $x = -1$.

EXAMPLE 2

If $h(x) = 4x^2 - 3x + 1$, find

a. $h(3)$
b. $h(-c)$

c. $h(a)$
d. $h(a) - h(b)$

Solution

a. $h(3) = 4(3)^2 - 3(3) + 1 = 28$
b. $h(-c) = 4(-c)^2 - 3(-c) + 1 = 4c^2 + 3c + 1$
c. $h(a) = 4(a)^2 - 3(a) + 1 = 4a^2 - 3a + 1$
d. $h(a) - h(b) = [4(a)^2 - 3(a) + 1] - [4(b)^2 - 3(b) + 1] = 4a^2 - 3a - 4b^2 + 3b$

EXAMPLE 3

Given $f(x) = 3x - 1$, find

a. $f(x + h)$
b. $f(x + h) - f(x)$
c. $\dfrac{f(x + h) - f(x)}{h}, \quad h \neq 0$

Solution

a. $f(x + h) = 3(x + h) - 1 = 3x + 3h - 1$
b. $f(x + h) - f(x) = (3x + 3h - 1) - (3x - 1) = 3h$
c. $\dfrac{f(x + h) - f(x)}{h} = \dfrac{3h}{h} = 3$

EXAMPLE 4

Given $g(x) = x^2 - 3x$, find

$$\dfrac{g(x + h) - g(x)}{h}, \quad \text{for } h \neq 0$$

Solution

$$\dfrac{g(x + h) - g(x)}{h} = \dfrac{[(x + h)^2 - 3(x + h)] - [x^2 - 3x]}{h}$$

$$= \dfrac{x^2 + 2xh + h^2 - 3x - 3h - x^2 + 3x}{h}$$

$$= \dfrac{2xh + h^2 - 3h}{h}$$

$$= 2x + h - 3$$

Exercise 9-3

1. If $f(x) = 3x - 1$, find
   a. $f(0)$
   b. $f(2)$
   c. $f(-3)$
   d. $f(a)$

2. If $g(x) = 1 - 4x$, find
   a. $g(1)$
   b. $g(3)$
   c. $g(-4)$
   d. $g(-b)$

3. If $f(x) = 4x^2 - 3$, find
   a. $f(0)$
   b. $f(-1)$
   c. $f(-2)$
   d. $f(-x)$

4. If $h(x) = 3x^2 - 2x$, find
   a. $h(3)$
   b. $h(-3)$
   c. $h(2)$
   d. $h(-x)$

5. If $g(x) = x^3 - 4x^2 + 3$, find
   a. $g(-1)$
   b. $g(2)$
   c. $g(1)$

6. If $f(x) = x^3 - 8x - 1$, find
   a. $f(0)$
   b. $f(2)$
   c. $f(-3)$

7. If $f(x) = x^2 - 3x$, find
   a. $f(x + 2)$
   b. $f(x + h)$
   c. $f(x + h) - f(x)$
   d. $\dfrac{f(x + h) - f(x)}{h}$

8. If $f(x) = x^2 - 8x$, find
   a. $f(x + 1)$
   b. $f(x + h)$
   c. $f(x + h) - f(x)$
   d. $\dfrac{f(x + h) - f(x)}{h}$

9. If $f(x) = 3x^2 + 2x + 3$, find $\dfrac{f(x+h) - f(x)}{h}$.

10. If $g(x) = 4x^2 + 3x + 1$, find $\dfrac{g(x+h) - g(x)}{h}$.

9-4 Graphing relations and functions

OBJECTIVE: After completing this section, you should be able to graph relations and functions.

We graphed linear equations in Chapter 7 by plotting two points and drawing the line through them or by using a point and the slope of the line. Graphing other relations and functions are much more difficult, however. We have seen in Example 2 of Section 9-1 that we can graph a relation by plotting a number of points and drawing a smooth curve connecting the points.

EXAMPLE 1

Graph the function $y = x^3 + 1$.

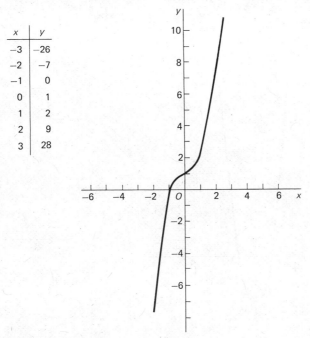

Figure 9-10

Solution

Figure 9-10 contains the table of selected values for *x* and their corresponding *y*-values along with the graph of the function.

Some properties of relations and functions are helpful in sketching their graphs.

INTERCEPTS

As with linear equations, we may find the *x*-intercept(s) of a relation from its equation by setting $y = 0$ and solving for *x*. The equation to be solved may be linear, quadratic, or even a higher-degree equation. If no real value for *x* exists, the curve does not intersect the *x*-axis. In the same manner, the *y*-intercept(s) of a relation may be found by setting $x = 0$ in the equation and solving for *y*. If the relation is a function, there will be no more than one *y*-intercept.

EXAMPLE 2

Find the *y*-intercept and the *x*-intercepts of $y = 3x^2 - 3$.

Solution

Setting $x = 0$ in the equation gives $y = 3(0)^2 - 3$ or $y = -3$ as the *y*-intercept. Thus the curve crosses the *y*-axis at $(0, -3)$.

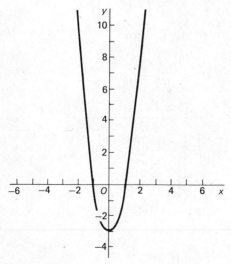

Figure 9-11

Setting $y = 0$ in the equation gives $0 = 3x^2 - 3$. Factoring gives $0 = 3(x-1)(x+1)$; so $x = 1$ and $x = -1$ are the x-intercepts. Thus the curve intersects the x-axis at $(1, 0)$ and $(-1, 0)$.

The graph of $y = 3x^2 - 3$ is given in Figure 9-11.

EXAMPLE 3

Find the x- and y-intercepts of the relation $x^2 + y^2 = 9$.

Solution

Setting $x = 0$ gives $y^2 = 9$. Solving for y gives y-intercepts $y = 3$ and $y = -3$. Setting $y = 0$ gives $x^2 = 9$. Solving for x gives x-intercepts $x = 3$ and $x = -3$. The graph of $x^2 + y^2 = 9$ is the circle in Figure 9-12.

SYMMETRY

If the part of a graph lying to the right of the y-axis is a mirror image of the part lying to the left of the y-axis, we say the graph is symmetric with respect to the y-axis. Figure 9-13 shows a curve which is symmetric with respect to the y-axis. The points (x, y) and $(-x, y)$ are symmetric with respect to the y-axis. Therefore a graph is symmetric with respect to the y-axis if $(-x, y)$ is on the graph whenever (x, y) is on the graph. **Thus, if substituting $-x$ in for x in an equation does not change the equation, the graph of the equation is symmetric with respect to the y-axis.** If the relation is a function, we may say that its graph is

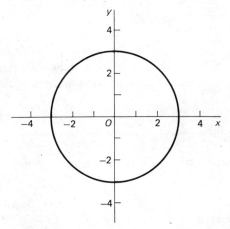

Figure 9-12

symmetric with respect to the y-axis if $f(-x) = f(x)$ for all x in the domain of f. Such a function is called an **even function**. [A function f such that $f(-x) = -f(x)$ is called an **odd function**.]

EXAMPLE 4

Is the graph of $y = 3x^2 - 3$ symmetric with respect to the y-axis?

Solution

Substitution of $-x$ for x in $y = 3x^2 - 3$ gives $y = 3(-x)^2 - 3$ or $y = 3x^2 - 3$. The equation was unchanged by substitution of $-x$ for x; so the graph is symmetric with respect to the y-axis. The graph of $y = 3x^2 - 3$ is given in Figure 9-13.

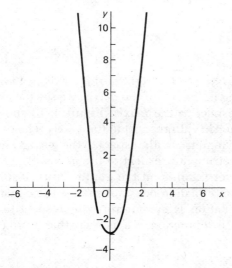

Figure 9-13

If substituting $-y$ in for y in an equation does not change the equation, the graph of the equation is symmetric with respect to the x-axis.

EXAMPLE 5

Is the graph of the relation $y^2 = 4x$ symmetric about the x-axis?

Solution

Substituting $-y$ in for y in $y^2 = 4x$ gives $(-y)^2 = 4x$ or $y^2 = 4x$. Thus the graph is symmetric with respect to the x-axis. The

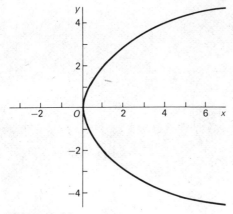

Figure 9-14

graph is in Figure 9-14. Note that the graph is not symmetric with respect to the y-axis. Substituting $-x$ in for x in the equation gives $y^2 = 4(-x)$ or $y^2 = -4x$, which is not the same as the original equation $y^2 = 4x$.

The graph of a relation is **symmetric with respect to the origin** if substituting $-x$ in for x and $-y$ in for y does not change its equation.

EXAMPLE 6

Is the graph of $x^2 + y^2 = 9$ symmetric with respect to the origin?

Solution

Substituting $-x$ in for x and $-y$ in for y gives $(-x)^2 + (-y)^2 = 9$, or $x^2 + y^2 = 9$. Thus the graph is symmetric with respect to the origin. The graph of this relation is given in Figure 9-15.

Note: The graph in Figure 9-15 is also symmetric with respect to the x- and y-axis. The next example illustrates a graph which is symmetric with respect to the origin, but not with respect to either axis.

EXAMPLE 7

Is the graph of $y = 1/x$ symmetric with respect to the origin?

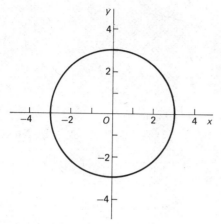

Figure 9-15

Solution

Substituting $-x$ in for x and $-y$ in for y gives $(-y) = 1/(-x)$, or $y = 1/x$. Thus the graph is symmetric with respect to the origin. Plotting some points with positive x-values and drawing a smooth curve give the curve in Figure 9-16(a). Using the prop-

x	y
1	1
2	$\frac{1}{2}$
3	$\frac{1}{3}$
4	$\frac{1}{4}$
5	$\frac{1}{5}$
$\frac{1}{5}$	5
$\frac{1}{4}$	4
$\frac{1}{3}$	3
$\frac{1}{2}$	2

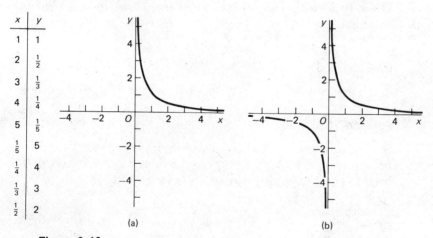

Figure 9-16

erty of symmetry about the origin gives the remainder of the graph [in Figure 9-16(b)]. Note that there are no x- or y-intercepts.

Exercise 9-4

Give the x- and y-intercepts, if any, for the graphs of the following equations.

1. $y = x^2 - 9$
2. $y = x^2 - 2x$
3. $y = \dfrac{1}{x} + 1$
4. $y = \dfrac{4}{x}$
5. $x^2 + y^2 = 4$
6. $x^2 - y^2 = 9$
7. $3x^2 - y = 3$
8. $x - 4y^2 = 16$
9. $x^2 + 4y^2 = 4$
10. $x - y = 4$

Discuss symmetry of the graphs of the following equations.

11. $y = x^2 - 9$
12. $y = x^2 - 2x$
13. $y = \dfrac{1}{x} + 1$
14. $y = \dfrac{4}{x}$
15. $x^2 + y^2 = 4$
16. $x^2 - y^2 = 9$
17. $3x^2 - y = 3$
18. $x - 4y^2 = 16$
19. $x^2 + 4y^2 = 4$
20. $x - y = 4$

Graph the following relations and functions.

21. $y = x^2 - 9$ (see Problems 1 and 11)
22. $y = x^2 - 2x$ (see Problems 2 and 12)
23. $y = \dfrac{1}{x} + 1$ (see Problems 3 and 13)
24. $y = \dfrac{4}{x}$ (see Problems 4 and 14)
25. $x^2 + y^2 = 4$ (see Problems 5 and 15)
26. $x^2 - y^2 = 9$ (see Problems 6 and 16)
27. $3x^2 - y = 3$ (see Problems 7 and 17)
28. $x - 4y^2 = 16$ (see Problems 8 and 18)
29. $x^2 + 4y^2 = 4$ (see Problems 9 and 19)
30. $x - y = 4$ (see Problems 10 and 20)
31. In which of the equations (Problems 1 through 10) is y a function of x?

9-5 Special functions

OBJECTIVES: After completing this section, you should be able to identify and sketch the graphs of linear, absolute value, and quadratic functions.

One of the simplest functions is the **identity function**, $f(x) = x$, which maps each element of the domain onto itself.

EXAMPLE 1

If $f(x) = x$, find:

a. $f(2)$
b. $f(-3)$
c. $f(a)$
d. $f(y)$

Solution

a. $f(2) = 2$
b. $f(-3) = -3$
c. $f(a) = a$
d. $f(y) = y$

EXAMPLE 2

Sketch the graph of $y = f(x) = x$.

Solution

$y = x$ is a linear equation with y-intercept 0 and slope 1. The graph is given in Figure 9-17.

Recall that any polynomial equation that has 1 as the highest degree of any term (that is, a first-degree equation) is a

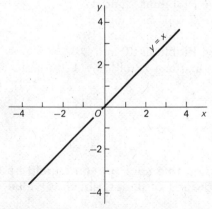

Figure 9-17

linear equation. All linear equations except those of the form $x = c$ express functional relationships; that is, any linear equation involving y defines a function. Thus equations such as $y = 3$, $y = 2x - 1$, and $3x + 2y = 4$ define linear functions on the real numbers.

Not all linear functions are expressed in the variables x and y. The following example uses a formula with C as the dependent variable and S as the independent variable.

EXAMPLE 3

A merchant used the formula $S = \frac{6}{5}C + 2$ to determine the selling price of a shipment of goods, where C is his cost. Make the graph of this function.

Solution

We construct a table of values, realizing that we would not have negative values for C (the cost). We see the points fall on a straight line, which is the graph. (See Figure 9-18.) But looking at the equation reveals that it is a linear equation, because the variables occur to the first degree.

The polynomial equation $y = ax^2 + bx + c$, where $a, b,$ and c are real numbers and $a \neq 0$, is the equation of a second-degree

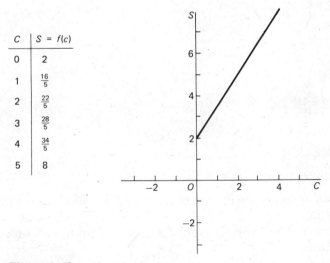

C	S = f(c)
0	2
1	$\frac{16}{5}$
2	$\frac{22}{5}$
3	$\frac{28}{5}$
4	$\frac{34}{5}$
5	8

Figure 9-18

function, or **quadratic function.** The graph of a quadratic function is a **parabola.**

If $a > 0$, the parabola opens up (see Figure 9-19). If $a < 0$, the parabola opens down (see Figure 9-20).

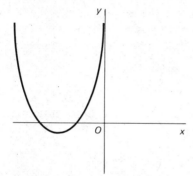

Figure 9-19

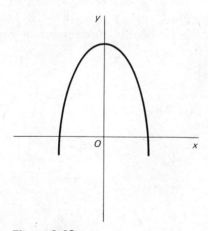

Figure 9-20

EXAMPLE 4

For the equation $y = x^2 - 4x + 3$:

a. Construct a table of values using integer values of x from -1 to 5.
b. Graph the equation using the points determined in a as a guide.

Solution

a.

x	x^2	$- 4x$	$+ 3$	$=$	y
-1	$(-1)^2$	$- 4(-1)$	$+ 3$	$=$	8
0	0^2	$- 4(0)$	$+ 3$	$=$	3
1	1^2	$- 4(1)$	$+ 3$	$=$	0
2	2^2	$- 4(2)$	$+ 3$	$=$	-1
3	3^2	$- 4(3)$	$+ 3$	$=$	0
4	4^2	$- 4(4)$	$+ 3$	$=$	3
5	5^2	$- 4(5)$	$+ 3$	$=$	8

b.

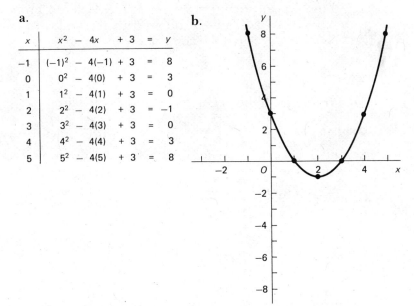

Figure 9-21

From the graph of the equation in Example 4 (Figure 9-21), we can make the following observations:

1. The graph opens up. (The coefficient of the squared term is $a = 1 > 0$).
2. The parabola has a **minimum** (lowest) point at $(2, -1)$. A parabola which opens down has a **maximum** (highest) point.
3. One-half of the parabola is a reflection of the other. We say that the parabola is **symmetric** about its **axis of symmetry**. In Example 4, the axis of symmetry is the vertical line through the minimum point $(2, -1)$. The axis of symmetry will always pass through the minimum or maximum of the parabola.
4. The parabola crosses the x-axis where the y-value is zero. Thus the x-values of the points where the parabola crosses the x-axis are the solutions to the equation

$$0 = x^2 - 4x + 3$$

The solutions (from the graph) are $x = 1$ and $x = 3$. Solving the equation $0 = x^2 - 4x + 3$ by factoring gives $0 = (x - 1)(x - 3)$; so $x = 1$ and $x = 3$ are the solutions. These values are called the **zeros** of the function.

EXAMPLE 5

Find the zeros, if they exist, and graph $y = x^2 - 6x + 8$.

Solution

Setting $y = 0$ and solving for x will tell where the curve crosses the x-axis.

$$0 = x^2 - 6x + 8$$

or

$$0 = (x - 2)(x - 4)$$

so

$$x = 2 \quad \text{or} \quad x = 4$$

x	x²	− 6x	+ 8	=	y
0	0²	− 6(0)	+ 8	=	8
1	1²	− 6(1)	+ 8	=	3
2	2²	− 6(2)	+ 8	=	0
3	3²	− 6(3)	+ 8	=	−1
4	4²	− 6(4)	+ 8	=	0
5	5²	− 6(5)	+ 8	=	3
6	6²	− 6(6)	+ 8	=	8

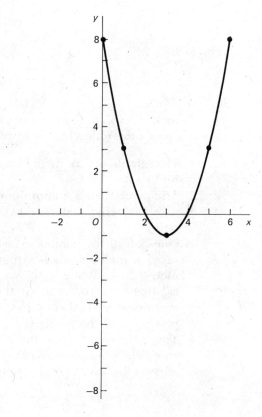

Figure 9-22

Thus the curve crosses the x-axis at $x = 2$ and at $x = 4$. Because the coefficient of x^2 is positive (+1), the parabola will open up.

Making a table of values near the two points we know are on the graph will help construct the graph (see Figure 9-22).

Note again that the "axis of symmetry" is a vertical line through the minimum point $(3, -1)$. Note also that we could have found the x-value of the minimum point by taking the value midway between the two x-values where the curve crosses the x-axis. Thus, since the curve crosses the x-axis at $x = 2$ and $x = 4$, the curve has its minimum value at $x = 3$.

EXAMPLE 6

Find the zeros, if any, and graph $f(x) = -2x^2 + 4x - 4$.

Solution

Setting $f(x) = 0$ gives $0 = -2x^2 + 4x - 4$. Solving by the quadratic formula gives

$$x = \frac{-4 \pm \sqrt{16 - 4(-2)(-4)}}{2(-2)}$$

or

$$x = \frac{-4 \pm \sqrt{-16}}{-4} = \frac{-4 \pm 4i}{-4} = \frac{-4(1 \pm i)}{-4} = 1 \pm i$$

Thus there are no real zeros; so the curve does not cross the x-axis. The parabola opens down, because the x^2 term has a negative coefficient. A table of values shows the axis of symmetry is the line $x = 1$; so the maximum is at $x = 1$. The graph is shown in Figure 9-23.

We shall discuss parabolas in more detail in Chapter 12, along with other related curves.

The **absolute-value function,** $f(x) = |x|$, maps each element of the domain onto its absolute value.

EXAMPLE 7

Sketch the graph of $y = |x|$.

Solution

The function $y = |x|$ is symmetric about the y-axis because substituting $-x$ for x gives $y = |-x|$ or $y = |x|$. Plotting points where $x \geq 0$ gives a straight line, as Figure 9-24(a) shows. Symmetry gives the left half of the graph, shown in Figure 9-24(b). Note that the graph satisfies the definition of $|x|$ given in Section 5-7,

$$|x| = \begin{cases} x & \text{if } x \geq 0 \\ -x & \text{if } x < 0 \end{cases}$$

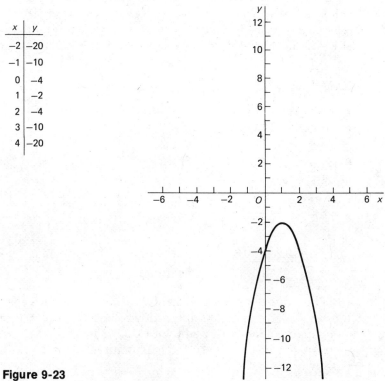

x	y
-2	-20
-1	-10
0	-4
1	-2
2	-4
3	-10
4	-20

Figure 9-23

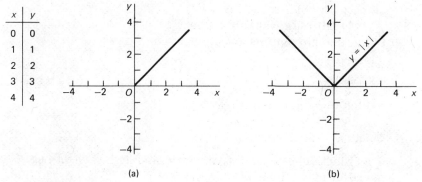

x	y
0	0
1	1
2	2
3	3
4	4

(a)   (b)

Figure 9-24

Exercise 9-5

Identify and sketch the graph of each of the following special functions.

1. $y = x$
2. $y = x - 3$
3. $y = 4x - 1$
4. $f(x) = x$
5. $y = \frac{1}{4}x + 1$
6. $y = 3x + 1$
7. $y = |2x|$
8. $y = |3x|$
9. $y = x^2 + 3$
10. $y = 4 - x^2$
11. $f = \frac{9}{5}c + 32$
12. $S = C^2 + 5$

Find the zeros, if any, for the following quadratic functions and sketch their graphs.

13. $y = x^2 - 4$
14. $y = x^2 + 5x + 4$
15. $y = -2x^2 + 18$
16. $y = 4 - x^2$
17. $y = x^2 + 4x + 4$
18. $y = x^2 + 6x - 9$
19. $y = x^2 + x + 2$
20. $y = -x^2 + 2x - 4$

9-6 Composite and inverse functions

OBJECTIVES: After completing this section, you should be able to:

a. Form the composite of two functions.
b. Show that two functions are inverses of each other.

We can combine functions in different ways to create new functions. For example, we can form the function $f + g$, where f and g are functions and $(f+g)(x) = f(x) + g(x)$. As the rule $(f+g)(x) = f(x) + g(x)$ indicates, $f + g$ is the **sum** of the two functions f and g. The **difference** of f and g, $f - g$, is defined by $(f - g)(x) = f(x) - g(x)$. We may also define the **product** of f and g by $(f \cdot g)(x) = f(x) \cdot g(x)$ and the **quotient** of f divided by g by $(f/g)(x) = f(x)/g(x)$.

There is another method of combining two functions to yield a new function. Just as we can substitute a number in for the independent variable in a function, we can substitute a second function in for the variable. When we do this, we get a new function, called a **composite function**. The composite function g of f is the function g ∘ f defined by

$$(g \circ f)(x) = g[f(x)]$$

The domain of g ∘ f is the subset of the domain of f for which g ∘ f is defined.

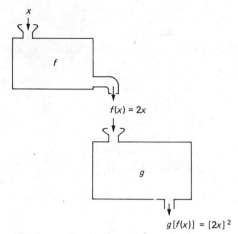

Figure 9-25

Figure 9-25 illustrates how the composite function $g \circ f$ is created from f and g, where $f(x) = 2x$ and $g(x) = x^2$. (See Example 1.)

EXAMPLE 1

If $f(x) = 2x$ and $g(x) = x^2$, find $g[f(x)]$.

Solution

Substituting $f(x) = 2x$ in for x in $g(x)$ gives

$$g[f(x)] = g[2x] = [2x]^2 = 4x^2$$

EXAMPLE 2

If $f(x) = 4x$ and $g(x) = \frac{1}{4}x$ find

a. $(g \circ f)(x) = g[f(x)]$
b. $(f \circ g)(x) = f[g(x)]$

Solution

a. $g[f(x)] = g[4x] = \frac{1}{4}[4x] = x$; so $(g \circ f)(x) = x$
b. $f[g(x)] = f[\frac{1}{4}x] = 4[\frac{1}{4}x] = x$; so $(f \circ g)(x) = x$

We see that both $g \circ f$ and $f \circ g$ in Example 2 are identity functions. This leads us to the definition of inverse functions.

If f and g are two functions such that $(g \circ f)(x) = x$ for every element in the domain of f and $(f \circ g)(x) = x$ for every element in the domain of g, then f and g are said to be the inverse functions of each other.

EXAMPLE 3

Show that $f(x) = x - 3$ and $g(x) = x + 3$ are inverse functions of each other.

Solution

$$(g \circ f)(x) = g[f(x)] = g[x - 3] = [x - 3] + 3 = x$$

and

$$(f \circ g)(x) = f[g(x)] = f[x + 3] = [x + 3] - 3 = x$$

Thus f and g are inverse functions.

We may denote the fact that f and g are inverse functions by $f = g^{-1}$ or by $g = f^{-1}$. Note that the -1 is not an exponent, but rather a means of indicating the inverse of a function.

EXAMPLE 4

Show that $g = h^{-1}$ if $h(x) = 4x - 1$ and $g(x) = \dfrac{x + 1}{4}$.

Solution

$$(h \circ g)(x) = h\left[\frac{x + 1}{4}\right] = 4\left[\frac{x + 1}{4}\right] - 1 = x + 1 - 1 = x$$

and

$$(g \circ h)(x) = g[4x - 1] = \frac{[4x - 1] + 1}{4} = \frac{4x}{4} = x$$

Thus $g = h^{-1}$. We may also write $h = g^{-1}$.

A function is a one-to-one function if and only if there is exactly one element in the range for every element in the domain. The function f in Figure 9-26(a) is a one-to-one function. Note that f maps each element of its domain to exactly one element of the range.

Because f is a one-to-one function, we can define a new

function which maps the range of f onto the domain of f. If the new function does this by reversing the correspondence determined by f, the new function is f^{-1}, the inverse function of f. Figure 9-26(b) shows the inverse of the function in Figure 9-26(a).

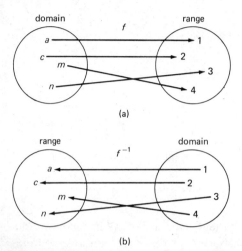

Figure 9-26

EXAMPLE 5

Which of the following functions are one-to-one functions?

a. $y = 2x$
b. $y = x^2$

Solution

a. There is exactly one value of y for each value of x and exactly one value of x for each value of y; so $y = 2x$ is a one-to-one function.
b. y is a function of x, but if $y = 4$, x can be either 2 or -2; so the function is not one to one.

It can be shown that a function has an inverse if and only if the function is a one-to-one function. If a function is one to one and is defined by the formula $y = f(x)$, then the range of f becomes the domain of its inverse function f^{-1}, and the domain of f becomes the range of f^{-1}. Because we have been letting x's represent the elements of the domain and y's represent the elements of the range, we can find the formula for f^{-1} by interchanging the x's and y's in the formula for f.

EXAMPLE 6

Find the formula for the inverse of the function defined by $f(x) = y = x - 3$.

Solution

Interchanging x and y gives the formula $x = y - 3$, which may be written as $y = x + 3$. Thus the inverse function is $f^{-1}(x) = y = x + 3$.
We have shown in Example 3 that $g(x) = x + 3$ is the inverse of $f(x) = x - 3$.

EXAMPLE 7

Find the inverse function of $h(x) = 4x - 1$.

Solution

Defining the function h by the formula $y = 4x - 1$ and interchanging x and y gives $x = 4y - 1$. Solving for y gives $y = (x + 1)/4$, which defines the inverse function of h. Note that in Example 4 we have shown that $g(x) = (x + 1)/4$ is the inverse of $h(x) = 4x - 1$.

Because the inverse function is found by interchanging the variables x and y in the formula for the original function, the graph of the inverse function is a reflection of the original about the line $y = x$; that is, the graphs of a function and its inverse are symmetric about the line $y = x$.

EXAMPLE 8

Graph $y = 4x - 1$ and its inverse $y = (x + 1)/4$.

Solution

Both equations are linear and are graphed in Figure 9-27. Note the symmetry about the line $y = x$ (dotted on graph).

EXAMPLE 9

Graph $y = x^3 + 1$ and its inverse on the same graph.

Solution

The inverse of $y = x^3 + 1$ is $x = y^3 + 1$, or $y = \sqrt[3]{x - 1}$. The graphs of the two functions are in Figure 9-28.

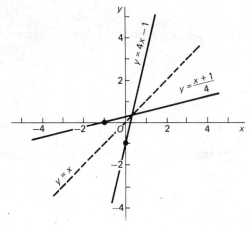

Figure 9-27

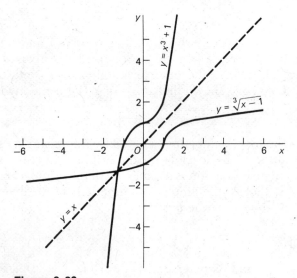

Figure 9-28

Exercise 9-6

1. If $f(x) = 3x$ and $g(x) = 5x$, find
   a. $(g \circ f)(x)$ b. $(f \circ g)(x)$
2. If $f(x) = x^2$ and $g(x) = 3x$, find
   a. $(g \circ f)(x)$ b. $(f \circ g)(x)$
3. If $h(x) = 5x$ and $g(x) = \frac{1}{5}x$, find
   a. $g[h(x)]$ b. $h[g(x)]$

4. If $f(x) = x + 2$ and $g(x) = x - 2$, find
   a. $g[f(x)]$ b. $f[g(x)]$
5. Are $h(x) = 5x$ and $g(x) = \frac{1}{5}(x)$ inverse functions (see Problem 3)?
6. Are $f(x) = x + 2$ and $g(x) = x - 2$ inverse functions (see Problem 4)?
7. Show that $f = g^{-1}$ if $f(x) = 2x - 1$ and $g(x) = \frac{x+1}{2}$ are inverse functions.
8. Show that $g = f^{-1}$ if $f(x) = 2x - 3$ and $g(x) = \frac{x+3}{2}$.
9. Is $y = f(x) = 3x - 1$ a one-to-one function?
10. Is $y = f(x) = x$ a one-to-one function?
11. Find the inverse function of $y = 3x - 1$.
12. Find the inverse function of $y = \frac{x+3}{4}$.
13. Graph $y = 3x + 4$ and its inverse.
14. Graph $y = \frac{x-1}{3}$ and its inverse.
15. Graph $y = \sqrt[3]{x+1}$ and its inverse.

CHAPTER TEST

Graph the following relations.

1. $\{(6, 2), (3, -1), (1, 3)\}$
2. $\{(x, y): y \leq x, x \in U, y \in U\}$ where $U = \{0, 1, 2, 3\}$
3. $y < 2x$
4. State the domain and range of $y = \sqrt{x-3}$.
5. Is $\{(1, 3), (3, 5), (4, 5)\}$ a function?
6. If $y^2 = 4x$, is y a function of x?
7. What is the domain of $y = \frac{4}{x+5}$?
8. If $f(x) = 5x^2 - 3x + 2$, what is $f(3)$?
9. If $g(x) = x^3 - 4x + 1$, what is $g(-2)$?

For the following relations and functions, give the x- and y-intercepts, if any, discuss symmetry, and sketch the graphs.

10. $y = x^2 - 3$
11. $y^2 = 16 - x^2$
12. $x - 4y^2 = 4$

Identify and sketch each of the following special functions.

13. $C = \dfrac{5f - 160}{9}$
14. $f(x) = x$
15. $y = |4x|$
16. $y = x^2 + 5x + 6$
17. Find the zeros of, and sketch the graph of, $y = -3x^2 + 27$.
18. Is $y = 5x + 3$ a one-to-one function?
19. If $f(x)$ and x^2 and $g(x) = 5x + 1$:
    a. find $(g \circ f)(x)$
    b. find $(f \circ g)(x)$
20. Are $f(x) = 3x + 1$ and $g(x) = \dfrac{x-1}{3}$ inverse functions?
21. Find the inverse function of $y = \dfrac{x-5}{2}$.
22. Graph $y = x^3 + 3$ and its inverse.

10

exponential and logarithmic functions

10-1 Exponential functions

OBJECTIVE: After completing this section, you should be able to graph exponential functions.

One special group of functions is called **exponential functions**. We define exponential functions as follows.

> *If a is a positive real number, then the function* $f(x) = a^x$ *is an exponential function with base* **a**.

An example of an exponential function is $y = 2^x$. A table of values satisfying this equation and its graph are shown in Figure 10-1.

As Figure 10-1 shows, the graph approaches, but never reaches, the x-axis. We say that the x-axis is an **asymptote** for the graph and that the graph approaches the x-axis asymptotically. Clearly, there is no value of x that will make $2^x = 0$ or 2^x equal a negative number. The domain of this function contains all real numbers, and the range contains all positive real numbers.

EXAMPLE 1

Graph $y = 10^x$.

Solution

A table of values and the graph are shown in Figure 10-2.
Note that the graphs of $y = 2^x$ and $y = 10^x$ are very similar, except that the graph of $y = 10^x$ (Figure 10-2) rises much more

272 Exponential and Logarithmic Functions

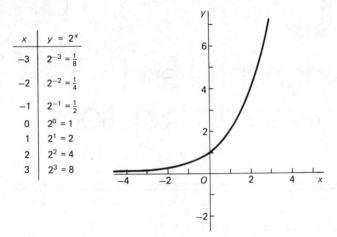

Figure 10-1

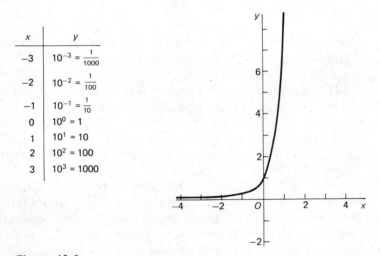

Figure 10-2

rapidly than the graph of $y = 2^x$ (Figure 10-1). Based on the observation of the graphs of $y = 2^x$ and $y = 10^x$, we can guess that the graph of $y = 3^x$ would rise more rapidly than that of $y = 2^x$, but not as rapidly as that of $y = 10^x$.

EXAMPLE 2

Graph $y = 3^x$.

Solution

A table of values and the graph are shown in Figure 10-3. Note that the graph of $y = 3^x$ fits "between" the graphs of $y = 2^x$ and $y = 10^x$.

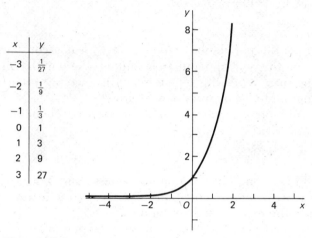

x	y
−3	$\frac{1}{27}$
−2	$\frac{1}{9}$
−1	$\frac{1}{3}$
0	1
1	3
2	9
3	27

Figure 10-3

A special function, which occurs frequently in economics and in biological growth, is $y = e^x$, where e is a certain irrational number (approximately 2.71828 . . .). Because e is between 2 and 3, the graph of $y = e^x$ can be sketched by referring to their graphs (see Figure 10-4).

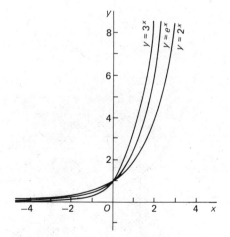

Figure 10-4

The number e is involved in many functions occurring in nature. For example, the **normal probability density function,** which is basic to the study of statistics, is an exponential function. The growth curve of a population (of insects, for example) in biology has equation

$$y = P_0 e^{h(t-t_0)}$$

where P_0 is the population size at time t_0, h is a constant depending on the type of population, and y is the population size at any instant t. When this equation applies, we say y increases according to the **exponential law of growth.**

The number of atoms, y, of a radioactive element at an instant t is

$$y = W_0 e^{-h(t-t_0)}$$

where W_0 is the number of atoms at time t_0, and h depends on the element. When the equation holds, we say y decreases according to the **exponential law of decay.**

We should note that the graph of an exponential function will be quite different if the base is between 0 and 1. The table of values and the graph for $y = (\tfrac{1}{2})^x$, or $y = 2^{-x}$, are shown in Figure 10-5.

The graphs of $y = (\tfrac{1}{3})^x$ and e^{-x} would be similar to the above graph.

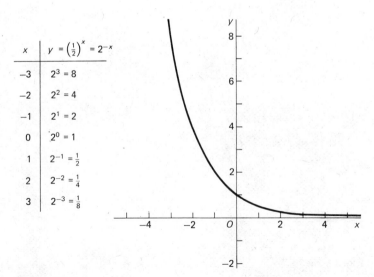

Figure 10-5

EXAMPLE 3

Graph $y = e^{-2x}$

Solution

The table of values and graph are shown in Figure 10-6. (Powers of e can be found in Table III of the Appendix.)

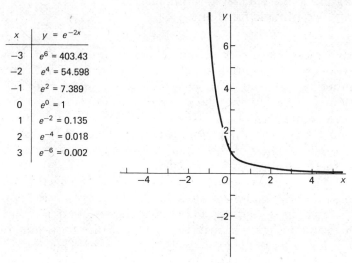

x	$y = e^{-2x}$
−3	e^6 = 403.43
−2	e^4 = 54.598
−1	e^2 = 7.389
0	e^0 = 1
1	e^{-2} = 0.135
2	e^{-4} = 0.018
3	e^{-6} = 0.002

Figure 10-6

Exercise 10-1

1. Graph the function $y = 4^x$.
2. Graph the function $y = 8^x$.
3. Graph the function $y = 2(3^x)$.
4. Graph the function $y = 3(2^x)$.
5. Graph $y = 2^{x+1}$.
6. Graph $y = 3^{x-1}$.
7. Graph $y = e^x$.
8. Graph $y = 2e^x$.
9. Graph $y = 3^{-x}$.
10. Graph $y = 3^{-2x}$.
11. Graph $y = e^{-x}$.
12. Graph $y = \frac{1}{3}e^x$.
13. A biologist has a population of 100 insects, which grow according to the formula $y = 100e^n$, where n is the number of days. How many insects are present after 4 days?

10-2 Logarithmic functions

OBJECTIVES: After completing this section, you should be able to:

a. Convert equations for logarithmic functions from logarithmic to exponential form and vice versa.
b. Evaluate some special logarithms.
c. Graph logarithmic functions.

Every exponential function, $y = a^x$, is a one-to-one function; so every exponential function has an inverse. As with other functions, we find the inverse function of an exponential function by interchanging the x's and y's in the equation. Thus the inverse function of $y = 10^x$ is $x = 10^y$. We may write the equation $x = 10^y$ in the **logarithmic form** $y = \log_{10} x$. In general, the inverse function of the exponential function $f(x) = a^x$, $a \neq 1$, is the logarithmic function $f^{-1}(x) = \log_a x$.

The equation $y = \log_a x$ means that y is the logarithm to the base a of the number x, or, stated differently, y is the power to which a must be raised to obtain x. As we mentioned earlier, a logarithmic function may be written in the logarithmic form, $y = \log_a x$, or in the exponential form, $x = a^y$. Thus if $y = \log_2 x$, then $x = 2^y$; so y is the power to which we must raise 2 to get x.

EXAMPLE 1

a. Write $2 = \log_4 16$ in exponential form.
b. Write $64 = 4^3$ in logarithmic form.

Solution

a. $2 = \log_4 16$ is equivalent to $16 = 4^2$.
b. $64 = 4^3$ is equivalent to $3 = \log_4 64$.

EXAMPLE 2

a. Evaluate $\log_2 8$.
b. Evaluate $\log_3 9$.

Solution

a. If $y = \log_2 8$, then $8 = 2^y$. But $2^3 = 8$; so $\log_2 8 = 3$.
b. If $y = \log_3 9$, then $9 = 3^y$. But $3^2 = 9$; so $\log_3 9 = 2$.

EXAMPLE 3

If $4 = \log_2 x$, what is x?

Solution

If $4 = \log_2 x$, then $2^4 = x$; so $x = 16$.

EXAMPLE 4

If $2 = \log_x 9$, what is x?

Solution

If $2 = \log_x 9$, then $x^2 = 9$; so $x = 3$.

The logarithmic function $y = \log_{10} x$ (frequently written $y = \log x$) is the inverse of the exponential function $y = 10^x$. Thus we may graph $y = \log_{10} x$ by graphing $y = 10^x$ and then graphing its reflection about the line $y = x$. The graph of the two functions is shown in Figure 10-7.

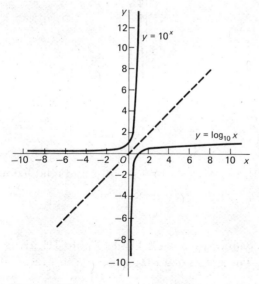

Figure 10-7

EXAMPLE 5

Graph $y = \log_2 x$.

Solution

We may graph $y = \log_2 x$ by graphing $x = 2^y$. The table of values (found by substituting values in for y and calculating x), along with the graph, are shown in Figure 10-8. Note that it is symmetric to the graph of $y = 2^x$ (in dotted lines) with respect to the line $y = x$.

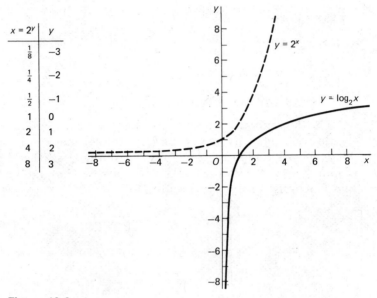

$x = 2^y$	y
$\frac{1}{8}$	-3
$\frac{1}{4}$	-2
$\frac{1}{2}$	-1
1	0
2	1
4	2
8	3

Figure 10-8

EXAMPLE 6

Graph $y = \log_e x$ (frequently written $y = \ln x$ and read y equals the natural log of x).

Solution

$y = \log_e x$ is equivalent to $x = e^y$. Using Table III gives the values for $x = e^y$. The graph of $x = e^y$ is shown in Figure 10-9.

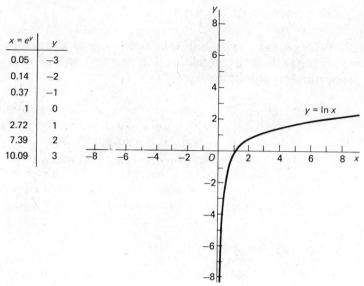

Figure 10-9

Exercise 10-2

1. Write $4 = \log_2 16$ in exponential form.
2. Write $3 = \log_3 27$ in exponential form.
3. Write $\frac{1}{2} = \log_4 2$ in exponential form.
4. Write $-2 = \log_3 \left(\frac{1}{9}\right)$ in exponential form.
5. Write $2^5 = 32$ in logarithmic form.
6. Write $5^3 = 125$ in logarithmic form.
7. Write $4^{-1} = \frac{1}{4}$ in logarithmic form.
8. Write $9^{1/2} = 3$ in logarithmic form.
9. Evaluate $\log_3 27$.
10. Evaluate $\log_4 16$.
11. Evaluate $\log_9 3$.
12. Evaluate $\log_5 \frac{1}{5}$.
13. Graph $y = \log_3 x$.
14. Graph $y = \log_4 x$.
15. Graph $y = \log_\pi x$.
16. Graph $y = \log_9 x$.
17. Graph $y = \log_2 (-x)$.
18. Graph $y = \log_e (-x) = \ln(-x)$.

10-3 Properties of logarithms

OBJECTIVES: After completing this section, you should be able to simplify logarithmic expressions and perform operations, using the properties of logarithms.

If $y = \log_a x$, then y is an exponent ($x = a^y$). The logarithm of a number x to the base a is the power (or exponent) to which the base must be raised to obtain x. For example, because $3^4 = 81$, the logarithm of 81 to the base 3 is 4; that is, $\log_3 81 = 4$.

Because logarithms are exponents, the properties of logarithms can be derived from the properties of exponents. Proofs of these properties may be found in the Appendix.

PROPERTIES OF LOGARITHMS

If $a > 0$, $a \neq 1$, and M and N are positive real numbers, the following properties exist.

Property I. $\log_a a^x = x$

Example: $\log_4 4^3 = 3$

Property II. $a^{\log_a x} = x$

Example: $2^{\log_2 4} = 4$

Property III. $\log_a (MN) = \log_a M + \log_a N$

Example: $\log_2 (16 \cdot 4) = \log_2 16 + \log_2 4$

Property IV. $\log_a \left(\dfrac{M}{N}\right) = \log_a M - \log_a N$

Example: $\log_3 \left(\tfrac{9}{27}\right) = \log_3 9 - \log_3 27$

Property V. $\log_a (M^N) = N \log_a M$

Example: $\log_3 (9^2) = 2 \log_3 9$

EXAMPLE 1

Evaluate the following.

a. $\log_2 2^4$
b. $3^{\log_3 9}$

Solution

a. $\log_2 2^4 = 4$, by Property I
b. $3^{\log_3 9} = 9$, by Property II

EXAMPLE 2

Find $\log_{10} (4 \cdot 5)$, if $\log_{10} 4 = 0.6021$ and $\log_{10} 5 = 0.6990$.

Solution

$$\log_{10} (4 \cdot 5) = \log_{10} 4 + \log_{10} 5 = 0.6021 + 0.6990 = 1.3011$$

EXAMPLE 3

Find $\log_{10} (\tfrac{16}{5})$, if $\log_{10} 16 = 1.2041$ and $\log_{10} 5 = 0.6990$.

Solution

$$\log_{10} (\tfrac{16}{5}) = \log_{10} 16 - \log_{10} 5 = 1.2041 - 0.6990 = 0.5051$$

EXAMPLE 4

If $\log_3 729 = 6$, find $\log_3 (729^4)$.

Solution

$$\log_3 (729^4) = 4 \cdot \log_3 729 = 4 \cdot 6 = 24$$

Exercise 10-3

Evaluate the following.

1. $\log_3 3^4$ 2. $\log_5 5^2$ 3. $\log_e e^x$ 4. $\log_\pi \pi^5$ 5. $\log_{10} 10$
6. $\log_e e$ 7. $\log 10$ 8. $\ln e$ 9. $\log_3 3$ 10. $\log_4 4$
11. $3^{\log_3 x}$ 12. $4^{\log_4 5}$ 13. $10^{\log 2}$ 14. $e^{\ln 3}$
15. Find $\log_{10} (3 \cdot 4)$, if $\log_{10} 3 = 0.4771$ and $\log_{10} 4 = 0.6021$.
16. Find $\log_{10} (12 \cdot 43)$, if $\log_{10} 12 = 1.0792$ and $\log_{10} 43 = 1.6335$.
17. Find $\log_3 (9 \cdot 27)$, if $\log_3 9 = 2$ and $\log_3 27 = 3$. Does the answer equal $\log_3 243$?
18. Find $\log_2 (8 \cdot 16)$.
19. Find $\log_{10} (\tfrac{13}{4})$, if $\log_{10} 13 = 1.1139$ and $\log_{10} 4 = 0.6021$.
20. Find $\log_3 (\tfrac{27}{9})$, if $\log_3 27 = 3$ and $\log_3 9 = 2$. Does the answer equal $\log_3 3$?
21. Find $\log_{10} (16^2)$, given that $\log_{10} 16 = 1.2041$.

22. Find $\log_4 (16^3)$.
23. Find $\log_{10} \sqrt{351}$, given $\log_{10} 351 = 2.5453$.
24. Find $\log_4 \sqrt[3]{256}$, given $\log_4 256 = 4$.
25. Find $\log_3 \sqrt{81}$.
26. Find $\log_2 16^2$.

10-4 Computations with logarithms

OBJECTIVES: After completing this section, you should be able to:

a. Use a table of logarithms to find logarithms of numbers.
b. Find products, quotients, powers, and roots of numbers, using logarithms.

We can use any positive number except 1 as a base for logarithms, but since our number system is based on 10, we shall begin our discussion with logarithms to the base 10, called **common** logarithms. It is customary to omit the symbol for the base when referring to common (or base 10) logarithms. So $\log 1000 = 3$ means $\log_{10} 1000 = 3$.

EXAMPLE 1

Evaluate

a. $\log_{10} 10$
b. $\log 100$
c. $\log 0.01$
d. $\log_{10} 1$

Solution

a. If $y = \log_{10} 10$, then $10^y = 10$; so $y = 1$. Thus $\log_{10} 10 = 1$.
b. If $y = \log_{10} 100$, then $10^y = 100$; so $y = 2$. Thus $\log_{10} 100 = 2$.
c. If $y = \log_{10} 0.01$, then $10^y = 0.01$ or $10^y = 10^{-2}$. Thus $\log_{10} 0.01 = -2$.
d. The exponent to which we must raise 10 to get 1 is 0. Thus $\log_{10} 1 = 0$.

We have seen in Example 1 how to find the common logarithm of a number if it is an integral power of 10. If the number is not a power of 10, we must (in most cases) refer to a table of common logarithms. (See Table II in the Appendix.)

In the Table of Four-Place Common Logarithms (Table II) we see the first number in the left-hand column is 10. The last number in the left-hand column is 99. This table gives us the common logarithms of numbers between 1 and 9.99. Since log 1 = 0 and log 10 = 1, the logarithms in the table are between 0 and 1. It is assumed, therefore, that every number has a decimal point after the first digit and that every logarithm value in the table is preceded by a decimal point. The logarithm values in the table are called mantissas.

Thus to find log 5.42, we look down the left-hand column until we reach 54. We then move to the right until we reach the column headed by the digit 2. The entry is 7340. Thus log 5.42 = .7340 = 0.7340.

EXAMPLE 2

Find the following common logarithms.

a. log 3.23
b. log 1.04

Solution

a. log 3.23 = 0.5092
b. log 1.04 = 0.0170

Now consider log 54.2. We know log 5.42 = 0.7340, and log 54.2 = log (5.42 · 10). But since logarithms are exponents, log (5.42 · 10) = log 5.42 + log 10^1 = 0.7340 + 1 = 1.7340. Note that the mantissa is the same for both log 5.42 and log 54.2. Only the whole number part, or **characteristic**, is different.

EXAMPLE 3

Find log 542.

Solution

$$\begin{aligned} \log 542 = \log (5.42 \cdot 100) &= \log 5.42 + \log 10^2 \\ &= 0.7340 + 2 \\ &= 2.7340 \end{aligned}$$

Thus, to find the common logarithm of a number N, we first write the number as the product of two factors, one of them a number between 1 and 10, and the other an integral power of 10. That is, $N = a \cdot 10^c$.

The **mantissa** which is found in the table is the logarithm of a (the number between 1 and 10), and the **characteristic** is c (the power to which 10 has been raised).

If the characteristic is negative, we do not combine the characteristic and the mantissa. Combining a negative characteristic with a positive mantissa would give a negative logarithm. If, for example, we have the characteristic -1 and the mantissa .4315, we may add and subtract 10 to combine the characteristic and mantissa. That is, $-1 = 9 - 10$, so

$$-1 + .4315 = 9.4315 - 10$$

EXAMPLE 4

Find log 0.00723.

Solution

$$\begin{align} \log 0.00723 &= \log (7.23 \cdot 10^{-3}) \\ &= \log 7.23 + (-3) \\ &= 0.8591 + (-3) \end{align}$$

Rather than adding -3 to the mantissa, we write -3 as $7 - 10$. Thus $0.859 + (-3) = 7.8591 - 10$, and the logarithm can be used as a positive number. This will make it unnecessary to add positive and negative logarithms in the computational work that follows.

EXAMPLE 5

Add log 0.166 to log 206.

Solution

$$\begin{array}{rl} \log 0.166 = \log (1.66 \cdot 10^{-1}) = & 9.2201 - 10 \\ + \log 206 = \log 2.06 \cdot 10^2 = & 2.3139 \\ \hline & 11.5340 - 10 \end{array}$$

The sum of the logs is $11.5340 - 10$ or 1.5340.

EXAMPLE 6

Find log $(0.166 \cdot 206)$.

Solution

$$\log (0.166 \cdot 206) = \log 0.166 + \log 206$$
$$= 1.5340 \quad \text{(from Example 5)}$$

EXAMPLE 7

Use logs to multiply 0.166 by 206.

Solution

From Example 6 we have the log of the product. We know the log of the product has mantissa .5340, which can be found in the table under the column heading 2 and in the row headed by 34. Because the characteristic is 1, the product is $3.42 \times 10^1 = 34.2$.

EXAMPLE 8

Use logarithms to find $0.0263 \cdot 3.19$.

Solution

Let $x = 0.0263 \cdot 3.19$.
Then
$$\log x = \log (0.0263 \cdot 3.19)$$
$$= \log 0.0263 + \log 3.19$$

$$\log 0.0263 = \log (2.63 \cdot 10^{-2}) = 8.4200 - 10$$
$$\underline{\log 3.19 = \log (3.19 \cdot 10^0) \quad = 0.5038}$$
$$\log (0.0263 \cdot 3.19) = \qquad\qquad\qquad 8.9238 - 10$$

Looking up the mantissa .9238 in the table entries for mantissas, we see that this mantissa corresponds to the number 8.39. The characteristic is -2; so

$$0.0263 \cdot 3.19 = 8.39 \cdot 10^{-2} = 0.0839$$

Division can be performed by using the subtraction property of logarithms.

EXAMPLE 9

Use logs to divide 82.9 by 25.3.

Solution

Let $y = 82.9 \div 25.3$.
Then
$$\log y = \log (82.9 \div 25.3) = \log 82.9 - \log 25.3$$

$$\begin{array}{r} \log 82.9 = 1.9186 \\ -\log 25.3 = 1.4031 \\ \hline \log (82.9 \div 25.3) = 0.5155 \end{array}$$

The mantissa entry in the table closest to 0.5155 is 0.5159, which corresponds to 3.28. Thus the quotient is (approximately) $3.28 \cdot 10^0 = 3.28$. The quotient is accurate to three significant digits.

EXAMPLE 10

Use logs to divide 16.1 by 432.

Solution

$$\log 16.1 = 1.2068$$
$$-\log 432 = 2.6355$$

Clearly we cannot subtract without getting a negative logarithm. So we rewrite log 16.1 by adding and subtracting 10:

$$\begin{array}{r} \log 16.1 = 11.2068 - 10 \\ \log 432 = 2.6355 \\ \hline \log (16.1 \div 432) = 8.5713 - 10 \end{array}$$

The mantissa closest to .5713 in the table is .5717; so

$$16.1 \div 432 = 3.73 \cdot 10^{-2} = 0.0373$$

to three significant figures.

We may use logarithms to find powers and roots of numbers by using

Property V. $\log (M^N) = N \log M$.

EXAMPLE 11

Find $\sqrt[3]{81.7}$.

Solution

$$\log (\sqrt[3]{81.7}) = \log (81.7^{1/3}) = \tfrac{1}{3} \log 81.7$$
$$= \tfrac{1}{3}(1.9122) = 0.6374$$

Thus $\sqrt[3]{81.7} = 4.34$ to three significant digits.

Some problems involving roots and characteristics containing -10 can be made easier by slightly altering the way the characteristic is written.

EXAMPLE 12

Find $\sqrt[4]{0.241}$.

Solution

$$\log (\sqrt[4]{0.241}) = \log (0.241^{1/4}) = \tfrac{1}{4}(\log 0.241)$$
$$= \tfrac{1}{4}(9.3820 - 10)$$

So that we can easily divide by 4, we rewrite the problem with -40 instead of with -10. (We do this by adding and subtracting 30.)

$$\log (\sqrt[4]{0.24}) = \tfrac{1}{4}(39.3820 - 40)$$
$$= 9.8455 - 10$$

Thus the log of the root has characteristic -1 and mantissa .8455. So $\sqrt[4]{0.241} = 0.701$ to three significant digits.

EXAMPLE 13

Use logarithms to compute $\sqrt{\dfrac{18^3}{54}}$.

Solution

Let

$$y = \sqrt{\dfrac{18^3}{54}}$$

288 Exponential and Logarithmic Functions

Then
$$\log y = \log \sqrt{\frac{18^3}{54}} = \tfrac{1}{2} \log \frac{18^3}{54}$$
$$= \tfrac{1}{2}(3 \log 18 - \log 54)$$

$$\begin{array}{rl}
3 \log 18 = 3(1.2553) = & 3.7659 \\
-\log 54 = & 1.7324 \\
\hline
3 \log 18 - \log 54 = & 2.0335
\end{array}$$

Thus

$$\tfrac{1}{2}(3 \log 18 - \log 54) = \tfrac{1}{2}(2.0335) = 1.0168$$

The mantissa entry in the table closest to .0168 is .0170. Thus

$$\frac{18^3}{54} = 1.04 \cdot 10^1 = 10.4$$

Exercise 10-4

1. Find log 324.
2. Find log 32,400.
3. Find log 0.00324.
4. Find log 65.3.

Find the following products, using logarithms (correct to three significant digits).

5. $124 \cdot 532$
6. $0.432 \cdot 4670$
7. $0.0021 \cdot 6670$
8. $0.421 \cdot 201 \cdot 3.16$
9. $0.0031 \cdot 15 \cdot 101$

Use logarithms to find (correct to three significant digits)

10. $763 \div 125$
11. $750 \div 15.0$
12. $17{,}600 \div 136$
13. $16.5 \div 1860$
14. 16.1^3
15. 0.031^2
16. $\sqrt[3]{291}$
17. $\sqrt{0.0365}$
18. $\sqrt{165 \cdot 13^3}$
19. $\dfrac{\sqrt{0.165 \cdot 128}}{16.2^2}$
20. $\sqrt[4]{\dfrac{16^2}{5.11}}$

CHAPTER TEST

1. Graph $y = 2^x$.
2. Graph $y = 3^{-x}$.
3. Graph $y = e^x$.
4. Write $\log_3 81 = 4$ in exponential form.
5. Write $4^3 = 64$ in logarithmic form.
6. Evaluate $\log_9 3$.
7. Evaluate $\log_5 \frac{1}{5}$.
8. Graph $\log_2 x$.
9. Graph $\log_e x$.
10. Evaluate $4^{\log_4 9}$.
11. Evaluate $\log_{10} 10$.
12. Find log 64.7.
13. Find log 0.00732.

Use logarithms to compute the following (correct to three significant figures).

14. $64.7 \cdot 0.00732$
15. $4.67 \div 18.6$
16. $\sqrt[4]{0.0361}$
17. $\sqrt{139 \cdot 15^3}$

11
higher-degree polynomial functions

11-1 Zeros of polynomials

OBJECTIVES: After completing this section, you should be able to:

a. Find zeros of polynomial functions.
b. Find the remainder if a polynomial is divided by a binomial of the form $x - r$.
c. Write a factor of the polynomial, given a zero of a polynomial.
d. Write a zero of the polynomial, given a factor of a polynomial.

A function f such that

$$f(x) = a_n x^n + a_{n-1} x^{n-1} + \ldots + a_1 x + a_0, \qquad a_n \neq 0$$

where the coefficients are constants and n is a nonnegative integer, is called a **polynomial function of degree n**. Polynomial functions are frequently denoted using P rather than f. For example,

$$P(x) = x^4 - 3x^2 + 3x - 4$$

is a polynomial function of degree 4. The domain of a polynomial function is usually assumed to be the set of all real numbers.

We have studied polynomial functions of degree 2 (quadratic functions) in Chapter 9. We wish to find the zeros of, and graph, polynomial functions of degree 3 or greater in this chapter.

If a number r is a solution of $P(x) = 0$, then r is called a **zero of the polynomial** $P(x)$. But if r is a solution of $P(x) = 0$, then $P(r) = 0$. That is, substituting r in for x will result in the value 0 if r is a zero of $P(x)$.

EXAMPLE 1

a. Is -3 a solution of $x^2 - x - 12 = 0$?
b. Is -3 a zero of $P(x)$ if $P(x) = x^2 - x - 12$?

Solution

a. Yes, because substituting -3 in for x gives
$(-3)^2 - (-3) - 12 = 0$.
b. Yes, because $P(-3) = (-3)^2 - (-3) - 12 = 0$.

Note that the two questions in Example 1 are asking the same thing in different ways. We can determine if a value r is a zero of a higher degree polynomial $P(x)$ by evaluating $P(r)$.

EXAMPLE 2

If $P(x) = 2x^3 - 5x^2 - 7x$:

a. Is -1 a zero of $P(x)$?
b. Is 2 a zero of $P(x)$?
c. Is 0 a zero of $P(x)$?

Solution

a. Yes, because $P(-1) = 2(-1)^3 - 5(-1)^2 - 7(-1) = 0$
b. No, because $P(2) = 2(2)^3 - 5(2)^2 - 7(2) = -18 \neq 0$.
c. Yes, because $P(0) = 2(0)^3 - 5(0)^2 - 7(0) = 0$.

If we divide any polynomial, $P(x)$, by $x - r$, we get a result of the following form:

$P(x) = (x - r)Q(x) + R,$

where $Q(x)$ is the quotient and R is the remainder resulting from the division.

EXAMPLE 3

Divide $x^3 - 6x^2 + 3x + 21$ by $x - 2$.

Solution

$$\begin{array}{r}
x^2 - 4x - 5 \\
x-2\overline{)x^3 - 6x^2 + 3x + 21}\\
\underline{x^3 - 2x^2}\\
-4x^2 + 3x\\
\underline{-4x^2 + 8x}\\
-5x + 21\\
\underline{-5x + 10}\\
11
\end{array}$$

The quotient is $x^2 - 4x - 5$ and the remainder is 11. The division may be written as $x^3 - 6x^2 + 3x + 21 = (x - 2)(x^2 - 4x - 5) + 11$.

Note that if $P(x) = (x - r)Q(x) + R$, then $P(r) = (r - r)Q(r) + R$, or $P(r) = R$. This result may be stated as the **remainder theorem**.

The Remainder Theorem. If a polynomial $P(x)$ is divided by $x - r$ until a constant remainder R is obtained, then $P(r) = R$.

EXAMPLE 4

Use the remainder theorem to determine the remainder if:

a. $P(x) = x^3 - 4x^2 + 6x - 1$ is divided by $x - 1$.
b. $Q(x) = x^4 + 3x^3 + x^2 - x + 3$ is divided by $x + 2$.

Solution

a. $P(1) = 1^3 - 4(1)^2 + 6(1) - 1 = 2$; the remainder is 2.
b. $Q(-2) = (-2)^4 + 3(-2)^3 + (-2)^2 - (-2) + 3 = 1$; the remainder is 1.

According to the remainder theorem, we may write the division by $x - r$ as $P(x) = (x - r)Q(r) + P(r)$. It is easily seen from this equation that $x - r$ is a factor of $P(x)$ if and only if $P(r) = 0$; that is, if and only if r is a zero of the polynomial. This result is the **factor theorem**.

The Factor Theorem. If r is a zero of the polynomial $P(x)$, then $x - r$ is a factor of $P(x)$; conversely, if $x - r$ is a factor of $P(x)$, then r is a zero of $P(x)$.

This theorem states that we can find a factor of a polynomial if we know a zero and that we can find a zero if we know a factor of the polynomial.

EXAMPLE 5

a. Find the zeros of $P(x) = (x-1)(x+2)(x-4)$.
b. Is $x-3$ a factor of $P(x) = x^3 - 3x^2 + 2x - 6$?
c. Does the graph of $y = x^4 - 4x^2 + 3x - 6$ cross the x-axis at $x - 2$?

Solution

a. $x-1$ is a factor of $P(x)$, so 1 is a zero of $P(x)$. $x+2$ is a factor of $P(x)$, so -2 is a zero of $P(x)$. $x-4$ is a factor of $P(x)$, so 4 is a zero of $P(x)$.
b. $P(3) = 3^3 - 3(3)^2 + 2(3) - 6 = 0$, so $x-3$ is a factor of $P(x)$.
c. If $P(x) = x^4 - 4x^2 + 3x - 6$, $P(2) = 2^4 - 4(2)^2 + 3(2) - 6 = 0$. Thus, 2 is a zero of the polynomial, so the graph crosses the x-axis at $(2, 0)$.

Exercise 11-1

1. Is 2 a solution of $x^2 - 3x + 2 = 0$?
2. Is 3 a solution of $x^2 + 6x - 9 = 0$?
3. Is -1 a solution of $x^2 + 3x - 4 = 0$?
4. Is 4 a solution of $x^2 - 8x + 16 = 0$?
5. Is -2 a zero of $P(x) = x^3 + 3x^2 + 2x - 8$?
6. Is 1 a zero of $P(x) = x^3 - 2x^2 + x$?
7. Is -3 a zero of $x^3 + 4x^2 - 8$?
8. Is -4 a zero of $x^2 - 4x + 4$?
9. What is the remainder if $x^2 - 3x + 2$ is divided by $x - 3$?
10. What is the remainder if $x^3 - 3x^2 + 2x + 1$ is divided by $x - 2$?
11. What is the remainder if $x^3 - 4x^2 + 2x - 3$ is divided by $x + 1$?
12. What is the remainder if $x^4 - x^3 + x^2 + x - 3$ is divided by $x + 2$?
13. Find the zeros of $P(x) = (x-3)(x+2)(x-6)$.
14. Find the zeros of $P(x) = (x-5)(x+2)(x-1)$.
15. Is $x - 5$ a factor of $x^3 - 5x^2 + 3x - 15$?
16. Is $x + 1$ a factor of $x^3 - 4x^2 - 3x + 2$?
17. Is $x - \frac{1}{2}$ a factor of $2x^2 + 3x - 2$?
18. Is $x - 3$ a factor of $x^3 - 2x^2 + 4x - 6$?
19. Does the graph of $y = x^2 - 4x + 3$ cross the x-axis at 3?
20. Does the graph of $y = x^3 - 3x^2 + x + 2$ cross the x-axis at 1?

11-2 Synthetic division

OBJECTIVE: After completing this section, you should be able to:
a. Use synthetic division to divide polynomials by binomials of the form $x - r$.
b. Use synthetic division to determine if $x - r$ is a factor of a polynomial.
c. Use synthetic division to determine if r is a zero of a polynomial.

In Chapter 2 we divided polynomials by binomials. In this chapter we will learn to divide polynomials by a binomial of the form $x - r$ by **synthetic division**. We will use an example to illustrate the method.

Dividing $x^4 + 6x^3 + 5x^2 - 4x + 2$ by $x - 3$ by long division gives:

$$
\begin{array}{r}
x^3 + 4x^2 - 3x + 2 \quad \text{[quotient]} \\
[\text{divisor}] \quad x+2 \overline{)x^4 + 6x^3 + 5x^2 - 4x + 2} \quad \text{[dividend]} \\
\underline{x^4 + 2x^3} \\
4x^3 + 5x^2 \\
\underline{4x^3 + 8x^2} \\
-3x^2 - 4x \\
\underline{-3x^2 - 6x} \\
2x + 2 \\
\underline{2x + 4} \\
-2 \quad \text{[remainder]}
\end{array}
$$

Note that we repeated the writing of many terms and that the only important divisions involved the coefficients. The powers of x in the quotient begin with one less than the highest power of the dividend and decrease by one for each term. If we rewrite the work without the terms that were recopied and without the x's, the process looks like this:

$$
\begin{array}{r}
\boxed{1} + \textcircled{4} - \textcircled{3} + \textcircled{2} \\
+2 \overline{)\boxed{1} + 6 + 5 - 4 + 2} \\
2 \\
\textcircled{4} \\
8 \\
\ominus 3 \\
-6 \\
\textcircled{2} \\
4 \\
-2
\end{array}
$$

Note that, with the exception of the first term, the coefficients of the quotient are the same as the differences after the subtraction has been performed (circled). The last difference is, of course, the remainder. The coefficient of the first term of the quotient is the same as the coefficient of the first term of the dividend (in box). Thus, if we write the differences on one line and write the coefficient of the first term of dividend on that line, we have the coefficients of the quotient all on one line, as follows:

$$+2 \overline{)1 + 6 + 5 - 4 + 2}$$
$$ 2 + 8 - 6 + 4$$
$$ \boxed{1} + ④ - ③ + ② - 2$$

Now we will make the process faster by changing the sign on the divisor from $+2$ to -2 and adding rather than subtracting. The procedure now looks as follows:

$$\overline{1 + 6 + 5 - 4 + 2} \, \rfloor -2$$
$$ -2 - 8 + 6 - 4$$
$$\underbrace{1 + 4 - 3 + 2}_{\text{coefficients of quotient}} \underbrace{- 2}_{\text{remainder}}$$

The quotient polynomial is $x^3 + 4x^2 - 3x + 2$, with remainder -2.

The procedure for dividing a polynomial by $x - r$ follows:

PROCEDURE	EXAMPLE
To divide $P(x)$ by $x - r$:	Divide $x^4 - 4x^3 + 3x + 10$ by $x - 3$
1. Arrange the coefficients of $P(x)$ in descending powers of x, with a 0 for any missing power.	1. $1 - 4 + 0 + 3 + 10 \quad \rfloor 3$
2. Bring down the first coefficient of $P(x)$ to the third line.	2. $\begin{array}{r} 1 - 4 + 0 + 3 + 10 \quad \rfloor 3 \\ \hline 1 \end{array}$
3. Multiply the last number in line 3 by r and write the product under the next term of line 1.	3. $\begin{array}{r} 1 - 4 + 0 + 3 + 10 \quad \rfloor 3 \\ +3 \\ \hline 1 \end{array}$

4. **Add** the last number in line 2 to the number above in line 1, and write the sum in line 3.

4. $1 - 4 + 0 + 3 + 10$ (3
    $\underline{+ 3}$
    $1 - 1$

5. Repeat steps 3 and 4 until all numbers of line 1 have been used.

5. $1 - 4 + 0 + 3 + 10$ (3
    $\underline{+ 3 - 3 - 9 - 18}$
    $1 - 1 - 3 - 6 - 8$

6. Write the quotient as a polynomial of degree 1 less than the dividend. The last number in line 3 is the remainder.

6. The quotient is $x^3 - x^2 - 3x - 6$, with remainder -8.

EXAMPLE 1

Use synthetic division to divide $3x^4 - 6x^2 - 8x + 2$ by $x + 1$.

Solution

With $r = -1$, we proceed as follows:

$3 + 0 - 6 - 8 + 2$ $(-1$
$\underline{- 3 + 3 + 3 + 5}$
$3 - 3 - 3 - 5 + 7$

The quotient is $3x^3 - 3x^2 - 3x - 5$, remainder 7.

EXAMPLE 2

Use synthetic division to divide $4x^3 - 7x^2 + 3x - 10$ by $x - 2$.

Solution

$4 - 7 + 3 - 10$ (2
$\underline{+ 8 + 2 + 10}$
$4 + 1 + 5 + 0$

Note: The remainder of the division in Example 2 is 0. Thus, 2 is a zero of $4x^3 - 7x^2 + 3x - 10$ and $x - 2$ is a factor of $4x^3 - 7x^2 + 3x - 10$.

EXAMPLE 3

Use synthetic division to determine if $x - 1$ is a factor of $x^4 - 3x^3 + 5x^2 - x - 2$?

Solution

$$\begin{array}{r} 1 - 3 + 5 - 1 - 2 \quad \underline{(1} \\ + 1 - 2 + 3 + 2 \\ \hline 1 - 2 + 3 + 2 + 0 \end{array}$$

The remainder is 0, so $x - 1$ is a factor.

Note: Synthetic division will work only when dividing by a number of the form $x - a$. For example, we cannot use synthetic division to divide a polynomial by $x^2 - 4$.

Exercise 11-2

Use synthetic division to perform the following divisions.

1. $3x^2 + 2x - 3 \div x - 3$
2. $2x^2 + 3x + 4 \div x - 2$
3. $4x^2 - 3x + 6 \div x - 1$
4. $5x^2 - 3x + 1 \div x - 4$
5. $5x^2 - 3x + 6 \div x + 2$
6. $2x^2 + 4x - 3 \div x + 3$
7. $x^3 + 4x^2 - 3x + 2 \div x - 1$
8. $x^3 - 4x^2 + 5x - 4 \div x + 2$
9. $2x^3 + 5x^2 - x + 3 \div x + 2$
10. $3x^3 + 2x^2 + 6x - 1 \div x + 1$
11. $4x^3 - 3x - 1 \div x - 1$
12. $x^3 - 4x^2 + 8 \div x - 2$
13. $x^3 + 4x - 3 \div x - \frac{1}{3}$
14. $4x^3 + 2x^2 - 2x - 1 \div x + \frac{1}{2}$

Use synthetic division to determine if

15. $x - 1$ is a factor of $3x^2 - 4x + 1$.
16. $x + 2$ is a factor of $x^2 - 3x + 2$.
17. $x + 3$ is a factor of $x^3 + 3x^2 + 4x + 12$.
18. $x - \frac{1}{2}$ is a factor of $4x^3 + x - 1$.
19. 2 a zero of $x^4 - 2x^2 - 7x + 6$.
20. -3 a zero of $x^3 - 4x^2 - 3x + 18$.
21. -1 a zero of $x^3 + 9x^2 + 24x + 16$.
22. $\frac{1}{2}$ is a zero of $4x^3 + 16x^2 + 9x - 9$.
23. 5 a solution of $2x^3 - 12x^2 - x + 30 = 0$.
24. -2 is a solution of $x^4 + 3x^2 + 4 = 0$.
25. $\frac{2}{3}$ a solution of $3x^3 - 8x^2 - 2x + 4 = 0$.
26. $\frac{1}{2}$ a solution of $1 - 4x^2 - 16x^4 = 0$.

11 - 3   Solutions of polynomial equations

OBJECTIVE: After completing this section, you should be able to find solutions to certain polynomial equations.

The **fundamental theorem of algebra** states that **every polynomial equation, $P(x) = 0$, has at least one (real or complex) solution.** If we find one such solution, say r_1, we know $x - r_1$ is a factor of $P(x)$. Then we may write

$$P(x) = (x - r_1)Q_1(x)$$

where $Q_1(x)$ is quotient polynomial when $P(x)$ is divided by $x - r_1$. But the fundamental theorem also applies to $Q_1(x)$. If we call this solution r_2, then $x - r_2$ is a factor of $Q_1(x)$ so

$$P(x) = (x - r_1)(x - r_2)Q_2(x).$$

Continuing this process until one of the quotients is a constant, say a, we have

$$P(x) = (x - r_1)(x - r_2) \ldots (x - r_n)a$$

Note that the degree of the polynomial was reduced by one each time it was divided by a linear factor. Thus, if the polynomial is of degree n, n divisions are necessary before the quotient is a constant. Thus, **every polynomial $P(x)$ of degree n can be expressed as the product of n linear factors, so $P(x)=0$ has exactly n solutions (not all distinct).** The nondistinct solutions are called **multiple solutions.** The solutions are complex numbers, and may not all be real.

EXAMPLE 1

If 2 and -2 are solutions of $x^4 + 6x^3 + 5x^2 - 24x - 36 = 0$, find the remaining solutions

Solution

If 2 and -2 are solutions of $P(x) = 0$, then $x - 2$ and $x + 2$ are factors of $P(x)$. Dividing the product of these two factors $x^2 - 4$, into the polynomial gives

$$x^4 + 6x^3 + 5x^2 - 24x - 36 = (x^2 - 4)(x^2 + 6x + 9)$$

But

$$x^2 + 6x + 9 = (x + 3)(x + 3)$$

Thus, the solutions are 2, −2, and −3. −3 is a **multiple solution, with multiplicity** 2, or a **double solution**.

We may use synthetic division to find the solutions of $P(x) = 0$, as the following example shows.

EXAMPLE 2

Solve the equation $x^4 - x^3 - 8x^2 + 5x + 3 = 0$, given that 1 and 3 are solutions.

Solution

One factor is $x - 1$, so using synthetic division with 1 gives

$$\begin{array}{r} 1 - 1 - 8 + 5 + 3 \quad (1 \\ + 1 + 0 - 8 - 3 \\ \hline 1 + 0 - 8 - 3 + 0 \end{array}$$

Thus, $x^4 - x^3 - 8x^2 + 5x + 3 = (x - 1)(x^3 + 0x^2 - 8x - 3)$.
Using synthetic division again, with 3, gives

$$\begin{array}{r} 1 + 0 - 8 - 3 \quad (3 \\ + 3 + 9 + 3 \\ \hline 1 + 3 + 1 + 0 \end{array}$$

Thus, $x^4 - x^3 - 8x^2 + 5x + 3 = (x - 1)(x - 3)(x^2 + 3x + 1)$.
The solutions related to the quotient polynomial after $(x - 1)$ and $(x - 3)$ have been divided out $[x^2 + 3x + 1]$ may be found by the quadratic formula:

$$x = \frac{-3 \pm \sqrt{9 - 4(1)(1)}}{2}$$

or

$$x = \frac{-3 \pm \sqrt{5}}{2}$$

Thus, the solution set for $x^4 - x^3 - 8x^2 + 5x + 3 = 0$ is

$$\left\{ 1, 3, \frac{-3 + \sqrt{5}}{2}, \frac{-3 - \sqrt{5}}{2} \right\}$$

Again, we have 4 solutions to the 4th degree polynomial equation.

As the following example will show, we can combine two or more synthetic divisions into one process.

EXAMPLE 3

Find the solutions of $x^4 - 9x^2 + 22x - 24 = 0$, given that 2 and -4 are solutions.

Solution

We may use synthetic division to find the quotient polynomial after $x - 2$ and $x + 4$ have been divided into $P(x)$:

$$
\begin{array}{r}
1 + 0 - 9 + 22 - 24 \quad (2 \\
\underline{+ 2 + 4 - 10 + 24} \\
1 + 2 - 5 + 12 \quad (-4 \\
\underline{- 4 + 8 - 12} \\
1 - 2 + 3
\end{array}
$$

Thus, the quotient polynomial is $x^2 - 2x + 3$.
The solutions to $x^2 - 2x + 3 = 0$ may be found by factoring (if possible) or by the quadratic formula. The quadratic formula gives

$$x = \frac{2 \pm \sqrt{4 - 12}}{2} = \frac{2 \pm i\sqrt{8}}{2} = \frac{2(1 \pm i\sqrt{2})}{2} = 1 \pm i\sqrt{2}$$

Thus, the solution set for $x^4 - 9x^2 + 22x - 24 = 0$ is

$$\{2, -4, 1 + i\sqrt{2}, 1 - i\sqrt{2}\}$$

Note that in Example 3 we found 2 complex solutions to $P(x) = 0$. We may state in general that **every polynomial with real coefficients which has $a + bi$ as a solution will also have $a - bi$ as a solution.** The two solutions, which differ only in the sign before the imaginary part, are called **conjugate solutions**. That complex solutions occur in pairs can easily be seen by looking at the quadratic formula, which gives two roots which differ only in the sign before the radical.

Exercise 11-3

1. Solve the equation $x^3 - 8x^2 + 21x - 18 = 0$, given that 3 is a solution.
2. Solve the equation $x^3 - 6x^2 + 11x - 6 = 0$, given that 2 is a solution.
3. Solve $x^3 - 7x + 6 = 0$, given that 2 is one solution.
4. Solve $x^3 - 3x^2 + x + 1 = 0$, given that 1 is a solution.

5. Solve $x^3 - 4x^2 + x + 6 = 0$, given that 2 is one solution.
6. Solve $x^3 - 2x^2 - 11x + 12 = 0$, given that 4 is one solution.
7. Solve $x^4 + 2x^3 - 7x^2 - 8x + 12 = 0$, given that 1 and 2 are solutions.
8. Solve $2x^4 - 11x^3 - 8x^2 + 11x + 6 = 0$, given that -1 and 6 are solutions.
9. Solve $x^3 - 6x^2 + 10x - 4 = 0$, given that 2 is a solution.
10. Solve $x^3 - 4x^2 + 6x - 3 = 0$, given that 1 is a solution.
11. Solve $x^3 - x^2 - 2x + 8 = 0$, given that -2 is a solution.
12. Solve $x^3 - x^2 - 4x - 6 = 0$, given that 3 is a solution.

11-4 Rational solutions of polynomial equations

OBJECTIVE: After completing this section, you should be able to find the rational solutions of polynomial equations.

If we look at the solutions to the polynomial equation in Example 1 of Section 11-2, we see that they are all factors of the constant term of the polynomial. In Example 2 and Example 3 of Section 11-3, the **integer** solutions to the polynomial equations are factors of the constant term of the respective polynomials. In general, if a polynomial equation $P(x) = 0$ has **integer** solutions $r_1, r_2, \ldots, r_k$, then $P(x) = (x - r_1)(x - r_2) \cdots (x - r_k)Q(x)$, and the constant term of $P(x)$ must have $r_1, r_2, \ldots, r_k$ as factors. Thus we have the following theorem.

If $P(x)$ has integer coefficients and a coefficient of 1 on its highest power of x, then all integer solutions of $P(x) = 0$ are factors of the constant term of $P(x)$.

If the coefficient of the highest power of x is not 1, we can factor this coefficient out of the polynomial. That is,

$$P(x) = a_n x^n + a_{n-1} x^{n-1} + \cdots + a_1 x + a_0$$

can be written as

$$P(x) = a_n \left(x^n + \frac{a_{n-1}}{a_n} x^{n-1} + \cdots + \frac{a_1 x}{a_n} + \frac{a_0}{a_n} \right)$$

Thus we may conclude the following:

The rational solutions of the polynomial equation $a_n x^n + a_{n-1} x^{n-1} + \cdots + a_1 x + a_0 = 0$, with integer coefficients, must be of the form **p/q**, where **p** *is a factor of* a_0 *and* **q** *is a factor of* a_n.

The discussion in this chapter indicates we may use the following procedure to find the solutions of some higher-degree polynomials equations.

PROCEDURE	EXAMPLE
To find solutions of higher-degree polynomial equations:	Find the solutions of $P(x) = 2x^4 + 9x^3 + 10x^2 - 3x - 6 = 0$
1. Use the degree of the polynomial to indicate how many solutions there are.	1. $P(x)$ is fourth degree; so there are four solutions.
2. Find the possible solutions, p/q, where p is a factor of the constant and q is a factor of the coefficient of the highest power of x.	2. Possible solutions are ± 1, $\pm\frac{1}{2}$, ± 2, $\pm\frac{2}{2}$, ± 3, $\pm\frac{3}{2}$, ± 6, $\pm\frac{6}{2}$. (Note that $\pm\frac{2}{2}$ and $\pm\frac{6}{2}$ are repetitions.)
3. Find $P(p/q)$ to determine if any of the possible solutions is a solution.	3. $P(1) = 2 + 9 + 10 - 3 - 6 \neq 0$ $P(-1) = 2 - 9 + 10 + 3 - 6 = 0$ -1 is a solution.
4. Use synthetic division to find a quotient polynomial which is 1 degree less than the polynomial.	4. $\begin{array}{r} 2 + 9 + 10 - 3 - 6 \\ \underline{ - 2 - 7 - 3 + 6} \\ 2 + 7 + 3 - 6 \end{array}$ $(-1$ The quotient polynomial is $2x^3 + 7x^2 + 3x - 6$.
5. Repeat steps 2 to 4 until the quotient polynomial is a quadratic. (Try integer values first, as they are easier.)	5. Possible solutions of quotient polynomial are still ± 1, $\pm\frac{1}{2}$, ± 2, ± 3, $\pm\frac{3}{2}$, and ± 6. $P(2) = 32 + 72 + 40 - 6 - 6 \neq 0$ $P(-2) = 32 - 72 + 40 + 6 - 6 = 0$ -2 is a solution. $\begin{array}{r} 2 + 7 + 3 - 6 \\ \underline{ - 4 - 6 + 6} \\ 2 + 3 - 3 \end{array}$ $(-2$ The quotient polynomial is $2x^2 + 3x - 3$.

6. Use the quadratic formula or factoring to find the remaining two solutions.

6. $x = \dfrac{-3 \pm \sqrt{9+24}}{4}$
$= \dfrac{-3 \pm \sqrt{33}}{4}$

7. List the solution set.

7. The solution set is
$\left\{ -1, -2, \dfrac{-3+\sqrt{33}}{4}, \dfrac{-3-\sqrt{33}}{4} \right\}$

EXAMPLE 1

Find the solutions of the equation $P(x) = x^3 - 2x^2 - 5x + 6 = 0$.

Solution

The polynomial is of degree 3, so there are three solutions. If we can find one of the roots, we can use the quadratic formula (or factoring) to find the other two.

The possible rational solutions are factors of 6: $\pm 1, \pm 2, \pm 3, \pm 6$. We can determine if 1 is a solution by evaluating $P(1)$: $P(1) = 1 - 2 - 5 + 6 = 0$. Thus 1 is a solution, and $x - 1$ is a factor. Synthetic division gives the quotient polynomial

$$\begin{array}{r} 1 - 2 - 5 + 6 \quad (1 \\ + 1 - 1 - 6 \\ \hline 1 - 1 - 6 \end{array}$$

The quotient polynomial is $x^2 - x - 6$, which factors into $(x - 3)(x + 2)$. Thus the solution set is $\{1, 3, -2\}$.

EXAMPLE 2

Find the zeros of $P(x) = 2x^3 + 3x^2 - 12x - 4$.

Solution

The possible factors are $\pm 1, \pm\tfrac{1}{2}, \pm 2, \pm\tfrac{2}{2}, \pm 4, \pm\tfrac{4}{2}$. Clearly $\pm\tfrac{2}{2}$ and $\pm\tfrac{4}{2}$ are accounted for by ± 1 and ± 2, respectively.

We may use synthetic division to find $P(a)$ [recall $P(a)$ is the remainder after $P(x)$ is divided by $x - a$.] The advantage to using synthetic division is that we find the quotient polynomial $Q(x)$ whenever we find a zero.

```
2 + 3 − 12 −  4    (1
  + 2 +  5 −  7
2 + 5 −  7 − 11      P(1) = −11

2 + 3 − 12 −  4    (2
  + 4 + 14 +  4
2 + 7 +  2 +  0      P(2) = 0
```

Thus 2 is a zero of $P(x)$ [that is, a solution to $P(x) = 0$]. The quotient polynomial is $2x^2 + 7x + 2$. The quadratic formula gives the remaining zeros.

$$x = \frac{-7 \pm \sqrt{49 - 16}}{4} = \frac{-7 \pm \sqrt{33}}{4}$$

The zeros are 2, $\dfrac{-7 + \sqrt{33}}{4}$, and $\dfrac{-7 - \sqrt{33}}{4}$.

The following observation will help isolate real zeros of polynomials with real coefficients.

Changing Signs. If $P(a)$ and $P(b)$ are of opposite sign, then there is an odd number (at least one) of zeros between a and b.

EXAMPLE 3

Is there a solution to $2x^3 - 9x^2 + x + 12 = 0$ that is between 1 and 2?

Solution

We shall find $P(1)$ and $P(2)$ by synthetic division.

```
2 − 9 +  1 + 12    (1
  + 2 −  7 −  6
2 − 7 −  6 +  6      P(1) = 6

2 − 9 +  1 + 12    (2
  + 4 − 10 − 18
2 − 5 −  9 −  6      P(2) = −6
```

Thus the graph of $P(x)$ crosses the x-axis [that is, $P(x) = 0$] at least once between $x = 1$ and $x = 2$. The solution between 1 and 2 is $\frac{3}{2}$.

EXAMPLE 4

Find the solutions to $2x^3 + 3x^2 - 5x - 6 = 0$.

11-4 Rational solutions of polynomial equations

Solution

The possible rational solutions are factors of 6 divided by factors of 2: $\pm 1, \pm\frac{1}{2}, \pm 2, \pm\frac{2}{2}, \pm 3, \pm\frac{3}{2}, \pm 6, \pm\frac{6}{2}$. Testing the integer values by synthetic division shows

$$
\begin{array}{rrrr}
2+3- & 5- & 6 & \quad (1 \\
+2+ & 5+ & 0 & \\
\hline
2+5+ & 0- & 6 & \quad P(1)=-6
\end{array}
$$

$$
\begin{array}{rrrr}
2+3- & 5- & 6 & \quad (2 \\
+4+ & 14+ & 18 & \\
\hline
2+7+ & 9+ & 12 & \quad P(2)=12
\end{array}
$$

Thus there is a solution between 1 and 2 (the signs changed). But the only possible rational solution between 1 and 2 is $\frac{3}{2}$; so $\frac{3}{2}$ is a good possibility for a zero of $P(x)$.

$$
\begin{array}{rrrr}
2+3- & 5- & 6 & \quad (\frac{3}{2} \\
+3+ & 9+ & 6 & \\
\hline
2+6+ & 4+ & 0 & \quad P(\frac{3}{2})=0
\end{array}
$$

The quotient polynomial is $2x^2 + 6x + 4$. Factoring gives $2(x^2 + 3x + 2) = 2(x + 1)(x + 2)$. The solutions set is $\{\frac{3}{2}, -1, -2\}$.

EXAMPLE 5

Find the solutions of $P(x) = 3x^4 - 5x^3 - 10x^2 + 2x + 4 = 0$.

Solution

There are four (complex) solutions. The possible rational solutions are factors of 4 divided by the factors of 3. Thus the possible rational solutions are $\pm 1, \pm\frac{1}{3}, \pm 2, \pm\frac{2}{3}, \pm 4, \pm\frac{4}{3}$. Let us begin by testing positive integer values:

$$
\begin{array}{rrrrr}
3- & 5-10+ & 2+ & 4 & \quad (1 \\
+ & 3- 2- & 12- & 10 & \\
\hline
3- & 2-12- & 10- & 6 & \quad P(1)=-6
\end{array}
$$

$$
\begin{array}{rrrrr}
3- & 5-10+ & 2+ & 4 & \quad (2 \\
+ & 6+ 2- & 16- & 28 & \\
\hline
3+ & 1- 8- & 14- & 24 & \quad P(2)=-24
\end{array}
$$

$$
\begin{array}{rrrrr}
3- & 5-10+ & 2+ & 4 & \quad (4 \\
+12+28+ & 72+ & 296 & \\
\hline
3+ 7+18+ & 74+ & 300 & \quad P(4)=300
\end{array}
$$

A real solution exists between 2 and 4.

Thus a real solution exists between 2 and 4 (the signs changed). There are no possible rational solutions between 2 and 4; so there is an irrational solution (which we cannot find at this time) between 2 and 4. Testing negative integer values gives

$$\begin{array}{r} 3 - 5 - 10 + 2 + 4 \\ -3 + 8 + 2 - 4 \\ \hline 3 - 8 - 2 + 4 - 0 \end{array} \quad \begin{array}{l}(-1 \\ \\ P(-1) = 0\end{array}$$

One solution is -1. Possible solutions of the quotient polynomial, $Q(x) = 3x^3 - 8x^2 - 2x + 4$, are the same as those for $P(x)$, because the coefficient of x^3 is 3, and the constant is 4. We can easily see that $P(0) = 4$; so there is at least one solution between 0 and 1 [recall $P(1) = -6$]. Checking $\frac{1}{3}$ and $\frac{2}{3}$ in the quotient polynomial $Q(x)$ gives

$$\begin{array}{r} 3 - 8 - 2 + 4 \\ + 1 - \frac{7}{3} - \frac{13}{9} \\ \hline 3 - 7 - \frac{13}{3} + \frac{23}{9} \end{array} \quad \begin{array}{l}(\frac{1}{3} \\ \\ Q(\frac{1}{3}) = \frac{23}{9}\end{array}$$

$$\begin{array}{r} 3 - 8 - 2 + 4 \\ + 2 - 4 - 4 \\ \hline 3 - 6 - 6 + 0 \end{array} \quad \begin{array}{l}(\frac{2}{3} \\ \\ Q(\frac{2}{3}) = 0\end{array}$$

The quadratic factor which remains is $3x^2 - 6x - 6 = 3(x^2 - 2x - 2)$. The quadratic formula gives solutions

$$x = \frac{2 \pm \sqrt{4+8}}{2} = 1 \pm \sqrt{3}.$$

The solution set is $\{-1, \frac{2}{3}, 1 + \sqrt{3}, 1 - \sqrt{3}\}$. Note that $1 + \sqrt{3}$ is the irrational solution between 2 and 4.

Exercise 11-4

Solve the following polynomial equations.

1. $x^3 - 4x^2 + 5x - 2 = 0$
2. $x^3 - 3x^2 - x + 3 = 0$
3. $x^3 - 3x^2 - 4x + 12 = 0$
4. $x^3 + 2x^2 - 5x - 6 = 0$
5. $x^3 - 7x - 6 = 0$
6. $x^3 - 12x - 16 = 0$
7. $x^3 - 5x^2 - 2x + 24 = 0$
8. $x^3 + 10x^2 + 31x + 30 = 0$
9. $x^3 - 5x^2 + 7x - 2 = 0$
10. $x^3 + 8x^2 + 18x + 9 = 0$
11. $x^3 + 2x^2 + 6x + 4 = 0$

12. $x^3 - 5x^2 + 10x - 8 = 0$
13. $2x^3 - 11x^2 + 17x - 6 = 0$
14. $6x^3 + 5x^2 - 2x - 1 = 0$
15. $3x^3 - 7x^2 - 22x + 8 = 0$
16. $3x^3 - 5x^2 + x + 1 = 0$
17. $2x^3 + x^2 - 8x - 2 = 0$
18. $3x^3 - x^2 - 5x - 1 = 0$
19. $4x^3 - 11x^2 + 8x - 4 = 0$
20. $2x^3 - 8x^2 + 11x - 6 = 0$
21. $x^4 - x^3 - 7x^2 + 17x + 6 = 0$
22. $x^4 - x^3 - 7x^2 + x + 6 = 0$
23. $x^4 - 7x^3 + 8x^2 + 28x - 48 = 0$
24. $x^4 - 4x^3 - 7x^2 + 22x + 24 = 0$
25. $x^4 + x^3 - 27x - 27 = 0$
26. $x^4 - 3x^3 - 11x^2 - 27x + 18 = 0$

CHAPTER TEST

1. Is 3 a solution of $x^2 - 6x + 9 = 0$?
2. Is -2 a zero of $x^3 - 4x^2 + 6x + 4$?
3. What is the remainder if $x^3 - 5x^2 + 3x + 2$ is divided by $x - 3$?
4. Find the zeros of $P(x) = (x - 1)(x + 3)(x + 6)$.
5. Is $x - 3$ a factor of $x^3 - 5x^2 + 3x + 9$?
6. Does the graph of $y = 2x^2 - 3x + 1$ cross the x-axis at $\frac{1}{2}$?

Use synthetic division to

7. Divide $x^3 + 4x^2 - 3x + 1$ by $x - 4$.
8. Divide $4x^3 + 3x + 6$ by $x + 2$.
9. Determine if $x + 3$ is a factor of $3x^3 + 9x^2 + 6x + 9$.
10. Determine if $\frac{1}{2}$ is a solution of $4x^2 - 4x + 1 = 0$.

Solve the following equations.

11. $x^3 + 6x^2 - x - 6 = 0$, given that 1 is a solution.
12. $x^4 + 3x^3 - 12x^2 + 8 = 0$, given that 1 and 2 are solutions.
13. $2x^3 + 7x^2 + 7x + 12 = 0$
14. $x^4 - 3x^3 - 2x^2 + 6x + 4 = 0$
15. Find the solution of $x^3 + 2x^2 - 2x - 1$ which lies between 0 and -1.

12

conic sections

12-1 Parabolas

OBJECTIVES: After completing this section, you should be able to:

a. Graph parabolas from their equations in standard form or general form.
b. Locate the focus of a parabola.
c. Change the equation of a parabola from the general form to the standard form.

An equation which has at least one second-degree term and no term with degree greater than two is a **quadratic equation**. The **general form** of the quadratic equation in two unknowns is

$$Ax^2 + Bxy + Cy^2 + Dx + Ey + F = 0$$

with A, B, and C not all 0. We shall limit our discussion to equations with $B = 0$. The graph of the quadratic function, $y = ax^2 + bx + c$, $a \neq 0$, is a parabola which opens up or down (see Section 9-5). A quadratic equation with a y^2 term and no x^2 term is a parabola which opens to the right or left.

EXAMPLE 1

Graph the equation $y^2 - 4x = 0$.

Solution

A table of values and the graph are shown in Figure 12-1(a). Note that this graph passes through the origin (x-intercept is 0 and y-intercept is 0) and is symmetric about the x-axis.

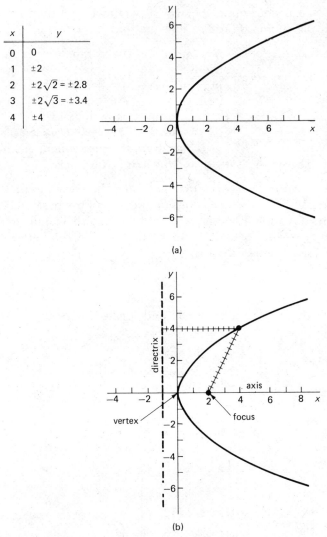

Figure 12-1

We may define a parabola as the locus of all points equidistant from a fixed point and a fixed line. The point is called the **focus** of the parabola, and the line is called the **directrix**. Figure 12-1(b) is the graph of $y^2 - 4x = 0$ with the focus and directrix drawn in. The parabola intersects its axis at the **vertex**.

The parabola in Example 1 opens to the right and has its vertex at (0, 0). The axis of this parabola is the x-axis.

We can determine the way a parabola will open by looking at the **standard form** of its equation:

1. If $y^2 = 4px$ and $\begin{cases} p \text{ is positive, the parabola opens to the right.} \\ p \text{ is negative, the parabola opens to the left.} \end{cases}$

   The vertex is (0, 0), and the axis is the x-axis.

2. If $x^2 = 4py$ and $\begin{cases} p \text{ is positive, the parabola opens up.} \\ p \text{ is negative, the parabola opens down.} \end{cases}$

   The vertex is (0, 0), and the axis is the y-axis.

The value of p is the distance from the vertex to the focus of the parabola. The focus lies on the axis of the parabola. Writing the equation of Example 1 in the form $y^2 = 4x$ indicates that $p = +1$; so the focus is 1 unit to the right of the vertex, or at (1, 0) [see Figure 12-1(b)].

EXAMPLE 2

Graph the equation $x^2 = -8y$ and locate the focus of the parabola.

Solution

The form of the equation indicates that the parabola will open down, the vertex is (0, 0), and the focus will be 2 units down from the vertex, at (0, 2). A table of values helps complete the graph. The graph is shown in Figure 12-2.

x	y
-4	-2
-3	$-\frac{9}{8}$
-2	$-\frac{1}{2}$
-1	$-\frac{1}{8}$
0	0
1	$-\frac{1}{8}$
2	$-\frac{1}{2}$
3	$-\frac{9}{8}$
4	-2

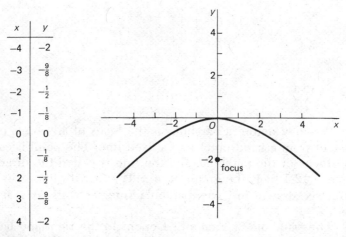

Figure 12-2

The vertex of the parabola is at the point (h, k) instead of at $(0, 0)$ if the standard form of its equation is $(y - k)^2 = 4p(x - h)$ or $(x - h)^2 = 4p(y - k)$.

1. If $(y - k)^2 = 4p(x - h)$, the axis of the parabola is parallel to the x-axis. The curve opens to the right if p is positive and to the left if p is negative.
2. If $(x - h)^2 = 4p(y - k)$, the axis of the parabola is parallel to the y-axis. The curve opens up if p is positive and down if p is negative.

EXAMPLE 3

Graph $(y - 3)^2 = -4(x - 1)$ and locate the focus.

Solution

The parabola opens to the left and has its vertex at $(1, 3)$. The focus is 1 unit to the left of $(1, 3)$, at $(0, 3)$. The table gives some additional points on the curve. The graph is shown in Figure 12-3.

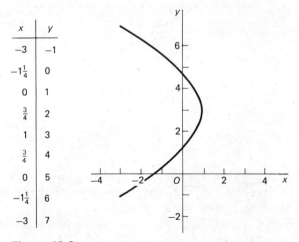

x	y
-3	-1
$-1\frac{1}{4}$	0
0	1
$\frac{3}{4}$	2
1	3
$\frac{3}{4}$	4
0	5
$-1\frac{1}{4}$	6
-3	7

Figure 12-3

If the equation of a parabola is in **general form**, we must complete the square to get the equation in standard form.

EXAMPLE 4

Write the standard form of the equation $x^2 + 2x - 4y - 3 = 0$ and sketch its graph.

Solution

This is the general form of a parabola because just the x term is squared. We complete the square just as we did in Chapter 6:

$$x^2 + 2x = 4y + 3$$
$$x^2 + 2x + 1 = 4y + 3 + 1$$
$$(x + 1)^2 = 4(y + 1)$$

The vertex is at $(-1, -1)$; the axis is parallel to the y-axis; and the curve opens up. The focus is 1 unit up from the vertex, at $(-1, 0)$. A table of values and the graph are shown in Figure 12-4.

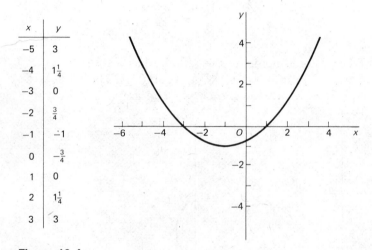

x	y
-5	3
-4	$1\frac{1}{4}$
-3	0
-2	$-\frac{3}{4}$
-1	-1
0	$-\frac{3}{4}$
1	0
2	$1\frac{1}{4}$
3	3

Figure 12-4

Remember that the equation of the parabola is characterized by the presence of the square of either (but not both) x or y and the first power of the other.

The uses of the parabolas in industry and science are many and varied. A partial list follows.

1. Some of the largest steel bridges are built with archs which are arcs of parabolas.
2. The path of a projectile, if air resistance is neglected, is a parabola.
3. The equation expressing the relation between the distance a freely falling body travels and the time required is the equation of a parabola.

Hollywood

HOLLYWOOD BRANDS CENTRALIA, ILL. 62801 U.S.A.
A CONSOLIDATED FOODS COMPANY • RESPONSIVE TO CONSUMER NEEDS

★ BUTTERNUT

BUTTERNUT

CARAMEL AND PEANUTS   NET WT 1.25 OZ (35g)

★ BUTTERNUT

STAR, BUTTERNUT, HOLLYWOOD REG. U.S. PAT. OFF. MADE IN U.S.A. PRINTED IN U.S.A.

INGREDIENTS: SUGAR, CORN SYRUP, PEANUTS, HYDROGENATED VEGETABLE OIL, WATER, SKIM MILK SOLIDS, NON FAT DRY MILK, WHEY SOLIDS, COCOA, CHOCOLATE, SALT, LECITHIN, MONO- AND DIGLYCERIDES.

4. The surface generated by revolving a parabola about it axis is called a paraboloid. This surface is used as a reflector in automobile headlights because a light at its focus causes the light rays that hit the reflector to reflect parallel to the axis.

Exercise 12-1

1. Graph $y^2 - 12x = 0$.
2. Graph $y^2 - 8x = 0$.
3. Graph $y^2 + 4x = 0$.
4. Graph $y^2 + 8x = 0$.
5. Graph $x^2 - 4y = 0$.
6. Graph $x^2 - 8y = 0$.
7. Graph $x^2 + 12y = 0$.
8. Graph $x^2 + 4y = 0$.
9. Graph $x^2 = -16y$ and locate the focus.
10. Graph $y^2 = -4x$ and locate the focus.
11. Graph $y^2 = -12x$ and locate the focus.
12. Graph $x^2 = 8x$ and locate the focus.
13. Graph $(x - 2)^2 = 4(y - 1)$ and locate the focus.
14. Graph $(y + 1)^2 = 8(x + 3)$ and locate the focus.
15. Graph $(y + 1)^2 = 6(x + 3)$ and locate the focus.
16. Graph $(x - 4)^2 = -10(y + 3)$ and locate the focus.
17. Write the standard form and sketch the graph of $x^2 - 6x - 4y + 13 = 0$.
18. Write the standard form and sketch the graph of $x^2 + 2x - 6y - 11 = 0$.
19. Write the standard form and sketch the graph of $y^2 + 4y - 4x - 16 = 0$.
20. Write the standard form and sketch the graph of $y^2 - 2y - 8x - 15 = 0$.
21. Write the standard form and sketch the graph of $y^2 - 6y + 4x + 5 = 0$.
22. Write the standard form and sketch the graph of $x^2 + 10x + 8y + 49 = 0$.
23. Write the standard form and sketch the graph of $4x^2 + 12x + 32y + 25 = 0$.
24. Write the standard form and sketch the graph of $4y^2 - 4y + 24x + 37 = 0$.

12-2 Circles

OBJECTIVES: After completing this section, you should be able to:

a. Find the center and radius of a circle from its equation.
b. Graph circles from their equations in either standard or general form.
c. Determine when a quadratic equation represents a circle (or its degenerate form).

The parabola is called a **conic section** because one curve, which results when a cone is cut by a plane, is a parabola (see Figure 12-5).

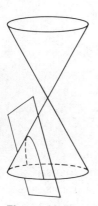

Figure 12-5

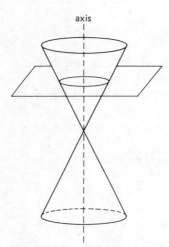

Figure 12-6

A second conic section is the **circle**. A circle is formed when a plane cuts the cone perpendicular to the axis of the cone (see Figure 12-6).

We may define a circle as the locus of all points equidistant from a fixed point.

The standard form of the equation of a circle is

$$(x - h)^2 + (y - k)^2 = r^2$$

where h and k represent the x- and y-coordinates, respectively, of the center of the circle, and r represents the radius of the circle.

EXAMPLE 1

Find the center and radius of each of the circles with the following equations.

a. $(x - 2)^2 + (y + 3)^2 = 16$
b. $x^2 + y^2 = 25$
c. $(x - 7)^2 + (y - 1)^2 = 12$

Solution

a. The center is $(2, -3)$; the radius is 4.
b. The center is $(0, 0)$; the radius is 5.
c. The center is $(7, 1)$; the radius is $\sqrt{12} = 2\sqrt{3}$.

The standard form of a circle can be easily written if its center and radius are known.

EXAMPLE 2

Write the equation of the circle

a. with center at $(0, 0)$ and radius 3;
b. with center at $(3, -1)$ and radius 2.

Solution

a. $(x - 0)^2 + (y - 0)^2 = 3^2$, or $x^2 + y^2 = 9$
b. $(x - 3)^2 + [y - (-1)]^2 = 2^2$, or $(x - 3)^2 + (y + 1)^2 = 4$

If we do the indicated squaring and combining of like terms in Example 2b, we get

$$x^2 + y^2 - 6x + 2y + 6 = 0$$

This is the **general form** of the equation of a circle which has its center at $(3, -1)$ and has radius 2.

To find the center and radius of a circle whose equation is given in general form, we complete the squares to transform the equation into standard form.

EXAMPLE 3

Find the center and radius and sketch the graph of the circle whose equation is $x^2 + y^2 - 6x + 2y + 6 = 0$.

Solution

We must complete the squares for both the x and y variables:

$$(x^2 - 6x) + (y^2 + 2y) = -6$$
$$(x^2 - 6x + 9) + (y^2 + 2y + 1) = -6 + 9 + 1$$
$$(x - 3)^2 + (y + 1)^2 = 4$$

The center is $(3, -1)$ and the radius is 2. The graph is shown in Figure 12-7.

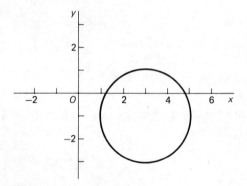

Figure 12-7

EXAMPLE 4

Find the center and radius of the circle whose equation is $x^2 + y^2 - 4x + 2y - 20 = 0$; sketch the circle.

Solution

Rewriting the equation and completing the squares give

$$(x^2 - 4x) + (y^2 + 2y) = 20$$
$$(x^2 - 4x + 4) + (y^2 + 2y + 1) = 20 + 4 + 1$$
$$(x - 2)^2 + (y + 1)^2 = 25$$

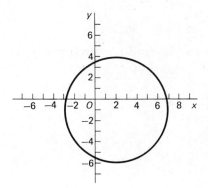

Figure 12-8

The center is at $(2, -1)$ and the radius is 5. The graph is shown in Figure 12-8.

We can determine when a quadratic equation represents a circle or its degenerate form (a point) by looking at the squared terms in the equation. **The equation**

$$Ax^2 + Cy^2 + Dx + Ey + F = 0$$

represents a circle if $A = C$. That is, if the equation has both variables squared and if the squared variables have the same coefficients, the graph of the equation is a circle.

EXAMPLE 5

Determine which of the following equations represent circles.

a. $x^2 - 8x + y^2 - 4y + 3 = 0$
b. $4x^2 + 8x + 4y^2 + 2y - 4 = 0$
c. $x^2 - y^2 = 4$
d. $4x^2 + 8x + y^2 - 6 = 0$
e. $x^2 + 4x + y - 4 = 0$

Solution

a. Is the equation of a circle.
b. Is the equation of a circle. (We would have to divide both sides of the equation by 4 to get it into standard form.)
c. Is not the equation of a circle. The coefficient of x^2 is 1, and the coefficient of y^2 is -1.
d. Is not the equation of a circle. The coefficient of x^2 is 4, and the coefficient of y^2 is 1.
e. Is not the equation of a circle. Only one variable is squared; so it is the equation of a parabola.

Exercise 12-2

Find the center and radius of each of the circles with the following equations; sketch their graphs.

1. $x^2 + y^2 = 36$
2. $x^2 + y^2 = 16$
3. $x^2 + y^2 - 49 = 0$
4. $x^2 + y^2 - 18 = 0$
5. $(x-1)^2 + (y-3)^2 = 81$
6. $(x-3)^2 + (y-4)^2 = 144$
7. $(x+2)^2 + (y+1)^2 = 46$
8. $(x+1)^2 + (y-3)^2 = 32$
9. $x^2 + y^2 - 2x - 2y - 2 = 0$
10. $x^2 + y^2 + 4x - 6y - 3 = 0$
11. $x^2 + y^2 - 6x + 8y = 0$
12. $x^2 + y^2 + 2x + 2y - 7 = 0$
13. $x^2 + y^2 - x + 3y - \frac{27}{2} = 0$
14. $x^2 + y^2 - 5x + y + \frac{5}{2} = 0$
15. $2x^2 + 2y^2 - 2x + 6y - 13 = 0$
16. $3x^2 + 3y^2 + 6y - 20 = 0$

Do each of the following equations have circles as their graphs?

17. $x^2 + y^2 - 6x + 4y - 6 = 0$
18. $2x^2 + 2y^2 - 4x + 8y - 10 = 0$
19. $3x^2 + 2y^2 + 4x - 6y + 1 = 0$
20. $x^2 + 3x + 4y - 6 = 0$

12-3 Ellipses

OBJECTIVES: After completing this section, you should be able to:

a. **Graph ellipses from their equations in standard or general form.**
b. **Determine if a quadratic equation is the equation of a parabola, circle, or ellipse.**

The equation of an ellipse is similar to the equation of a circle. The graph of $3x^2 + 3y^2 = 6$ is a circle, but the graph of $2x^2 + 3y^2 = 6$ is an ellipse (see Figure 12-9). If the equation has two variables squared whose coefficients are different but have the same sign, the graph of the equation is an ellipse. The ellipse has two axes of symmetry, with one longer than the other. The

12-3 Ellipses

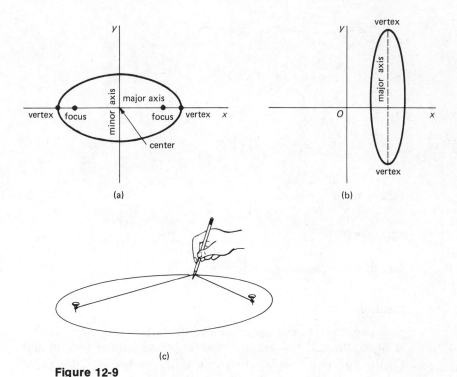

Figure 12-9

longer axis is called the **major axis**, and the shorter is the **minor axis**.

We may define an ellipse as the locus of all points such that the sum of the undirected distances from two fixed points to the points is always constant. The two fixed points are called the **foci** of the ellipse. We can construct an ellipse by fastening the ends of a string to two points (the foci) and moving a pencil against the string. The pencil will trace out an ellipse whose major axis will pass through the foci. [See Figure 12-9 (c).] We call the points where the ellipse intersects the major axis the **vertices** of the ellipse.

Figure 12-10 shows how an ellipse is formed as a conic section. **The standard form of the equation of an ellipse with center at the origin and major axis on the x-axis is**

$$\frac{x^2}{a^2} + \frac{y^2}{b^2} = 1, \quad \text{where } a > b$$

The graph of this equation will intersect the x-axis at a and $-a$, and the y-axis at b and $-b$. The length of the longer axis of the ellipse (the major axis) has length $2a$. The shorter (minor) axis (along the y-axis for this equation) has length $2b$.

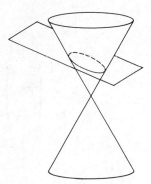

Figure 12-10

EXAMPLE 1

Graph the equation $\dfrac{x^2}{25} + \dfrac{y^2}{9} = 1$.

Solution

Because $25 > 9$, the major axis is along the x-axis. The center is at $(0, 0)$, and the graph intersects the x-axis at $(-5, 0)$ and $(5, 0)$. The two points where the ellipse intersects the major axis are called the vertices of the ellipse. The ellipse intersects the y-axis at $(0, -3)$ and $(0, 3)$. The graph of the equation is shown in Figure 12-11.

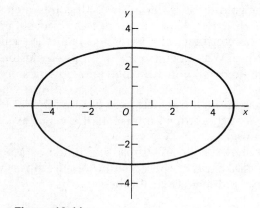

Figure 12-11

If the equation of an ellipse has the standard form

$$\frac{x^2}{b^2} + \frac{y^2}{a^2} = 1, \quad a > b$$

the center of the ellipse is at the origin, and the major axis is along the y-axis.

EXAMPLE 2

Graph the equation $\dfrac{x^2}{9} + \dfrac{y^2}{25} = 1$.

Solution

The graph is an ellipse with center at the origin and major axis along the y-axis. The ellipse will intersect the y-axis at $(0, -5)$ and $(0, 5)$ and the x-axis at $(3, 0)$ and $(-3, 0)$. The graph is shown in Figure 12-12.

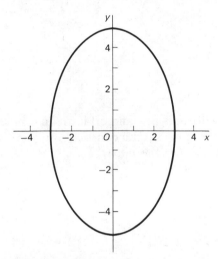

Figure 12-12

The standard form of the equation of an ellipse with center at (h, k) and major axis parallel to the x-axis is

$$\frac{(x - h)^2}{a^2} + \frac{(y - k)^2}{b^2} = 1$$

EXAMPLE 3

Graph the equation $\dfrac{(x - 1)^2}{9} + \dfrac{(y + 2)^2}{4} = 1$.

Solution

The graph is an ellipse with center at $(1, -2)$. The major axis intersects the ellipse 3 units to the right and left of $(1, -2)$, at

(4,−2) and (−2,−2). The minor axis intersects the ellipse 2 units above and below the center, at (1, 0) and (1, −4). The graph is shown in Figure 12-13.

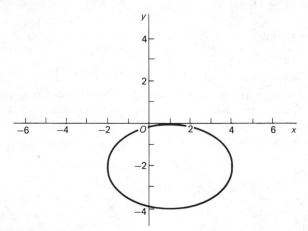

Figure 12-13

If an equation of an ellipse is in standard form and the constant under the $(y-k)^2$ is larger than the constant under $(x-h)^2$, the major axis will be parallel to the y-axis.

EXAMPLE 4

Graph the equation $\dfrac{(x+2)^2}{1} + \dfrac{(y-2)^2}{4} = 1$.

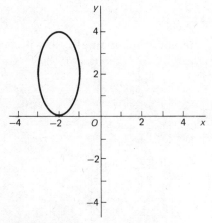

Figure 12-14

Solution

The graph is an ellipse with center $(-2, 2)$ and major axis parallel to the y-axis. The ellipse intersects the major axis 2 units up and down from the center, at $(-2, 4)$ and $(-2, 0)$. The ellipse intersects the minor axis 1 unit to the right and left of the center, at $(-1, 2)$ and $(-3, 2)$. The graph is shown in Figure 12-14.

The equation $Ax^2 + Cy^2 + Dx + Ey + F = 0$, with A and C not equal but with the same sign, is the **general form of the equation of an ellipse**.

EXAMPLE 5

Identify each of the following as equations of ellipses, circles, parabolas, or none of these.

a. $x^2 + 3y^2 + 6x + 12y + 12 = 0$
b. $4x^2 + 4y^2 + 8x - 4y + 4 = 0$
c. $x^2 - 4y^2 = 8$
d. $x^2 + 4x + 6y - 3 = 0$

Solution

a. ellipse
b. circle
c. none of these
d. parabola

We can convert the general form of the equation of an ellipse to the standard form by completing the squares and dividing both sides of the equation to make the constant on the right side of the equation equal to 1.

EXAMPLE 6

Change the equation $4x^2 + y^2 + 16x - 4y + 16 = 0$ to standard form and sketch its graph.

Solution

The graph is an ellipse. In completing the square, note that we factor 4 out of the x terms so that we get the form $(x - h)^2$. Completing the square gives

$$(4x^2 + 16x \quad) + (y^2 - 4y \quad) = -16$$

$$4(x^2 + 4x \quad) + (y^2 - 4y \quad) = -16$$
$$4(x^2 + 4x + 4) + (y^2 - 4y + 4) = -16 + 16 + 4$$

[Note that we added 16 to both sides to complete the square for $4(x^2 + 4x)$.]

$$4(x + 2)^2 + (y - 2)^2 = 4$$

Dividing both sides by 4 gives

$$\frac{(x + 2)^2}{1} + \frac{(y - 2)^2}{4} = 1$$

The ellipse has center at $(-2, 2)$, intersects the major axis at $(-2, 4)$ and $(-2, 0)$, and intersects the minor axis at $(-1, 2)$ and $(-3, 2)$. The graph is shown in Figure 12-15.

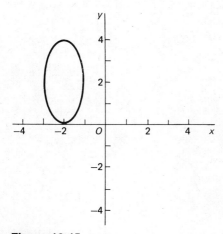

Figure 12-15

EXAMPLE 7

Change the equation $x^2 + 2y^2 + 2x - 12y + 1 = 0$ to standard form and sketch its graph.

Solution

The graph is an ellipse. Completing the square gives

$$(x^2 + 2x \quad) + (2y^2 - 12y \quad) = -1$$

or

$$(x^2 + 2x + 1) + 2(y^2 - 6y + 9) = -1 + 1 + 18$$

or

$$(x + 1)^2 + 2(y - 3)^2 = 18$$

The standard form is

$$\frac{(x+1)^2}{18} + \frac{(y-3)^2}{9} = 1$$

The graph is shown in Figure 12-16.

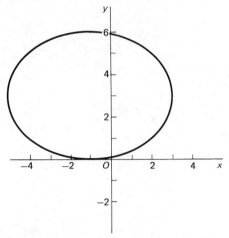

Figure 12-16

The ellipse has many applications in science. For example:
1. Elliptic gears, used in power punches and presses, produce a slow, powerful stroke with a fast return.
2. The planets travel in elliptic orbits with the sun at one focus.
3. Any ray originating at one focus will strike the ellipse, reflect, and pass through the other focus; whispering galleries usually have elliptical ceilings arranged so that one may stand at the focus. Thus situated, he can hear a slight noise made at the other focus, while a person standing between the foci hears nothing.

Exercise 12-3

Graph the following equations.

1. $\frac{x^2}{16} + \frac{y^2}{4} = 1$
2. $\frac{x^2}{9} + \frac{y^2}{4} = 1$
3. $\frac{x^2}{9} + \frac{y^2}{5} = 1$
4. $\frac{x^2}{49} + \frac{y^2}{6} = 1$

5. $\dfrac{x^2}{4}+\dfrac{y^2}{16}=1$ 6. $\dfrac{x^2}{9}+\dfrac{y^2}{25}=1$

7. $\dfrac{x^2}{4}+\dfrac{y^2}{18}=1$ 8. $\dfrac{x^2}{8}+\dfrac{y^2}{25}=1$

9. $\dfrac{(x-1)^2}{4}+\dfrac{(y+2)^2}{1}=1$ 10. $\dfrac{(x+1)^2}{9}+\dfrac{(y-4)^2}{4}=1$

11. $\dfrac{(x-\frac{1}{3})^2}{16}+\dfrac{(y+4)^2}{9}=1$ 12. $\dfrac{(x+4)^2}{49}+\dfrac{(y-5)^2}{25}=1$

13. $\dfrac{(x-3)^2}{4}+\dfrac{(y+1)^2}{25}=1$ 14. $\dfrac{(x+2)^2}{1}+\dfrac{(y-2)^2}{9}=1$

15. $\dfrac{(x-\frac{1}{2})^2}{9}+\dfrac{(y-\frac{3}{2})^2}{16}=1$ 16. $\dfrac{(x+1)^2}{5}+\dfrac{(y-6)^2}{8}=1$

Convert the following equations to standard form and sketch their graphs.

17. $x^2+4y^2-2x+16y+13=0$
18. $9x^2+y^2+36x-4y+31=0$
19. $16x^2+9y^2-32x-54y-47=0$
20. $4x^2+9y^2+24x-36y+36=0$
21. $3x^2+4y^2-6x-8y-5=0$
22. $9x^2+4y^2-6x-12y-26=0$
23. $2x^2+y^2-4+4y=0$
24. $2x^2+y^2-8x=0$

Identify the graphs of the following quadratic equations.

25. $x^2+4y^2+x-y-4=0$
26. $4x^2+4y^2-6x+5y-2=0$
27. $3x^2+4x+4y+6=0$
28. $5x^2+5y^2-6=0$
29. $5x^2+6y^2-7=0$
30. $5x^2+6y^2-5x+6y-3=0$

12-4 Hyperbolas

OBJECTIVES: After completing this section, you should be able to:

a. Determine if a quadratic equation is the equation of a hyperbola, an ellipse, a circle, or a parabola.
b. Graph a hyperbola from its equation in standard or general form.

12-4 Hyperbolas

The remaining conic section is a hyperbola. It is formed when a plane intersects the cone parallel to the axis of the cone (see Figure 12-17). If the plane intersects the cone on the axis of the cone, we get two intersecting lines, which is a degenerate case of the hyperbola.

If the quadratic equation $Ax^2 + Cy^2 + Dx + Ey + F = 0$ has two variables squared and their coefficients (A and C) have opposite signs, the graph of the equation is a hyperbola.

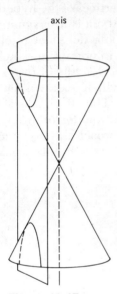

Figure 12-17

EXAMPLE 1

Which of the following equations are equations of hyperbolas?

a. $x^2 - y^2 = 4$
b. $x^2 + 3y^2 + 6x + 6y - 3 = 0$
c. $x^2 - 3y^2 + 6x + 6y - 8 = 0$
d. $x^2 - 3y = 6$

Solution

a. Is the equation of a hyperbola.
b. Is not the equation of a hyperbola. Is the equation of an ellipse.
c. Is the equation of a hyperbola.
d. Is not the equation of a hyperbola. Is the equation of a parabola.

A hyperbola has two branches, which open away from its center. The **standard form of the equation of a hyperbola** that has its center at the origin and opens right and left is

$$\frac{x^2}{a^2} - \frac{y^2}{b^2} = 1$$

The graph of this equation will intersect the x-axis at the distance a to the right and left of the origin (center of the hyperbola). The axis which joins these **vertices** of the hyperbola is called the **transverse axis**. This hyperbola is also symmetric about the y-axis. Although this hyperbola does not intersect this axis (called the **conjugate axis**), it is important in sketching the graph of the hyperbola. We use the lengths of these two axes as the dimensions of a rectangle which is drawn with its center at the center of the hyperbola. The diagonals of this **central rectangle**, extended, form the **asymptotes** of the hyperbola. The branches of the hyperbola approach the asymptotes, but do not intersect them. Use of these asymptotes make it much easier to graph the hyperbola.

EXAMPLE 2

Graph the equation $\frac{x^2}{4} - \frac{y^2}{9} = 1$.

Solution

The graph is a hyperbola. Because the $x^2/4$ term is positive, the hyperbola has its center at $(0, 0)$ and opens left and right. The hyperbola will intersect the transverse axis (along the x-axis) 2 units to the right and left of $(0, 0)$, at $(-2, 0)$ and $(2, 0)$. The hyperbola may be sketched by

a. drawing the central rectangle and its diagonals extended;
b. drawing the hyperbola through its vertices and approaching the asymptotes.

The graph is shown in Figure 12-18.

Note that the size of the transverse axis is not necessarily longer than the conjugate axis. It may be the same length, longer, or shorter. The direction the hyperbola opens will be affected, however, by which term is positive when the equation is in standard form.

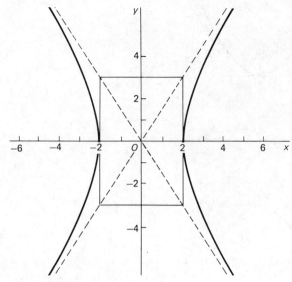

Figure 12-18

1. An equation of the form

$$\frac{x^2}{(\ )^2} - \frac{y^2}{(\ )^2} = 1$$

   always opens left and right.
2. An equation of the form

$$\frac{y^2}{(\ )^2} - \frac{x^2}{(\ )^2} = 1$$

   always opens up and down.

EXAMPLE 3

Graph the equation $\frac{y^2}{36} - \frac{x^2}{9} = 1$.

Solution

The graph is a hyperbola with its center at (0, 0). Because the $y^2/36$ term is positive, it opens up and down. Its vertices are 6 units up and down from (0, 0), at (0, 6) and (0, −6).

The graph is shown in Figure 12-19.

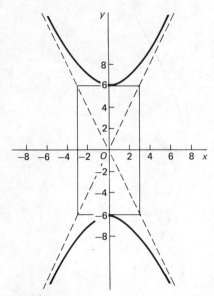

Figure 12-19

If the equation has the standard form

$$\frac{(x-h)^2}{a^2} - \frac{(y-k)^2}{b^2} = 1 \quad \text{or} \quad \frac{(y-k)^2}{a^2} - \frac{(x-h)^2}{b^2} = 1$$

the center of the hyperbola is at (h, k).

EXAMPLE 4

Graph the equation $\frac{(y-1)^2}{4} - \frac{(x+2)^2}{16} = 1$.

Solution

The graph is a hyperbola with its center at $(-2, 1)$. It opens up and down with its vertices 2 units up and down from $(-2, 1)$, at $(-2, 3)$ and $(-2, -1)$. The graph is shown in Figure 12-20.

As with other conic sections, we complete the square to go from the general form to the standard form of the equation of a hyperbola.

EXAMPLE 5

Find the standard form of $3x^2 - y^2 + 12x + 2y + 2 = 0$ and graph the equation.

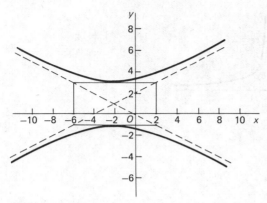

Figure 12-20

Solution

Factoring and completing the square gives
$$3(x^2 + 4x \quad) - (y^2 - 2y \quad) = -2$$
$$3(x^2 + 4x + 4) - (y^2 - 2y + 1) = -2 + 12 - 1$$
$$3(x + 2)^2 - (y - 1)^2 = 9$$
So
$$\frac{(x + 2)^2}{3} - \frac{(y - 1)^2}{9} = 1$$

is the standard form.

The hyperbola has center at $(-2, 1)$, opens right and left, and has vertices at $(-2 + \sqrt{3}, 1)$ and $(-2 - \sqrt{3}, 1)$. The graph is shown in Figure 12-21.

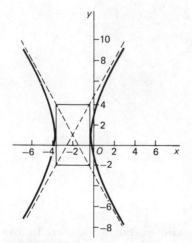

Figure 12-21

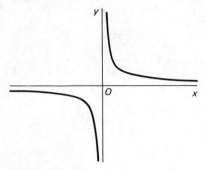

Figure 12-22

Although we have not discussed quadratic equations containing xy terms, there is one other important equation form which has the hyperbola as its graph. This equation has the form $xy = c$.

If $c > 0$, the graph has the form shown in Figure 12-22. If $c < 0$, the graph is as shown in Figure 12-23. Note that the asymptotes are the x- and y-axis.

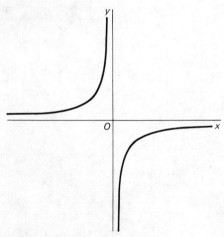

Figure 12-23

EXAMPLE 6

Graph $xy = 4$.

Solution

The table of sample values and graph are shown in Figure 12-24. Note the graph is symmetric about the origin.

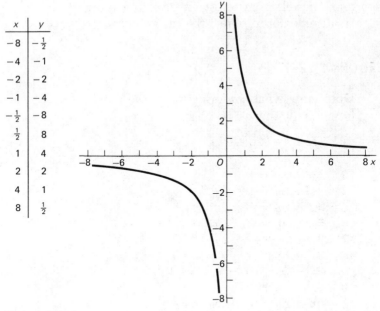

Figure 12-24

The hyperbola appears many places in nature. A relatively fast-moving body coming under the influence of an inverse-square attractive or repulsive force will travel in a hyperbolic orbit. The motion of some comets is an example of this. The path of an alpha particle shot near an atomic nucleus is hyperbolic. The hyperbola also has applications in electrostatic force repulsion and magnetic repulsion forces. The equation $PV = $ constant, expressing Boyle's Law, has a hyperbolic graph whose asymptotes are the coordinate axes. As with the ellipse, there are two foci associated with the hyperbola. The location of the foci of the hyperbola are important for the following reasons.

In military applications, the hyperbola is useful in locating the source of sound (such as enemy gun position). Two observers at points that are the foci of a family of hyperbolas record differences in time of arrival of the same sound and thus determine a certain hyperbola in the family. If two other foci are chosen and another hyperbola is determined, the intersection of these hyperbolas sketched on a map locates the source of the sound. The hyperbola is basic to LORAN, a system of radio location and guidance of planes and ships based on the time of delay of radio signals received from two points. A

334 Conic Sections

type of reflecting telescope uses a hyperbolic mirror as the small reflector to reflect the image to the eyepiece.

Exercise 12-4

Determine what conic section is the graph of each of the following equations.

1. $x^2 - y^2 = 4$
2. $x^2 + y^2 = 4$
3. $4x^2 - 4y^2 = 9$
4. $4x^2 + 5y^2 = 9$
5. $4x^2 + 4y = 9$
6. $5x^2 - 4y = 9$
7. $3x^2 - y^2 + 4x + 2y - 1 = 0$
8. $3x^2 + y^2 + 5x + y - 4 = 0$
9. $3x^2 - 5y^2 + 6x - y - 5 = 0$
10. $6x^2 - 5y + 6x = 8$
11. $6x^2 + 6y^2 + 3x - 6y + 1 = 0$
12. $2x^2 - 3y^2 + 6x - 2y - 3 = 0$

Graph the following equations.

13. $\dfrac{x^2}{9} - \dfrac{y^2}{8} = 1$
14. $\dfrac{x^2}{16} - \dfrac{y^2}{9} = 1$
15. $\dfrac{x^2}{4} - \dfrac{y^2}{25} = 1$
16. $\dfrac{x^2}{1} - \dfrac{y^2}{4} = 1$
17. $\dfrac{y^2}{4} - \dfrac{x^2}{1} = 1$
18. $\dfrac{y^2}{16} - \dfrac{x^2}{9} = 1$
19. $\dfrac{y^2}{5} - \dfrac{x^2}{16} = 1$
20. $\dfrac{y^2}{16} - \dfrac{x^2}{25} = 1$
21. $\dfrac{(y-1)^2}{4} - \dfrac{(x+3)^2}{9} = 1$
22. $\dfrac{(y+3)^2}{16} - \dfrac{(x-1)^2}{9} = 1$
23. $\dfrac{(x-3)^2}{16} - \dfrac{(y+2)^2}{4} = 1$
24. $\dfrac{(x-4)^2}{9} - \dfrac{(y-2)^2}{25} = 1$
25. $\dfrac{(y-2)^2}{9} - \dfrac{(x+5)^2}{1} = 1$
26. $\dfrac{(x+1)^2}{25} - \dfrac{(x-4)^2}{16} = 1$

Write the standard form of the following equations and sketch their graphs.

27. $x^2 - 4y^2 - 2x + 8y - 7 = 0$
28. $4x^2 - 9y^2 + 8x - 36y - 68 = 0$

29. $9x^2 - y^2 + 54x + 8y + 56 = 0$
30. $4y^2 - 9x^2 - 16y - 18x - 29 = 0$

Graph the following equations.

31. $xy = 9$
32. $xy = 16$
33. $xy = -4$
34. $xy = -25$

CHAPTER TEST

Determine what conic section is the graph of each of the following equations.

1. $x^2 + 4y^2 + 4x - 4y + 2 = 0$
2. $x^2 - y^2 + 6x - 4y - 1 = 0$
3. $x^2 - 4x + 5y - 6 = 0$
4. $4x^2 + 4y^2 - 8x + 10y - 4 = 0$

Graph the following functions.

5. $x^2 - 8y = 0$
6. $(x - 1)^2 = -4(y + 3)$
7. $y^2 - 6y - 4x + 13 = 0$
8. $x^2 + y^2 - 25 = 0$
9. $(x + 3)^2 + (y - 4)^2 = 16$
10. $2x^2 + 2y^2 - 2x + 2y - 17 = 0$
11. $\dfrac{x^2}{16} + \dfrac{y^2}{4} = 1$
12. $\dfrac{x^2}{9} + \dfrac{y^2}{25} = 1$
13. $\dfrac{(x - 1)^2}{9} + \dfrac{(y + 3)^2}{4} = 1$
14. $9x^2 + 25y^2 - 54x + 100y - 74 = 0$
15. $\dfrac{x^2}{8} - \dfrac{y^2}{16} = 1$
16. $\dfrac{(y + 2)^2}{4} - \dfrac{(x + 2)^2}{1} = 1$
17. $16x^2 - 9y^2 - 32x - 36y - 164 = 0$
18. $xy = 25$
19. $xy = -16$

13

sequences and series

13-1 Sequences

OBJECTIVES: After completing this section, you should be able to:

a. Write a specified number of terms of a sequence.
b. Find specified terms of arithmetic sequences.
c. Find specified terms of geometric sequences.

A sequence function is a function whose domain is the set of positive integers.

Because these functions are very special, it is customary to use the notation a_n to represent the functional value $f(n)$. The n is used to remind us that the domain contains only positive integers. The set of functional values $a_1, a_2, a_3, \ldots$ of the sequence function is called a **sequence**. The values $a_1, a_2, a_3, \ldots$ are called **terms** of the sequence, with a_1 the first term, a_2 the second term, and so on.

EXAMPLE 1

Find the first ten terms of the sequence defined by the formula $a_n = 1/2n$.

Solution

The first ten terms of the sequence are $a_1, a_2, a_3, a_4, a_5, a_6, a_7, a_8, a_9, a_{10}$ or $\frac{1}{2}, \frac{1}{4}, \frac{1}{6}, \frac{1}{8}, \frac{1}{10}, \frac{1}{12}, \frac{1}{14}, \frac{1}{16}, \frac{1}{18}, \frac{1}{20}$.

EXAMPLE 2

Write the first four terms of the sequence whose nth term is $a_n = (-1)^n/n$.

Solution

The first four terms of the sequence are a_1, a_2, a_3, a_4 or $-1, \frac{1}{2}, -\frac{1}{3}, \frac{1}{4}$.

EXAMPLE 3

If the nth term of a sequence is $a_n = (n-2)/n(n+1)$, write the first five terms and the eighth term of the sequence.

Solution

The first five terms of the sequence are $-\frac{1}{2}, 0, \frac{1}{12}, \frac{1}{10}, \frac{1}{10}$. The eighth term is $\frac{1}{12}$.

If each term of a sequence after the first can be found by adding the same number to the preceding term, the sequence is called an **arithmetic sequence**. That is, a sequence is called an arithmetic sequence if there exists a number d, called the **common difference**, such that

$$a_n = a_{n-1} + d \quad \text{for } n > 1.$$

EXAMPLE 4

Write three additional terms of the following arithmetic sequences.

a. $1, 3, 5, \ldots$
b. $9, 6, 3, \ldots$
c. $\frac{1}{2}, \frac{5}{6}, \frac{7}{6}, \ldots$

Solution

a. The difference is 2; so the next three terms are 7, 9, 11.
b. The difference is -3; so the next three terms are $0, -3, -6$.
c. The difference is $\frac{1}{3}$; so the next three terms are $\frac{3}{2}, \frac{11}{6}, \frac{13}{6}$.

Because each term after the first in an arithmetic sequence is gotten by adding d, the second term is $a_1 + d$, the third is $a_1 + 2d, \ldots$ and the nth term is $a_1 + (n-1)d$. That is, $a_n = a_1 + (n-1)d$ in an arithmetic sequence.

EXAMPLE 5

Find the eleventh term of the arithmetic sequence with first term 3 and common difference -2.

Solution

The 11th term is $a_{11} = 3 + (10)(-2) = -17$.

EXAMPLE 6

If the first term of an arithmetic sequence is 4 and the ninth term is 20, find the fifth term.

Solution

Substituting the values in $a_n = a_1 + (n-1)d$ gives $20 = 4 + 8d$. Solving gives $d = 2$. The fifth term is $4 + 4(2) = 12$.

If each term of a sequence after the first can be found by multiplying the preceding term by the same number, the sequence is called a **geometric sequence**. That is, a sequence is called a **geometric sequence** if there exists a number r, called the **common ratio**, such that

$$a_n = r a_{n-1} \quad \text{for } n > 1$$

EXAMPLE 7

Write three additional terms of the following geometric sequences.

a. 1, 3, 9
b. 4, 2, 1
c. 3, -6, 12

Solution

a. The common ratio is 3; so the next three terms are 27, 81, 243.
b. The common ratio is $\frac{1}{2}$; so the next three terms are $\frac{1}{2}, \frac{1}{4}, \frac{1}{8}$.
c. The common ratio is -2; so the next three terms are -24, 48, -96.

Because each term after the first in a geometric sequence is gotten by multiplying by r, the second term is $a_1 r$; the third is a_1r^2; ... and the nth term is a_1r^{n-1}. That is, $a_n = a_1r^{n-1}$ in a **geometric sequence.**

EXAMPLE 8

Find the seventh term of the geometric sequence with first term 5 and common ratio -2.

Solution

The seventh term is $a_7 = 5(-2)^6 = 5(64) = 320$.

EXAMPLE 9

Find the fifth term of the geometric sequence with first term 8 and tenth term $\frac{1}{64}$.

Solution

Substituting values into the formula $a_n = a_1r^{n-1}$ gives $\frac{1}{64} = 8 \cdot r^9$. Solving gives $r = \frac{1}{2}$. The fifth term is $a_5 = 8(\frac{1}{2})^4 = \frac{1}{2}$.

EXAMPLE 10

A ball is dropped from 125 feet. If it rebounds $\frac{3}{5}$ of the height from which it falls every time it hits the ground, how high will it bounce after it strikes the ground for the fifth time?

Solution

The first rebound is $\frac{3}{5}(125) = 75$ feet; the second rebound is $\frac{3}{5}(75) = 45$ feet. The heights of the rebounds form a geometric sequence with first term 75 and common ratio $\frac{3}{5}$. Thus the fifth term is

$$a_5 = 75(\tfrac{3}{5})^4 = 75(\tfrac{81}{625}) = \tfrac{243}{25} = 9\tfrac{18}{25} \text{ feet}$$

Recall that the domain of a sequence function is the set of positive integers. Thus a sequence has an infinite number of terms. We have been finding the nth terms of sequences. Many sequences (with an infinite number of terms) will approach a number L as n gets very large. We call the number L the **limit** of the sequence, and say that **the sequence approaches** L **as** n **approaches infinity.** For example, the sequence

$$\tfrac{1}{2}, \tfrac{1}{4}, \tfrac{1}{8}, \tfrac{1}{16}, \ldots, (\tfrac{1}{2})^n, \ldots$$

has nth term $(\frac{1}{2})^n$. This sequence approaches 0 as n gets very large (that is, as n approaches infinity). We write

$$\lim_{n \to \infty} (\tfrac{1}{2})^n = 0$$

to indicate this and say that the sequence whose nth term is $(\frac{1}{2})^n$ **converges** to 0.

EXAMPLE 11

To what does the sequence converge?

$$\tfrac{1}{2}, \tfrac{3}{4}, \tfrac{7}{8}, \tfrac{15}{16}, \ldots, 1 - (\tfrac{1}{2})^n, \ldots$$

Solution

The nth term of this sequence is $1 - (\frac{1}{2})^n$. As n gets large, $(\frac{1}{2})^n$ approaches 0, so $1 - (\frac{1}{2})^n$ approaches 1.

EXAMPLE 12

Does the sequence $2, 4, 6, 8, \ldots, 2n, \ldots$ converge to a number or become infinitely large as n approaches infinity?

Solution

As n becomes very large, the sequence whose nth term is $2n$ becomes infinitely large. We say that this sequence **diverges**.

Exercise 13-1

1. Write the first ten terms of the sequence defined by $a_n = 3n$.
2. Find the first six terms of the sequence defined by $a_n = 4n$.
3. Write the first eight terms of the sequence defined by $a_n = n/3$.
4. Write the first seven terms of the sequence defined by $a_n = 2/n$.
5. Write the first six terms of the sequence whose nth term is $(-1)^n/4$.
6. Write the first five terms of the sequence whose nth term is $(-1)^n/3$.
7. Write the first six terms of the sequence whose nth term is $(-1)^n/2n$.

8. Write the first five terms of the sequence whose nth term is $(-1)^n/(n+1)$.
9. If the nth term of a sequence is $a_n = (n-4)/n(n+2)$, write the first four terms and the tenth term.
10. If the nth term of a sequence is $a_n = n(n-1)/(n+3)$, write the sixth term.
11. Write three additional terms of the arithmetic sequence $2, 5, 8, \ldots$.
12. Write three additional terms of the arithmetic sequence $3, 9, 15, \ldots$.
13. Write four additional terms of the arithmetic sequence $3, \frac{9}{2}, 6, \ldots$.
14. Write four additional terms of the arithmetic sequence $2, 2.75, 3.5, \ldots$.
15. Find the eighth term of the arithmetic sequence with first term -3 and common difference 4.
16. Find the eleventh term of the arithmetic sequence with first term 7 and common difference 4.
17. Find the eighth term of the arithmetic sequence with first term 6 and common difference $-\frac{1}{2}$.
18. Find the sixth term of the arithmetic sequence with first term $\frac{1}{2}$ and common difference $-\frac{1}{3}$.
19. Find the sixth term of the arithmetic sequence with first term 5 and eighth term 19.
20. Find the tenth term of the arithmetic sequence with first term 20 and tenth term 47.
21. Find the fourth term of the arithmetic sequence with first term 16 and ninth term 0.
22. Find the sixth term of the arithmetic sequence with first term $\frac{1}{2}$ and eleventh term 3.
23. Write four additional terms of the geometric sequence $3, 6, 12, \ldots$.
24. Write four additional terms of the geometric sequence $4, 12, 36, \ldots$.
25. Write three additional terms of the geometric sequence $81, 54, 36, \ldots$.
26. Write three additional terms of the geometric sequence $32, 48, 72, \ldots$.
27. Find the sixth term of the geometric sequence with first term 10 and common ratio 2.
28. Find the fifth term of the geometric sequence with first term 6 and common ratio 3.
29. Find the fourth term of the geometric sequence with first term 4 and common ratio $\frac{3}{2}$.

30. Find the fifth term of the geometric sequence with first term 3 and common ratio -2.
31. Find the fifth term of the geometric sequence with first term 6 and eighth term 768.
32. Find the third term of the geometric sequence with first term 2 and fourth term 54.
33. Find the third term of the geometric sequence with first term 6 and sixth term $\frac{3}{16}$.
34. Find the sixth term of the geometric sequence with first term 18 and fifth term $\frac{2}{9}$.
35. A ball is dropped from 128 feet. If it rebounds $\frac{3}{4}$ of the height from which it falls every time it hits the ground, how high will it bounce after it strikes the ground for the fourth time?
36. A pump removes $\frac{1}{3}$ of the water in a container with every stroke. What amount of water is still in an 81-cm^3 container after three strokes?
37. A machine is valued at $10,000. If the depreciation at the end of each year is 20 percent of its value at the beginning of the year, find its value at the end of four years.
38. To what limit does the sequence

$$1, \tfrac{1}{3}, \tfrac{1}{9}, \tfrac{1}{27}, \ldots, 1/n^3, \ldots$$

converge?

39. To what limit does the sequence

$$\tfrac{5}{3}, \tfrac{17}{9}, \tfrac{53}{27}, \ldots, 2 - 1/3^n, \ldots$$

converge?

13-2 Summation notation

OBJECTIVE: After completing this section, you should be able to:

a. Evaluate sums as indicated in summation notation.
b. Find the sum of a fixed number of terms of an arithmetic sequence.
c. Find the sum of a fixed number of terms of a geometric sequence.

We frequently desire to find the sum of some or all terms of a sequence. We use **summation notation** to express the sums. The symbol

$$\sum_{i=1}^{n} a_i$$

represents the sum of the first n terms of the sequence a_1, $a_2, \ldots, a_n, \ldots$; that is,

$$\sum_{i=1}^{n} a_i = a_1 + a_2 + a_3 + \ldots + a_n$$

The a_i represents a functional value, the **index** i represents the domain, and the n represents the maximum value of i used in the sum. For example,

$$\sum_{i=1}^{5} \frac{1}{2i}$$

represents the sum of the first five terms of the sequence $\frac{1}{2}, \frac{1}{4}, \frac{1}{6}, \frac{1}{8}, \frac{1}{10}, \ldots$; that is

$$\sum_{i=1}^{5} \frac{1}{2i} = \frac{1}{2} + \frac{1}{4} + \frac{1}{6} + \frac{1}{8} + \frac{1}{10}.$$

EXAMPLE 1

Evaluate $\sum_{i=1}^{6} \frac{i-1}{i^2}.$

Solution

$$\sum_{i=1}^{6} \frac{i-1}{i^2} = \frac{0}{1} + \frac{1}{4} + \frac{2}{9} + \frac{3}{16} + \frac{4}{25} + \frac{5}{36} = \frac{3451}{3600}$$

EXAMPLE 2

Evaluate $\sum_{k=1}^{5} \frac{k-3}{k+1}.$

Solution

Note that k is now used as the index instead of i.

$$\sum_{k=1}^{5} \frac{k-3}{k+1} = \frac{-2}{2} + \frac{-1}{3} + \frac{0}{4} + \frac{1}{5} + \frac{2}{6} = \frac{-4}{5}$$

EXAMPLE 3

Evaluate $\sum_{k=2}^{6} \frac{k^2}{k-1}.$

Solution

In this example, the index is initiated at $k = 2$. Thus

$$\sum_{k=2}^{6} \frac{k^2}{k-1} = \frac{4}{1} + \frac{9}{2} + \frac{16}{3} + \frac{25}{4} + \frac{36}{5} = \frac{1637}{60}$$

EXAMPLE 4

Find the sum of the first three terms of the sequence defined by $a_n = (-1)^n/3n$.

Solution

$$\sum_{k=1}^{3} \frac{(-1)^k}{3k} = \frac{-1}{3} + \frac{1}{6} + \frac{-1}{9} = -\frac{5}{18}$$

Because the first n terms of an arithmetic sequence can be written as

$$a_1, a_1 + d, a_1 + 2d, a_1 + 3d, \ldots, a_1 + (n-1)d$$

or as

$$a_1, a_1 + d, a_1 + 2d, \ldots, a_n - 2d, a_n - d, a_n$$

we can find the sum of the n terms as follows:

$$s_n = a_1 + (a_1 + d) + (a_1 + 2d) + \ldots + (a_n - 2d) + (a_n - d) + a_n$$

and (writing the sum backward)

$$s_n = a_n + (a_n - d) + (a_n - 2d) + \ldots + (a_1 + 2d) + (a_1 + d) + a_1$$

Adding (term by term) gives

$$2s_n = (a_1 + a_n) + (a_1 + a_n) + (a_1 + a_n) + \ldots + (a_1 + a_n) + (a_1 + a_n) + (a_1 + a_n)$$

So $2s_n = n(a_1 + a_n)$ because there are n terms. Thus the sum of the first n terms of an arithmetic sequence is

$$s_n = \frac{n}{2}(a_1 + a_n)$$

where a_1 is the first term and a_n is the nth term.

EXAMPLE 5

Find the sum of the first ten terms of the arithmetic sequence with first term 2 and common difference 4.

Solution

First we find the tenth term. $a_{10} = 2 + 9 \cdot 4 = 38$. The sum is

$$s_{10} = \tfrac{10}{2}(2 + 38) = 200$$

EXAMPLE 6

Find the sum of the first seven terms of the arithmetic sequence $\tfrac{1}{4}, \tfrac{7}{12}, \tfrac{11}{12}, \ldots$.

Solution

The first term is $\tfrac{1}{4}$, and the common difference is $\tfrac{1}{3}$. The seventh term is $a_7 = \tfrac{1}{4} + 6(\tfrac{1}{3}) = 2\tfrac{1}{4}$. The sum is

$$s_7 = \tfrac{7}{2}(\tfrac{1}{4} + \tfrac{9}{4}) = \tfrac{35}{4} = 8\tfrac{3}{4}$$

The first n terms of a geometric sequence can be written as

$$a_1, a_1r, a_1r^2, \ldots, a_1r^{n-1}$$

so the sum of the first n terms is

$$s_n = a_1 + a_1r + a_1r^2 + \cdots + a_1r^{n-1}$$

Then

$$rs_n = a_1r^2 + a_1r^3 + \cdots + a_1r^n$$

so

$$s_n - rs_n = a_1 + (a_1r - a_1r) + (a_1r^2 - a_1r^2) + \cdots + (a_1r^{n-1} - a_1r^{n-1}) - a_1r^n$$

Thus

$$s_n(1 - r) = a_1 - a_1r^n, \quad \text{so} \quad s_n = \frac{a_1 - a_1r^n}{1 - r}$$

That is, the sum of the first n terms of the geometric sequence is

$$s_n = \frac{a_1(1 - r^n)}{1 - r}$$

where a_1 is the first term of the sequence and r is the common ratio.

EXAMPLE 7

Find the sum of the first five terms of the geometric series with first term 4 and common ratio -3.

Solution

$$S_5 = \frac{4[1-(-3)^5]}{1-(-3)} = \frac{4[1-(-243)]}{4} = 244$$

EXAMPLE 8

Find the sum of the first six terms of the geometric sequence $\frac{1}{4}, \frac{1}{8}, \frac{1}{16}, \ldots$

Solution

$a_1 = \frac{1}{4}$ and $r = \frac{1}{2}$

so

$$S_6 = \frac{\frac{1}{4}[1-(\frac{1}{2})^6]}{1-\frac{1}{2}} = \frac{\frac{1}{4}(1-\frac{1}{64})}{\frac{1}{2}} = \frac{1-\frac{1}{64}}{2} = \frac{64-1}{128} = \frac{63}{128}$$

Exercise 13-2

1. Evaluate $\sum_{k=1}^{4} \frac{1}{k^2}$.

2. Evaluate $\sum_{i=1}^{6} \frac{2i}{i+1}$.

3. Evaluate $\sum_{i=1}^{5} \frac{i^2}{i+3}$.

4. Evaluate $\sum_{i=1}^{4} \frac{4}{i(i+1)}$.

5. Find the sum of the first four terms of the sequence defined by $a_n = 3/n$.
6. Find the sum of the first five terms of the sequence defined by $a_n = n/4$.
7. Find the sum of the first five terms of the sequence whose nth term is $(-1)^n/n$.
8. Find the sum of the first four terms of the sequence whose nth term is $(-1)^n/2n$.

9. Find the sum of the first eight terms of the arithmetic sequence with first term 2 and the common difference 3.
10. Find the sum of the first seven terms of the arithmetic sequence with first term 6 and the common difference 4.
11. Find the sum of the first six terms of the arithmetic sequence with first term 10 and common difference $\frac{1}{2}$.
12. Find the sum of the first eight terms of the arithmetic sequence with first term 12 and common difference -3.
13. Find the sum of the first five terms of the arithmetic sequence 2, 4, 6, ...
14. Find the sum of the first six terms of the arithmetic sequence 6, 9, 12, ...
15. Find the sum of the first eight terms of the arithmetic sequence 6, $\frac{9}{2}$, 3, ...
16. Find the sum of the first seven terms of the arithmetic sequence 12, 9, 6, ...
17. Find the sum of the first six terms of the geometric sequence with first term 6 and common ratio 3.
18. Find the sum of the first four terms of the geometric sequence with first term 3 and common ratio 4.
19. Find the sum of the first six terms of the geometric sequence with first term 4 and common ratio $-\frac{1}{2}$.
20. Find the sum of the first five terms of the geometric sequence with first term 9 and common ratio $\frac{1}{3}$.
21. Find the sum of the first five terms of the geometric sequence 1, 3, 9, ...
22. Find the sum of the first four terms of the geometric sequence 16, 64, 256, ...
23. Find the sum of the first five terms of the geometric sequence 6, 4, $\frac{8}{3}$, ...
24. Find the sum of the first five terms of the geometric sequence 9, -6, 4, ...

13-3 Series

OBJECTIVES: After completing this section, you should be able to:

a. **Find the sum of certain (infinite) series.**
b. **Find the sum of geometric series whose common ratio is between -1 and 1.**
c. **Write repeating decimals as the ratios of two integers.**

Sequences and Series

We have used sigma notation to find the sum of the first n terms of sequences. For example,

$$\sum_{k=1}^{4} \frac{1}{k^2} = \frac{1}{1} + \frac{1}{4} + \frac{1}{9} + \frac{1}{16}$$

We may write an infinite sum in the form

$$\sum_{k=1}^{\infty} \frac{1}{k^2} = \frac{1}{1} + \frac{1}{4} + \frac{1}{9} + \cdots + \frac{1}{n^2} \cdots$$

In general, **an expression of the form**

$$\sum_{k=1}^{\infty} a_k = a_1 + a_2 + a_3 + \cdots + a_n + \cdots$$

is called an (infinite) series. Thus, a series is an infinite sum. The value of this infinite sum (if it exists) is found by using a new sequence, called the **sequence of partial sums.** This sequence of partial sums is a function whose domain is the set of positive integers. For example,

$$s_1 = \sum_{k=1}^{1} a_k = a_1$$

$$s_2 = \sum_{k=1}^{2} a_k = a_1 + a_2$$

$$s_3 = \sum_{k=1}^{3} a_k = a_1 + a_2 + a_3$$

.
.
.

$$s_n = \sum_{k=1}^{n} a_k = a_1 + a_2 + a_3 + \cdots + a_n$$

.
.
.

The sum of the series, denoted

$$s = \sum_{k=1}^{\infty} a_k$$

is defined as the value of s_n as n approaches infinity (as n becomes very large), if the sum exists.

EXAMPLE 1

Find the sum of the series $\sum_{k=1}^{\infty} (\frac{1}{2})^k = \frac{1}{2} + \frac{1}{4} + \frac{1}{8} + \cdots$.

Solution

Looking at the first few terms of the sequence of sums helps us see what value the sum is approaching.

$s_1 = (\frac{1}{2})^1 = \frac{1}{2}$

$s_2 = (\frac{1}{2})^1 + (\frac{1}{2})^2 = \frac{1}{2} + \frac{1}{4} = \frac{3}{4}$

$s_3 = (\frac{1}{2})^1 + (\frac{1}{2})^2 + (\frac{1}{2})^3 = \frac{1}{2} + \frac{1}{4} + \frac{1}{8} = \frac{7}{8}$

$s_4 = (\frac{1}{2})^1 + (\frac{1}{2})^2 + (\frac{1}{2})^3 + (\frac{1}{2})^4 = \frac{1}{2} + \frac{1}{4} + \frac{1}{8} + \frac{1}{16} = \frac{15}{16}$

.
.
.

The sequence of sums seems to be approaching 1. Note that we can write the sums in an alternate form, which will permit us to find the general term s_n.

$s_1 = 1 - \frac{1}{2} = \frac{1}{2}$

$s_2 = 1 - (\frac{1}{2})^2 = \frac{3}{4}$

$s_3 = 1 - (\frac{1}{2})^3 = \frac{7}{8}$

$s_4 = 1 - (\frac{1}{2})^4 = \frac{15}{16}$

.
.
.

$s_n = 1 - (\frac{1}{2})^n$

.
.
.

Clearly the sum can get no larger than 1, and it is approaching 1 as n gets larger. Thus the sum of the infinite series is 1.

Not all series will have a finite sum. For example, $\sum_{k=1}^{\infty} 2n$ will get large without bound as n gets very large. An **arithmetic** series is the sum of the terms of an arithmetic sequence. Because the same number is added to each term of an arithmetic series, the sum of the series is never finite. (It increases without bound, either positively or negatively.)

EXAMPLE 2

Find the sum of the terms of geometric sequence $1, \frac{1}{2}, \frac{1}{4}, \frac{1}{8}, \ldots$

Solution

Listing the first few terms of the sequence of sums gives (using the formula for sums of geometric sequences):

$$s_1 = \frac{1(1-\frac{1}{2})}{1-\frac{1}{2}} = 1$$

$$s_2 = \frac{1[1-(\frac{1}{2})^2]}{1-\frac{1}{2}} = \frac{3}{2}$$

$$s_3 = \frac{1[1-(\frac{1}{2})^3]}{1-\frac{1}{2}} = \frac{7}{4}$$

$$\vdots$$

$$s_n = \frac{1[1-(\frac{1}{2})^n]}{1-\frac{1}{2}}$$

$$\vdots$$

As n gets larger, $(\frac{1}{2})^n$ gets smaller, approaching 0. Thus s_n approaches $1[1-0]/(1-\frac{1}{2})$ as n gets very large. Thus the sum of the terms is $1/\frac{1}{2} = 2$.

The sum of the terms of a geometric sequence is called a **geometric series**. If the common ratio r is such that $|r| < 1$, the sum of the geometric series is

$$s = \frac{a_1}{1-r}$$

We can see this by considering

$$S_n = \frac{a_1(1-r^n)}{1-r} = \frac{a_1}{1-r} - \frac{a_1 r^n}{1-r}$$

If r is between -1 and 1, r^n will approach 0 as n increases without bound. If $|r| \geq 1$, the geometric series will have no finite sum.

EXAMPLE 3

Find the sum of the geometric series with first term 5 and common ratio $-\frac{1}{3}$.

Solution

$$s = \frac{5}{1-(-\frac{1}{3})} = \frac{5}{\frac{4}{3}} = \frac{15}{4} = 3\frac{3}{4}$$

Note: The series of Example 1 is a geometric series with first term $\frac{1}{2}$ and common ratio $\frac{1}{2}$.

We stated in Chapter 1 that repeating decimals are rational numbers. Recall that a rational number is a number that can be written as the ratio of two integers. The following examples will illustrate how a repeating decimal can be written as the ratio of two integers.

EXAMPLE 4

Write $0.44444\ldots$ as the ratio of two integers.

Solution

$0.44444\ldots$ can be written as $0.4 + 0.04 + 0.004 + 0.0004 + 0.00004 + \cdots$, which is a geometric series with first term 0.4 and common ratio 0.1. Thus

$$s = \frac{0.4}{1-0.1} = \frac{4}{9}$$

so $0.4444\ldots = \frac{4}{9}$. We can check this result by dividing 4 by 9.

EXAMPLE 5

Write the fractional equivalent of 0.231231231. . . .

Solution

0.231231231 . . . can be written as $0.231 + 0.000231 + 0.000000231 + \cdots$, which is a geometric series with first term 0.231 and common ratio 0.001.

$$s = \frac{0.231}{1 - 0.001} = \frac{0.231}{0.999} = \frac{231}{999} = \frac{77}{333}$$

so

$$0.231231231\ldots = \frac{77}{333}$$

We can check these results by dividing 77 by 333.

Exercise 13-3

1. Does the sum of the series $\sum_{k=1}^{\infty} (\frac{1}{3})^k$ equal $\frac{1}{2}$ (do the partial sums approach $\frac{1}{2}$)?

2. Does the sum of the series $\sum_{k=1}^{\infty} (\frac{1}{5})^k$ equal $\frac{1}{4}$?

3. Does $\sum_{k=1}^{\infty} k$ have a finite sum?

4. Does any arithmetic series have a finite sum?

5. Do all geometric series have finite sums?

6. When do geometric series have finite sums?

7. Find the sum of the geometric series with first term 6 and common ratio $\frac{1}{3}$.

8. Find the sum of the geometric series with common ratio $\frac{3}{4}$ and first term 4.

9. Find the sum of the geometric series with common ratio $-\frac{1}{4}$ and first term 6.

10. Find the sum of the geometric series with first term 10 and common ratio $-\frac{1}{2}$.

11. Find the sum of the geometric series $6 + 2 + \frac{2}{3} + \frac{2}{9} + \cdots$, if it exists (that is, is finite).

12. Find the sum of the geometric series $4 + 2 + 1 + \frac{1}{2} + \cdots$, if it exists.

13. Find the sum of the geometric series $8 + (-4) + 2 + (-1) + \cdots$, if it exists.

14. Find the sum of the geometric series $12 + 18 + 27 + \cdots$, if it exists.
15. Write $0.5555\ldots$ as the ratio of two integers.
16. Write $0.6666\ldots$ as the ratio of two integers.
17. Write $0.676767\ldots$ as the ratio of two integers.
18. Write $8.383838\ldots$ as the ratio of two integers.

13-4 The binomial formula

OBJECTIVE: After completing this section, you should be able to use the binomial formula to expand a positive integral power of a binomial.

If we perform the multiplications of binomials indicated by the exponents, we obtain the following:

$(a + b)^1 = a + b$

$(a + b)^2 = a^2 + 2ab + b^2$

$(a + b)^3 = a^3 + 3a^2b + 3ab^2 + b^3$

$(a + b)^4 = a^4 + 4a^3b + 6a^2b^2 + 4ab^3 + b^4$

Note that if n is a positive power of $(a + b)$, the expansion has the following properties.

PROPERTIES	EXAMPLE
I. There are $n + 1$ terms in the expansion.	$n = 4$ $n + 1 = 5$ terms $(a + b)^④ = a^4 + 4a^3b + 6a^2b^2$ $\qquad + 4ab^3 + b^4$
II. The sum of the exponents (powers) of a and b in any term is n.	$(a + b)^④ = a^④ + 4a^③b^①$ $\qquad + 6a^②b^② + 4a^①b^③$ $\qquad + b^④$
III. The first term is a^n, and the power of a decreases by 1 in each succeeding term until a^1 occurs in the next-to-last term.	$(a + b)^4 = a^④ + 4a^③b + 6a^②b^2$ $\qquad + 4a^①b^3 + b^4$
IV. The second term contains b^1, and the power of b increases by 1 in each succeeding term.	$(a + b)^4 = a^4 + 4a^3b^① + 6a^2b^②$ $\qquad + 4ab^③ + b^④$

V. The coefficients of the first and last terms are 1, and the coefficient of the second term is n.

VI. If the coefficient of any term is multiplied by the exponent of a in that term and divided by the number of that term, the result will be the coefficient of the next term.

$$(a+b)^4 = \boxed{1}a^4 + \boxed{4}a^3b + 6a^2b^2 + 4ab^3 + \boxed{1}b^4$$

$$(a+b)^4 = 1a^4 + \left(\frac{1 \cdot 4}{1}\right)a^3b \quad \text{1st term, 2nd term}$$
$$+ \left(\frac{4 \cdot 3}{2}\right)a^2b^2 \quad \text{3rd term}$$
$$+ \left(\frac{6 \cdot 2}{3}\right)ab^3 \quad \text{4th term}$$
$$+ \left(\frac{4 \cdot 1}{4}\right)b^4 \quad \text{5th term}$$

The properties of the binomial expansion permit us to write it as a formula. The binomial formula is

$$(a+b)^n = a^n + \frac{n}{1}a^{n-1}b + \frac{n(n-1)}{1 \cdot 2}a^{n-2}b^2$$
$$+ \frac{n(n-1)(n-2)}{1 \cdot 2 \cdot 3}a^{n-3}b^3 + \cdots + b^n$$

EXAMPLE 1

Write the expansion of $(x+y)^5$.

Solution

$$(x+y)^5 = x^5 + \frac{5}{1}x^4y + \frac{5 \cdot 4}{1 \cdot 2}x^3y^2 + \frac{5 \cdot 4 \cdot 3}{1 \cdot 2 \cdot 3}x^2y^3$$
$$+ \frac{5 \cdot 4 \cdot 3 \cdot 2}{1 \cdot 2 \cdot 3 \cdot 4}xy^4 + \frac{5 \cdot 4 \cdot 3 \cdot 2 \cdot 1}{1 \cdot 2 \cdot 3 \cdot 4 \cdot 5}y^5$$
$$= x^5 + 5x^4y + 10x^3y^2 + 10x^2y^3 + 5xy^4 + y^5$$

Note the symmetry of the coefficients; that is, note that the coefficients of terms at fixed distances from opposite ends of the expansion are equal.

EXAMPLE 2

Write the first three terms of the expansion of $(x + y)^{20}$.

Solution

The first three terms are

$$x^{20} + \frac{20}{1}x^{19}y + \frac{20 \cdot 19}{1 \cdot 2}x^{18}y^2$$

Although the pattern of the exponents in the binomial expansion is easy to follow, the method we have used to find the coefficients of the terms is somewhat cumbersome. A device we may use to obtain the coefficients easily is Pascal's triangle, named for a French mathematician Blaise Pascal (1623–1662). The triangle is

```
                    1              n = 0
                  1   1            n = 1
                1   2   1          n = 2
              1   3   3   1        n = 3
            1   4   6   4   1      n = 4
          1   5  10  10   5   1    n = 5
        1   6  15   .   .   .      n = 6
```

The triangle is bordered by 1's. Each row after the second can be found by adding two adjacent numbers in the row above and writing the sum on the given row between the adjacent numbers. For example, the next number in the row associated with $n = 6$ is the sum of 10 and 10, or 20.

The 1 at the top of the triangle is associated with $n = 0$ because $(a + b)^0 = 1$. The second row is associated with $n = 1$ because $(a + b)^1 = 1a + 1b$. The third row is associated with $n = 2$ because $(a + b)^2 = 1a^2 + 2ab + 1b^2$, and so on.

EXAMPLE 3

a. What are the coefficients of each of the terms of $(a + b)^6$?
b. Write the expansion of $(a + b)^6$.

Solution

a. Completing the row associated in the triangle associated with $n = 6$ gives

  1   6   15   20   15   6   1

b. $(a + b)^6 = a^6 + 6a^5b + 15a^4b^2 + 20a^3b^3 + 15a^2b^4 + 6ab^5 + b^6$

EXAMPLE 4

Expand $(3u + 2v)^4$.

Solution

$$(3u + 2v)^4 = 1(3u)^4 + 4(3u)^3(2v) + 6(3u)^2(2v)^2 + 4(3u)(2v)^3 + 1(2v)^4$$
$$= 81u^4 + 4(27u^3)(2v) + 6(9u^2)(4v^2) + 4(3u)(8v^3) + 16v^4$$
$$= 81u^4 + 216u^3v + 216u^2v^2 + 96uv^3 + 16v^4$$

EXAMPLE 5

Expand $(2w - v)^3$.

Solution

$$(2w - v)^3 = [2w + (-v)]^3 = 1(2w)^3 + 3(2w)^2(-v) + 3(2w)(-v)^2 + 1(-v)^3$$
$$= 8w^3 + 3(4w^2)(-v) + 3(2w)(v^2) + 1(-v^3)$$
$$= 8w^3 - 12w^2v + 6wv^2 - v^3$$

As Example 5 illustrates, the expansion of the difference of two binomials has alternating signs.

Exercise 13-4

1. Write the first three terms of $(x + y)^{18}$.
2. Write the first three terms of $(y - 3)^{15}$.
3. Construct Pascal's triangle for $n = 0$ to $n = 8$.
4. Write the row associated with $n = 9$ in Pascal's triangle.

Expand by the binomial formula and simplify.

5. $(x + y)^3$
6. $(m + n)^5$
7. $(x - y)^4$
8. $(m - n)^5$
9. $(2a + b)^4$
10. $(v + 3w)^6$
11. $(3p - q)^5$
12. $(v - 3w)^6$
13. $(2x + 3)^4$
14. $(4z + 2)^8$
15. $(3v - 5)^6$
16. $(2u - 3)^7$
17. $(2x - 3y)^5$
18. $(5v - 4u)^3$
19. $(x^{1/3} - 1)^3$
20. $(x^{1/2} + 4)^4$
21. $(x^{-2} - y)^4$
22. $(w^{-1} + y^2)^3$

CHAPTER TEST

1. Write the first six terms of the sequence defined by $a_n = 3/n$.
2. Write the first six terms of the sequence whose nth term is $(-1)^n/(n+1)$.
3. Write three additional terms of the arithmetic sequence $4, 5\frac{1}{2}, 7, \ldots$.
4. Find the ninth term of the arithmetic sequence with first term 6 and common difference -2.
5. Find the sixth term of an arithmetic sequence with first term 18 and ninth term 22.
6. Write three additional terms of the geometric sequence $4, 16, 64, \ldots$.
7. Find the fifth term of the geometric sequence with first term 8 and common ratio $-\frac{1}{2}$.
8. Find the third term of the geometric sequence with first term 12 and fifth term 972.
9. A ball is dropped from 81 feet. If it rebounds $\frac{2}{3}$ of the height from which it falls every time it hits the ground, how high will it bounce after it strikes the ground for the third time?
10. Evaluate $\sum_{i=1}^{5} \dfrac{4i}{i+2}$.
11. Find the sum of the first four terms of the sequence defined by $a_n = (-1)^n/4n$.
12. Find the sum of the first seven terms of the arithemetic sequence with first term 3 and common difference 3.
13. Find the sum of the first six terms of the arithmetic sequence $4, 5\frac{1}{2}, 7, \ldots$.
14. Find the sum of the first five terms of the geometric sequence with first term 10 and common ratio $\frac{1}{2}$.
15. Find the sum of the first five terms of the geometric sequence $3, -6, 12, \ldots$.
16. Find the sum of the geometric series with first term 4 and common ratio $\frac{1}{4}$.
17. Find the sum of the geometric series $4 + (-2) + 1 + (-\frac{1}{2}) + \cdots$, if it exists.
18. Write $0.363636\ldots$ as the ratio of two integers.
19. Do all geometric series have finite sums?
20. Expand $(p + q)^4$ using the binomial formula.
21. Expand $(2m - n)^5$ using the binomial formula.

Appendix

$\sqrt{2}$ Is not rational (Proof)

We shall show that the $\sqrt{2}$ is not rational by an indirect proof; that is, we shall assume that $\sqrt{2}$ is rational and show that this assumption leads to a contradiction.

Proof: Assume that $\sqrt{2}$ is rational. Then $\sqrt{2}$ may be written as the ratio of two integers. Suppose that this ratio is written as a fraction reduced to lowest terms. If a/b is this reduced fraction, then 2 cannot be a factor of both a and b. But if $\sqrt{2} = a/b$, then $2 = a^2/b^2$, or $2b^2 = a^2$. Thus a^2 is even (divisible by 2).

But an odd number squared is always odd; so if a^2 is even, then a **must be even.** Since a is an even number, we may write $a = 2c$. Then we may write

$$2b^2 = (2c)^2 \quad \text{or} \quad 2b^2 = 4c^2$$

Dividing both sides of this equation by 2 gives $b^2 = 2c^2$. But then b^2 is even, and by the same argument as was used above for a, b **must be even.** Thus both a and b are divisible by 2, which is a contradiction of our earlier statement that 2 cannot be a factor of both a and b.

Thus $\sqrt{2}$ is not a rational number, and we say that $\sqrt{2}$ **is an irrational number.**

Properties of logarithms (Proofs)

Property I. $\log_a a^x = x$

Proof: If $v = \log_a a^x$, then the exponential form of the equation is $a^v = a^x$. Thus $v = x$; so $x = \log_a a^x$.

Property II. $a^{\log_a x} = x$

Proof: If $u = \log_a x$, then the exponential form of the equation is $a^u = x$. Thus $a^{\log_a x} = x$.

Property III. $\log_a (MN) = \log_a M + \log_a N$

Proof: Let $u = \log_a M$ and $v = \log_a N$. Then $a^u = M$ and $a^v = N$; so $MN = a^u a^v = a^{u+v}$. But $MN = a^{u+v}$ may be written as $\log_a (MN) = u + v$, or $\log_a (MN) = \log_a M + \log_a N$.

Property IV. $\log_a \left(\dfrac{M}{N}\right) = \log_a M - \log_a N$

Proof: Let $u = \log_a M$ and $v = \log_a N$. Then $a^u = M$ and $a^v = N$; so $\dfrac{M}{N} = \dfrac{a^u}{a^v} = a^{u-v}$. But $M/N = a^{u-v}$ may be written in the form $\log_a \left(\dfrac{M}{N}\right) = u - v$ or $\log_a \left(\dfrac{M}{N}\right) = \log_a M - \log_a N$.

Property V. $\log_a (M^N) = N \log_a M$

Proof: Let $u = \log_a M$. Then $a^u = M$; so $M^N = (a^u)^N = a^{Nu}$. But $M^N = a^{Nu}$ may be written in the form $\log_a (M^N) = Nu$, or $\log_a (M^N) = N \log_a M$.

Table I Squares, cubes, and square roots

n	n^2	n^3	$\sqrt{n}$	$\sqrt{10n}$
1	1	1	1.0000	3.1623
2	4	8	1.4142	4.4721
3	9	27	1.7321	5.4772
4	16	64	2.0000	6.3246
5	25	125	2.2361	7.0711
6	36	216	2.4495	7.7460
7	49	343	2.6458	8.3666
8	64	512	2.8284	8.9443
9	81	729	3.0000	9.4868
10	100	1000	3.1623	10.0000
11	121	1331	3.3166	10.4887
12	144	1728	3.4641	10.9545
13	169	2197	3.6056	11.4018
14	196	2744	3.7417	11.8322
15	225	3375	3.8730	12.2475
16	256	4096	4.0000	12.6491
17	289	4913	4.1231	13.0384
18	324	5832	4.2426	13.4164
19	361	6859	4.3589	13.7840
20	400	8000	4.4721	14.1421
21	441	9261	4.5826	14.4914
22	484	10648	4.6904	14.8324
23	529	12167	4.7958	15.1658
24	576	13824	4.8990	15.4919
25	625	15625	5.0000	15.8114
26	676	17576	5.0990	16.1245
27	729	19683	5.1962	16.4317
28	784	21952	5.2915	16.7332
29	841	24389	5.3852	17.0294
30	900	27000	5.4772	17.3205
31	961	29791	5.5678	17.6068
32	1024	32768	5.6569	17.8885
33	1089	35937	5.7446	18.1659
34	1156	39304	5.8310	18.4391
35	1225	42875	5.9161	18.7083
36	1296	46656	6.0000	18.9737
37	1369	50653	6.0828	19.2354
38	1444	54872	6.1644	19.4936
39	1521	59319	6.2450	19.7484
40	1600	64000	6.3246	20.0000

Table II Four-place logarithms

N	0	1	2	3	4	5	6	7	8	9
10	0000	0043	0086	0128	0170	0212	0253	0294	0334	0374
11	0414	0453	0492	0531	0569	0607	0645	0682	0719	0755
12	0792	0828	0864	0899	0934	0969	1004	1038	1072	1106
13	1139	1173	1206	1239	1271	1303	1335	1367	1399	1430
14	1461	1492	1523	1553	1584	1614	1644	1673	1703	1732
15	1761	1790	1818	1847	1875	1903	1931	1959	1987	2014
16	2041	2068	2095	2122	2148	2175	2201	2227	2253	2279
17	2304	2330	2355	2380	2405	2430	2455	2480	2504	2529
18	2553	2577	2601	2625	2648	2672	2695	2718	2742	2765
19	2788	2810	2833	2856	2878	2900	2923	2945	2967	2989
20	3010	3032	3054	3075	3096	3118	3139	3160	3181	3201
21	3222	3243	3263	3284	3804	3324	3345	3365	3385	3404
22	3424	3444	3464	3483	3502	3522	3541	3560	3579	3598
23	3617	3635	3655	3674	3692	3711	3729	3747	3766	3784
24	3802	3820	3838	3856	3874	3892	3909	3927	3945	3962
25	3979	3997	4014	4031	4048	4065	4082	4099	4116	4133
26	4150	4166	4183	4200	4216	4232	4249	4265	4281	4298
27	4314	4330	4346	4362	4378	4393	4409	4425	4440	4456
28	4472	4487	4502	4518	4533	4548	4564	4579	4594	4609
29	4624	4639	4654	4669	4683	4698	4713	4728	4742	4757
30	4771	4786	4800	4814	4829	4843	4857	4871	4886	4900
31	4914	4928	4942	4955	4969	4983	4997	5011	5024	5038
32	5051	5065	5079	5092	5105	5119	5132	5145	5159	5172
33	5185	5198	5211	5224	5237	5250	5263	5276	5289	5302
34	5315	5328	5340	5353	5366	5378	5391	5403	5416	5428
35	5441	5453	5465	5478	5490	5502	5514	5527	5539	5551
36	5563	5575	5587	5599	5611	5623	5635	5647	5658	5670
37	5682	5694	5705	5717	5729	5740	5752	5763	5775	5786
38	5798	5809	5821	5832	5843	5855	5866	5877	5888	5899
39	5911	5922	5933	5944	5955	5966	5977	5988	5999	6010
40	6021	6031	6042	6053	6064	6075	6085	6096	6107	6117
41	6128	6138	6149	6160	6170	6180	6191	6201	6212	6222
42	6232	6243	6253	6263	6274	6284	6294	6304	6314	6325
43	6335	6345	6355	6365	6375	6385	6395	6405	6415	6425
44	6435	6444	6454	6464	6474	6484	6493	6503	6513	6522
45	6532	6542	6551	6561	6571	6580	6590	6599	6609	6618
46	6628	6637	6646	6656	6665	6675	6684	6693	6702	6712
47	6721	6730	6739	6749	6758	6767	6776	6785	6794	6803
48	6812	6821	6830	6839	6848	6857	6866	6875	6884	6893
49	6902	6911	6920	6928	6937	6946	6955	6964	6972	6981
50	6990	6998	7007	7016	7024	7033	7042	7050	7059	7067
51	7076	7084	7093	7101	7110	7118	7126	7135	7143	7152
52	7160	7168	7177	7185	7193	7202	7210	7218	7226	7235
53	7243	7251	7259	7267	7275	7284	7292	7300	7308	7316
54	7324	7332	7340	7348	7356	7364	7372	7380	7388	7396

Appendix 363

(continued)

N	0	1	2	3	4	5	6	7	8	9
55	7404	7412	7419	7427	7435	7443	7451	7459	7466	7474
56	7482	7490	7497	7505	7513	7520	7528	7536	7543	7551
57	7559	7566	7574	7582	7589	7597	7604	7612	7619	7627
58	7634	7642	7649	7657	7664	7672	7679	7686	7694	7701
59	7709	7716	7723	7731	7738	7745	7752	7760	7767	7774
60	7782	7789	7796	7803	7810	7818	7825	7832	7839	7846
61	7853	7860	7868	7875	7882	7889	7896	7903	7910	7917
62	7924	7931	7938	7945	7952	7959	7966	7973	7980	7987
63	7993	8000	8007	8014	8021	8028	8035	8041	8048	8055
64	8062	8069	8075	8082	8089	8096	8102	8109	8116	8122
65	8129	8136	8142	8149	8156	8162	8169	8176	8182	8189
66	8195	8202	8209	8215	8222	8228	8235	8241	8248	8254
67	8261	8267	8274	8280	8287	8293	8299	8306	8312	8319
68	8325	8331	8338	8344	8351	8357	8363	8370	8376	8382
69	8388	8395	8401	8407	8414	8420	8426	8432	8439	8445
70	8451	8457	8463	8470	8476	8482	8488	8494	8500	8506
71	8513	8519	8525	8531	8537	8543	8549	8555	8561	8567
72	8573	8579	8585	8591	8597	8603	8609	8615	8621	8627
73	8633	8639	8645	8651	8657	8663	8669	8675	8681	8686
74	8692	8698	8704	8710	8716	8722	8727	8733	8739	8745
75	8751	8756	8762	8768	8774	8779	8785	8791	8797	8802
76	8808	8814	8820	8825	8831	8837	8842	8848	8854	8859
77	8865	8871	8876	8882	8887	8893	8899	8904	8910	8915
78	8921	8927	8932	8938	8943	8949	8954	8960	8965	8971
79	8976	8982	8987	8993	8998	9004	9009	9015	9020	9025
80	9031	9036	9042	9047	9053	9058	9063	9069	9074	9079
81	9085	9090	9096	9101	9106	9112	9117	9122	9128	9133
82	9138	9143	9149	9154	9159	9165	9170	9175	9180	9186
83	9191	9196	9201	9206	9212	9217	9222	9227	9232	9238
84	9243	9248	9253	9258	9263	9269	9274	9279	9284	9289
85	9294	9299	9304	9309	9315	9320	9325	9330	9335	9340
86	9345	9350	9355	9360	9365	9370	9375	9380	9385	9390
87	9395	9400	9405	9410	9415	9420	9425	9430	9435	9440
88	9445	9450	9455	9460	9465	9469	9474	9479	9484	9489
89	9494	9499	9504	9509	9513	9518	9523	9528	9533	9538
90	9542	9547	9552	9557	9562	9566	9571	9576	9581	9586
91	9590	9595	9600	9605	9609	9614	9619	9624	9628	9633
92	9638	9643	9647	9652	9657	9661	9666	9671	9675	9680
93	9685	9689	9694	9699	9703	9708	9713	9717	9722	9727
94	9731	9736	9741	9745	9750	9754	9759	9763	9768	9773
95	9777	9782	9786	9791	9795	9800	9805	9809	9814	9818
96	9823	9827	9832	9836	9841	9845	9850	9854	9859	9863
97	9868	9872	9877	9881	9886	9890	9894	9899	9903	9908
98	9912	9917	9921	9926	9930	9934	9939	9943	9948	9952
99	9956	9961	9965	9969	9974	9978	9983	9987	9991	9996

Table III Exponential functions

x	e^x	e^{-x}	x	e^x	e^{-x}
0.0	1.000	1.000	3.0	20.09	.0498
0.1	1.105	.9048	3.1	22.20	.0450
0.2	1.221	.8187	3.2	24.53	.0408
0.3	1.350	.7408	3.3	27.11	.0369
0.4	1.492	.6703	3.4	29.96	.0334
0.5	1.649	.6065	3.5	33.12	.0302
0.6	1.822	.5488	3.6	36.60	.0273
0.7	2.014	.4966	3.7	40.45	.0247
0.8	2.226	.4493	3.8	44.70	.0224
0.9	2.460	.4066	3.9	49.40	.0202
1.0	2.718	.3679	4.0	54.60	.0183
1.1	3.004	.3329	4.1	60.34	.0166
1.2	3.320	.3012	4.2	66.69	.0150
1.3	3.669	.2725	4.3	73.70	.0136
1.4	4.055	.2466	4.4	81.45	.0123
1.5	4.482	.2231	4.5	90.02	.0111
1.6	4.953	.2019	4.6	99.48	.0101
1.7	5.474	.1827	4.7	109.9	.0091
1.8	6.050	.1653	4.8	121.5	.0082
1.9	6.686	.1496	4.9	134.3	.0074
2.0	7.389	.1353	5.0	148.4	.0067
2.1	8.166	.1225	5.1	164.0	.0061
2.2	9.025	.1108	5.2	181.3	.0055
2.3	9.974	.1003	5.3	200.3	.0050
2.4	11.02	.0907	5.4	221.4	.0045
2.5	12.18	.0821	5.5	244.7	.0041
2.6	13.46	.0743	5.6	270.4	.0037
2.7	14.88	.0672	5.7	298.9	.0033
2.8	16.44	.0608	5.8	330.3	.0030
2.9	18.17	.0550	5.9	365.0	.0027
			6.0	403.4	.0025
			7.0	1097	.0009
			8.0	2981	.0003
			9.0	8103	.0001

Answers to odd-numbered exercises and chapter tests

Exercise 1-1

1. $x \in \{x, y, z, a\}$ 3. $3 \notin \{1, 2, 4, 5, 6\}$ 5. $12 \in \{1, 2, 3, 4, \ldots\}$
7. $5 \notin \{x: x \text{ is a natural number greater than } 5\}$ 9. $3 \notin \emptyset$ 11. $\{1, 2, 3, 4, 5, 6, 7\}$ 13. $\{11, 12, 13, 14, \ldots\}$ 15. $\emptyset$ 17. $\{x: x \text{ is a natural number less than } 10\}$ 19. $\{x: x \text{ is a natural number greater than } 6\}$
21. $\{x: x = 0\}$ 23. yes 25. no 27. $D \subseteq C$ 29. $D \subseteq A$
31. $A \subseteq B, B \subseteq A$ 33. $A = B$ 35. $D \neq E$ 37. A, and B, B and D, C and D 39. $\{a, d, e\}$ 41. $\emptyset$ 43. $\emptyset$ 45. $\{2, 1, 3, 6, 4, 5, 7, 8\}$
47. $\{a, e, i, o, u, b, c, d\}$ 49. $\{x: x \text{ is a natural number}\}$

Exercise 1-2

1. yes
5. $2 \leftrightarrow 6$
   $4 \leftrightarrow 12$
   $6 \leftrightarrow 14$
   $8 \leftrightarrow 18$

3. no
7. $1 \leftrightarrow 3$
   $2 \leftrightarrow 6$
   $3 \leftrightarrow 9$
   $4 \leftrightarrow 12$
    $\vdots$ $\vdots$

9.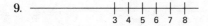

11.
$-8\ -7\ -6\ -5\ -4\ -3\ -2\ -1\ 0\ 1\ 2\ 3\ 4\ 5\ 6\ 7\ 8$

13. $-8 < -2$ 15. $-6 < -1$ 17. $-3 < 0$ 19. $-2 > -6$
21. $-8 < -4 < -2$ 23. $-6 < 2 < 4$ 25. $-8 < -6 < -2$
27. $-9 < -8 < -5$

Exercise 1-3

1.

3.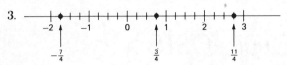

5. yes 7. no 9. no 11. 0.5 13. 0.4 15. 0.6666 ...
17. 0.5555 ... 19. $\frac{1}{8}$ 21. $\frac{123}{200}$ 23. $\frac{469}{100}$ 25. $\frac{33}{100}$

Exercise 1-4

1. rational 3. rational 5. rational 7. rational 9. rational
11. irrational 13. rational 15. rational 17. irrational 19. rational
21. 4.899 23. 37.947 25. 19.365 27. 5.732 29. −0.899
31. −0.748 33. 3.310

Exercise 1-5

1. commutative property of addition 3. associative property of addition
5. associative property of multiplication 7. additive inverse
9. multiplicative inverse 11. additive identity 13. multiplicative identity 15. commutative property of addition 17. additive identity
19. multiplicative inverse 21. additive inverse 23. distributive property 25. $(5 + 6) + 7$ 27. $6 + (3 + 4)$ 29. $(7 \cdot 3)2$
31. $4(2 \cdot 3)$ 33. $6 \cdot 3 + 6 \cdot 5 = 48$ 35. $5 \cdot 5 + 5 \cdot 8 = 65$

Exercise 1-6

1. 20 3. 19 5. 7 7. 33 9. 36 11. 304 13. 180 15. 4
17. 21 19. 6 21. 4 23. 4 25. 51 27. 26 29. 24 31. 4
33. 39 35. 3 37. 3 39. 1 41. 2 43. 3 45. 4 47. 0
49. $|-3|$ 51. $|-6|$

Exercise 1-7

1. −7 3. −11 5. 11 7. 1 9. −2 11. −2 13. −3 15. 0
17. −1 19. 6 21. −3 23. −16 25. −2 27. −6 29. 6
31. 2 33. 4 35. 2 37. −2 39. 1 41. −4.6 43. −13
45. −4 47. 7.03 49. 21 51. 29 53. −17 55. $340
57. 20,550

Exercise 1 - 8

1. 6 3. 20 5. −42 7. 4 9. 64 11. −58.8 13. −6
15. −30 17. −12 19. −96 21. −112 23. 24 25. 3 27. 2
29. 2 31. −2 33. −0.21 35. −11 37. −24 39. −4
41. −2 43. 2 45. −59 47. −4 49. 128 51. 0 53. 32
55. 0 57. undefined 59. 0 61. 0.4

CHAPTER 1 TEST

1. yes 2. no 3. {1, 2, 3, 4, 5, 6, 7, 8} 4. {x: x is an odd natural number less than 10} 5. no 6. yes 7. yes 8. {π}
9. {1, 2, 3, 4, 5, 6, 7, 11} 10. yes 11. one correspondence is
$$3 \leftrightarrow \pi$$
$$6 \leftrightarrow 6$$
$$8 \leftrightarrow 5$$
$$11 \leftrightarrow 4$$
$$\pi \leftrightarrow 3$$

12.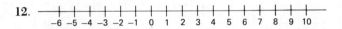

13. $-6 > -8$ 14. $-4 < -1 < 3$

15.

16. no 17. natural, integer, rational number 18. irrational number
19. rational number 20. 84 21. 244 22. 24 23. 6 24. 4
25. 10 26. 6 27. 32 28. $(5 \cdot 2)7$ 29. $4 \cdot 7$ 30. 6 31. 5
32. −8 33. −19 34. 9 35. 60 36. −36 37. −22 38. −4
39. 16

Exercise 2 - 1

1. binomial; 3 is the coefficient in $3x$; 2 is the coefficient in $2y$
3. monomial; 15 is the coefficient in $15x^2y$ 5. trinomial;
6 is the coefficient in $6x^2$; −3 is the coefficient in $-3x$; 4 is a constant
7. binomial; 46 is the coefficient in $46x^3$; −12 is a constant 9. 1. first;
2. second; 3. third; 4. third 11. $8x$ 13. $17y$ 15. $11y^3$ 17. $22z$
19. $4x^2 + 5x$ 21. $17x + 2y$ 23. $7x + 8a$ 25. $11t^2 + 5t^3$
27. $-4x^2y + 4xy$ 29. $9a^2 + 6b^2$ 31. $2x$ 33. $5xy$ 35. $10ab - 6a$
37. $13mn + 4n$ 39. $2xy$ 41. $-3pq$ 43. $2c - b$
45. $18x^2y - xy + 3y^2$

Exercise 2-2

1. $5x + 7y$ 3. $8m + 10n$ 5. $6x^2 + 6y$ 7. $5x + 5y + 5m$
9. $5m - x - y$ 11. $12a - b - 3$ 13. $4r + 7x$
15. $x^2 - 7x^2y - 2xy + 4$ 17. $x - 5y - 2$ 19. $4a + 6x$ 21. $3x - 9y$
23. $3a - 14b$ 25. $7 - 11ax^2y + 5axy - 2x + 4x^2y$

Exercise 2-3

1. x^6 3. m^9 5. 5^6 7. 4^5 9. a^3b^3 11. l^3m^5 13. $10x^5$
15. $12m^6$ 17. $2b^7$ 19. $24x^5$ 21. $18x^3y^4$ 23. $-12x^3yt$
25. $24x^4y^2$ 27. $-30m^4n^4$ 29. $2x^3y^3$ 31. $27x^4y^2$ 33. $-5r^3$
35. $3m$ 37. $4rs$ 39. $-8m$ 41. $6r^2$ 43. $1/x^4$ 45. $1/r$
47. $2x^2/y$ 49. $x^5/2m^3$ 51. $3x/y$ 53. $xy^2/2$ 55. y 57. $-3xy$
59. $3ab^2c^2$ 61. $10x^2$ 63. $11m$

Section 2-4

1. $5x - 8$ 3. $14x - 5$ 5. $4p + 2$ 7. $-2x$ 9. $2x^2 - x$
11. $2x^3 - 3$ 13. $2x - (-3y - 2z + 6)$
15. $4ab - (-3abx + 3x^2 - 2ax)$ 17. 22 19. $4x + 12$ 21. $6x + 12y$
23. 4 25. $7y + 3y^2$ 27. $7z$ 29. $2x$

Exercise 2-5

1. $5x + 5y$ 3. $a^2bx + ab^2y$ 5. $2ax^3 + 4ax^2y$
7. $-ax - bx^2 - cx^3$ 9. $4ax^3y^3 + 3ax^4y^2 + 4ax^2y^2$
11. $x^3 + 2x^2 + xy^2 - x^2y - 2xy - y^3$ 13. $x^2 + x^2y - xy^2 - y^3$
15. $y^3 - 4y^2z + x^2y + 4yz^2 - 2x^2z$ 17. $a^2z + abz + ab^2z - a^2b - ab^2 - ab^3$
19. $2a^3 - 7a^2b + 5ab^2 - b^3$ 21. $a^4 + 2a^3b + a^3b^2 - a^2b^2 - 2ab^3 - ab^4$
23. $a^3 - 2a^2b + a^2b^2 - 3ab^2 - 3ab^3$ 25. $y + 2 + 3xy^2$ 27. $2m - n - 4n^2$
29. $x - 3 - 2/y$ 31. $4 - a - 2/c$ 33. $3x + 2$ 35. $a - 3$
37. $y^2 + 2$ 39. $x - 2 - \dfrac{1}{x-2}$ 41. $4x^2 + 8x + 11 + \dfrac{15}{x-2}$
43. $x + 5y + \dfrac{4y^2}{x-y}$

CHAPTER 2 TEST

1. binomial 2. monomial 3. trinomial 4. $6ab, 6; 4a^2, 4$
5. $3a^2, 3; 4ax, 4; 13x^2, 13$ 6. $11x^2 + 19x - 6$ 7. $21pq - 2p^2$
8. $7x^3 + 4x^2 + 23x$ 9. $-x^2y^2 - 2x^3$ 10. $m^2 - 7n^2 - 3$ 11. $a + 8b - c$
12. $35x^2$ 13. $-24x^5y^6$ 14. $3rs$ 15. $-3m^2/n^3$ 16. $2m^2x^3$
17. $-3y^2$ 18. $9 + q$ 19. $6x$ 20. $5y$ 21. $2x + 2y$

22. $2ax^4 + a^2x^3 + a^2bx^2$ 23. $6y^2 - y - 12$ 24. $x^3 + 3x^2 + 6x + 4$
25. $a^3 - 5a^2b + 7ab^2 - 2b^3$ 26. $3 + m + 2m^2n$
27. $4x + y + 2/y$ 28. $3x^2 + 4x + 5$ 29. $a^2 + a + 4$

Exercise 3 - 1

1. $4(x^2 + 2)$ 3. $x(x + 2y)$ 5. $-a(x^2 - y)$ or $a(-x^2 + y)$
7. $6x(y + 2)$ 9. $3a(3b - 1)$ 11. $6b(a - 2a^2 + 3b)$
13. $4(2a^2b - 4ax + bx^2)$ 15. $x^2y(16y^2 - 8 - 9x)$ 17. $4(9x^2 - y^2)$
19. $4(x^2 + 2xy + y^2)$

Exercise 3 - 2

1. $x^2 + 2x + 1$ 3. $x^2 - 4x + 4$ 5. $y^2 + 5y + 6$ 7. $x^2 + 6x + 8$
9. $x^2 + x - 2$ 11. $x^2 - 2x - 3$ 13. $-a^2 - 5a - 6$ 15. $x^2 - 4$
17. $a^2 - 9$ 19. $x^2 - 6x + 9$ 21. $a^2 + 2a + 1$ 23. $y^2 - 16y + 64$
25. $x^2 - y^2$ 27. $2x^2 + 3xy + y^2$ 29. $4x^2 - 2x - 2$ 31. $12x^2 - 13x + 3$
33. $-15x^2 + 16x - 4$ 35. $9x^2 + 12x + 4$ 37. $6x^2 + 9xy + 2x + 3y$
39. $15x^2 + 26xy + 8y^2$ 41. $4x^2 - y^2$

Exercise 3 - 3

1. $(x + 2)(x + 1)$ 3. $(x + 1)(x - 2)$ 5. $(x + 5)(x + 1)$
7. $(x - 5)(x + 1)$ 9. $(x + 4)(x + 2)$ 11. $(x + 3)(x - 4)$
13. $(y - 9)(y + 1)$ 15. $(y + 3)(y - 12)$ 17. $(2x + 1)(x - 1)$
19. $(7x + 1)(x + 2)$ 21. $(3y + 2)(y - 3)$ 23. $(7a + 4)(a - 2)$
25. $(3x + 1)(x + 1)$ 27. $(5x + 7)(3x + 1)$ 29. $(2x - 1)(4x - 3)$
31. $(6x - 1)(x - 1)$ 33. $(5x - 1)(2x + 3)$ 35. $(4x + 1)(3x - 2)$
37. $(3x + 2)(2x + 7)$ 39. $(6x - 5)(2x - 3)$ 41. $(6x - 7)(x - 5)$
43. $(2x + 3)(2x - 5)$ 45. $(3x + 4)(2x + 1)$ 47. $(2x - 3)(4x - 7)$
49. $(x + 3y)(x - 2y)$ 51. $(2x - 5y)(2x + y)$ 53. $(5y - z)(2y + z)$

Exercise 3 - 4

1. $x^2 + 4x + 4$ 3. $x^2 + 8x + 16$ 5. $4x^2 + 12x + 9$ 7. $36x^2 - 60x + 25$
9. $9x^2 - 12xy + 4y^2$ 11. $16x^2 + 24xy + 9y^2$ 13. yes   15. no
17. no   19. yes   21. $(x - 2)^2$ 23. $(x - 5y)^2$ 25. $(2x - 2)^2$
27. $4(x + 6)^2$ or $(2x + 12)^2$ 29. $(2x + 5y)^2$

Exercise 3 - 5

1. $a^2 - b^2$ 3. $a^2 - 9$ 5. $x^2 - 25$ 7. $x^2 - y^2$ 9. $9x^2 - 4$
11. $25x^2 - 16$ 13. $9x^2 - 4y^2$ 15. $49x^2 - 4y^2$ 17. $(a - b)(a + b)$

19. $(x-y)(x+y)$ **21.** $(2x+1)(2x-1)$ **23.** $(6a-5)(6a+5)$
25. $(a+7)(a-7)$ **27.** $(2a-y)(2a+y)$ **29.** $(4x-5y)(4x+5y)$

Exercise 3 - 6

1. $x^3 + 3x^2 + 3x + 1$ **3.** $a^3 - 15a^2 + 75a - 125$
5. $8x^3 + 12x^2y + 6xy^2 + y^3$ **7.** $8x^3 - 48x^2 + 96x - 64$ **9.** $x^3 + 1$
11. $x^3 - 8$ **13.** $(x+1)^3$ **15.** $(x-4)^3$ **17.** $(x-4)(x^2 + 4x + 16)$
19. $(y+3)(y^2 - 3y + 9)$ **21.** $(x+3y)(x^2 - 3xy + 9y^2)$
23. $(m-4n)(m^2 + 4mn + 16n^2)$ **25.** $(2x+3y)(4x^2 - 6xy + 9y^2)$
27. $(2x-4m)(4x^2 + 8xm + 16m^2)$

Exercise 3 - 7

1. $5(x+y)$ **3.** $5(x+y)(x-y)$ **5.** $9(x-2y)(x+2y)$ **7.** $4(x+4)(x+2)$
9. $3(a-7)(a+3)$ **11.** $a(a-5)(a-3)$ **13.** $a(x+6)(x-2)$
15. $4(x+3)(x+1)$ **17.** $-a(y-5)(y+20)$ **19.** $a(a+8)^2$
21. $3(x^2 + 2x + 3)$ **23.** $4(4m+5)^2$ or $4(5+4m)^2$ **25.** $4a^2x(x+1)$

Exercise 3 - 8

1. $(x^2 + 3)^2$ **3.** $(p+3)^2(p-3)^2$ **5.** $(p^2 - q)(p^2 + q)$
7. $(x^2 + 9)(x-3)(x+3)$ **9.** $(x^4 + 16)(x^2 + 4)(x+2)(x-2)$
11. $(4+B)(x+y)$ **13.** $(a+4)(x+y)$ **15.** $(x-y)(3-a)$
17. $(5+q)(p-m)$ **19.** $(3a-x)(m+4x)$ **21.** $(x-4)(x^2+5)$
23. $(x-y)(x+y+4)$ **25.** $(x+2)(x-1)$ **27.** $(x+4+y)(x+4-y)$
29. $(x+y+4)(x+y-4)$ **31.** $(x-y)(x^2 + xy + y^2 + 4)$

CHAPTER 3 TEST

1. $x^2 - 6x + 8$ **2.** $6x^3 - 7x^2 - 3x$ **3.** $x^2 - 8x + 16$
4. $4a^2 + 12ab + 9b^2$ **5.** $b^2 - 9$ **6.** $25y^4 - 4$ **7.** $x^4 + 2x^2y^3 + y^6$
8. $5(x^2 - 2)$ **9.** $6ab(b - 2a + 4ab)$ **10.** $(x+1)(x-6)$
11. $(y-5)(y-2)$ **12.** $(5x-9)(x+3)$ **13.** $(3x-5)(2x-3)$
14. $(2x-3y)(4x+5y)$ **15.** $(x-5)^2$ **16.** $(2y+3)^2$
17. $(4x-1)(4x+1)$ **18.** $(2x-5y)(2x+5y)$ **19.** $4(x-2)^2$
20. $15a^2(y-b)(y+b)$ **21.** $(x^2 + 3y^2)^2$ **22.** $(x^2 + 9)(x-3)(x+3)$
23. $(x^3 + p^2)(x^3 - p^2)$ **24.** $(x-m)(6+y)$ **25.** $(x+2-y)(x+2+y)$

Exercise 4 - 1

1. $\dfrac{x-3}{x+2}$ **3.** $\dfrac{3x-1}{3x+1}$ **5.** $\dfrac{xy}{3}$ **7.** $2z^2$ **9.** $\dfrac{1}{2m}$ **11.** $\tfrac{1}{3}$ **13.** $\dfrac{x-2}{x+4}$

15. $\dfrac{x-1}{x-3}$ 17. $\dfrac{3}{x+y}$ 19. $\dfrac{a+1}{2-a}$ or $-\dfrac{a+1}{a-2}$ 21. $\dfrac{x+2}{x}$ 23. $x+1$
25. $3x+1$ 27. $x+1$ 29. $3x+2$ 31. $x+2$

Exercise 4 - 2

1. $\tfrac{1}{7}$ 3. $\tfrac{2}{5}$ 5. $1/x$ 7. b/a^2 9. 3 11. $\tfrac{15}{4}$ 13. $\dfrac{7(x+5)}{6(x-5)}$
15. 1 17. $\dfrac{x+1}{x+2}$ 19. $\dfrac{(x+5)(x-2)^2}{(x-5)(x+2)(x+3)}$ 21. $\dfrac{x+3}{x-2}$
23. $-\dfrac{x+3}{(x-5)(x+4)}$ or $\dfrac{x+3}{(5-x)(x+4)}$ 25. $-\dfrac{x-3}{(x-4)(x-1)}$ or $\dfrac{3-x}{(x-4)(x-1)}$
27. $\tfrac{3}{2}$ 29. $\tfrac{14}{15}$ 31. $2a/3d$ 33. $2x/3$ 35. $2x+4$ 37. $\tfrac{9}{2}$
39. $\dfrac{5y}{y-3}$ 41. $\dfrac{(x-1)(x-3)}{(x-2)(x-5)}$ or $\dfrac{x^2-4x+3}{x^2-7x+10}$ 43. $\dfrac{x+7}{x-2}$ 45. $r-s$

Exercise 4 - 3

1. $\tfrac{4}{3}$ 3. $\tfrac{7}{9}$ 5. $\dfrac{2x+3}{x-2}$ 7. 1 9. $\tfrac{1}{2}$ 11. $\dfrac{16+15x}{20x^2}$ 13. $\dfrac{a^2c+b^2}{ab^2c}$
15. $\dfrac{2x^2-9x+6}{(x-5)(x-4)}$ 17. $\dfrac{8a-3}{3(a-3)}$ 19. $\dfrac{x^2+4x-8}{2x(x+3)}$ 21. $\dfrac{2m+3}{(m-2)(m+2)}$
23. $\dfrac{2x^2-6}{(x-2)(x+1)(x+3)}$ 25. $\dfrac{1}{2x-1}$ 27. $\tfrac{2}{7}$ 29. $\dfrac{2x}{5a}$ 31. $\dfrac{-7n+1}{6n^2}$
or $\dfrac{1-7n}{6n^2}$ 33. $\dfrac{9}{2(x-4)}$ 35. $\dfrac{16a-15a^2}{12(x+2)}$ 37. $\dfrac{2a+27}{(a-4)(a+3)}$
39. $\dfrac{8-2x}{(x-1)(x-2)}$ 41. $\dfrac{y^2+8y+6}{(y-2)(y+1)}$ 43. $\dfrac{y^2+10y+1}{(y-5)(y+5)(y+2)}$
45. $\dfrac{85x-39}{30(x-2)}$ 47. $\dfrac{2x^2+9x^2-7}{2x(2x-1)}$ 49. $\dfrac{16x^2+30x+4}{(2x+3)(2x-3)(x-1)(x+1)}$

Exercise 4 - 4

1. 1 3. $\dfrac{6x}{5y^3}$ 5. $\dfrac{ab+b}{2ab+2a}$ 7. $\dfrac{a}{b}$ 9. $\dfrac{y+x}{y-x}$ 11. ab 13. $\tfrac{35}{36}$
15. $\dfrac{9b+8a}{4b+a}$

CHAPTER 4 TEST

1. $(x+2)/x$ 2. $2x^2/3y$ 3. $5x/6$ 4. $(x+3)/(x+1)$ 5. $x+2$
6. yes   7. $(x-4)/(x+4)$ 8. $4x/3y^2$ 9. $(x-4)/(x-1)$

10. $3ac/2bd$ 11. $(z-2)/4(z+1)$ 12. $x/(3x-2)$
13. $\dfrac{3x^2 + 10x + 5}{(x-4)(x+3)(x+2)}$ 14. $\dfrac{-y^2 + 7y + 13}{(2y+3)(y-2)(y+2)}$ 15. $\dfrac{4x^2 - 2x - 5}{2(3x-2)(x-3)}$
16. $\dfrac{6x + 18}{3x + 2}$ 17. $\dfrac{9b - 6a}{3 + 2a}$

Exercise 5-1

1. conditional   3. conditional   5. conditional   7. identity
9. identity   11. conditional   13. identity   15. yes   17. no
19. yes   21. $4 + 3 = 7$ 23. $3 \cdot 4 = 12$ 25. $x + 3 = 8$ 27. $3x = 6$
29. $3x - 4 = 8$ 31. $x + 7 = 16$ 33. $3x = 15$ 35. $2x + 4 = 24$
37. $2x - 6 = 0$ 39. $3x + 4 = 6$ 41. $19 - 4x = 7$ 43. $C = \pi d$
45. $P = a + b + c$ 47. $I = Prt$ 49. $P = 2l + 2w$ 51. $V = 160$ ft/sec
53. $A = 12$ sq. in.   55. $I = \$54$

Exercise 5-2

1. $x = 9$ 3. $x = -1$ 5. $x = 5$ 7. $x = -4$ 9. $y = 10$ 11. $y = 14$
13. $z = -1$ 15. $x = 2$ 17. $x = -7$ 19. $x = -17$ 21. $x = 4$
23. $x = \frac{3}{4}$ 25. $y = 4$ 27. $z = -1$ 29. $x = \frac{7}{2}$ 31. $x = 2$ 33. $x = 5$
35. $x = -\frac{12}{7}$ 37. $x = -\frac{5}{4}$ 39. $x = -2$ 41. $x = \frac{9}{2}$ 43. $x = 2$
45. $x = \frac{18}{7}$ 47. $x = \frac{15}{6}$ 49. $x = -\frac{22}{3}$ 51. $x = -6$ 53. $x = 0$
55. $y = \frac{1}{8}$ 57. $y = -\frac{17}{8}$ 59. $x = 5$ 61. $x = \frac{5}{28}$ 63. $x = 40$
65. $x = 10$ 67. $x = 1.7$ 69. $x = 1.2$ 71. $x = 2$ 73. $x = 19.2$
75. $x = -0.6$ 77. $x = -2$

Exercise 5-3

1. $x = (2 + b)/5$ 3. $x = (4 + a)/3$ 5. $x = -2b$ 7. $x = b/5$
9. $x = m/2$ 11. $x = (b + 8)/2$ 13. $y = -n$ 15. $y = n^2/(n - 3)$
17. $y = 4a$ 19. $y = b/a$ 21. $t = d/r$ 23. $P = I/rt$
25. $C = (5F - 160)/9$ 27. $b = 2A/h$

Exercise 5-4

1. $x = 6$ 3. $x = 53$ 5. $x = 7$ 7. $x = 24$ 9. length is 40
11. John has 9   13. Al is 8; Bob is 10   15. \$42,000 to one; \$76,500 to the other   17. 4 ft; 12 ft   19. \$10,000 at 6% = \$600; \$13,000 at 5% = \$675

Exercise 5-5

1. 2.5 pints   3. 15 quarts   5. 20 quarts   7. 50 pounds
9. 12 gallons   11. $1\frac{1}{3}$ pounds at 98¢ and $2\frac{2}{3}$ pounds at \$1.28
13. 7 pounds   15. 2 quarts

Exercise 5-6

1. $x < 2$ 3. $x \leq -4$ 5. $x < 2$ 7. $x < -3$ 9. $x < \frac{3}{2}$
11. $x \leq -1$

13. , $x < -4$

15. [number line], $x < 3$

17. [number line], $x \geq -\frac{7}{2}$

19. closed intervals   21. half open   23. closed   25. open
27. $\{x: 4 < x < 7\}$ 29. $\{x: -4 < x \leq 3\}$

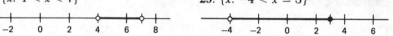

31. $\{x: x > 1 \text{ and } x > -1\} = \{x: x > 1\}$ 33. $\{x: x \leq 1 \text{ and } x \geq 1\} = \{1\}$

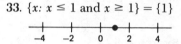

35. $\{x: x \leq 2 \text{ and } x \geq \frac{1}{3}\} = \{x: \frac{1}{3} \leq x \leq 2\}$

[number line]

Exercise 5-7

1. 3   3. 16   5. 0   7. x 9. $-(x-3)$ 11. $x = 4$ or $x = -2$
13. $y = -3$ or $y = 9$ 15. $x = -10$ or $x = 0$ 17. $x = \frac{2}{3}$ 19. no solution   21. $x > 4$ or $x < -4$ 23. $-2 < x < 2$ 25. $x < -2$ or $x > 3$
27. $x < 2$ 29. $x \leq \frac{2}{3}$

CHAPTER 5 TEST

1. is an identity   2. not an identity   3. $2x + 6 = 14$ 4. $2x - 5 = 15$
5. $A = 16$ sq. ft   6. $x = 9$ 7. $z = 4$ 8. $y = -4$ 9. $x = 3$
10. $z = -64$ 11. $x = 3$ 12. $x = -5$ 13. $x = -3$ 14. $x = -\frac{38}{3}$
15. $x = -40$ 16. $x = 8.8$ 17. $x = 5b/2$ 18. $x = (3b - 2)/28$
19. $t = (V - K)/32$ 20. $x = 13$ 21. 185 lbs and 210 lbs
22. Sam has $16, Jack has $21   23. 2.5 liters   24. $x > -7$
25. $x \geq -\frac{1}{7}$ 26. $-2 < x \leq 0$ 27. $x = \frac{2}{3}$ or $x = 2$ 28. $-\frac{2}{3} < x < 2$

Exercise 6-1

1. x^{12} 3. x^8 5. $30x^4y$ 7. $12y^4$ 9. x^{12} 11. 1   13. $1/x^2$
15. $1/y^2$ 17. $27m^3$ 19. m^4n^4 21. x^3/y^3 23. $-8/w^3$ 25. x^9

27. x^{15} 29. $27a^6$ 31. $216a^6b^3$ 33. $16a^2b^4$ 35. a^6/b^3 37. $16/a^4$
39. $36a^4b^2/25x^4$

Exercise 6-2

1. 1   3. 1   5. 1   7. 1   9. $2x$ 11. $16y^2$ 13. $1/x^2$ 15. $6/x^3$
17. $1/(3x)^2 = 1/9x^2$ 19. $3/xy$ 21. $4/(x^2)^2 = 4/x^4$ 23. $6/x^3$ 25. $2x$
27. $m^4/9n^2$ 29. $1/x^2$ 31. $1/x$ 33. $6/x$ 35. x^8 37. $2y^3/x$
39. b^8/a^4 41. a^4/b^2 43. a. $5/mn$; b. $5m^{-1}n^{-1}$ 45. a. $-7x/my$;
b. $-7m^{-1}xy^{-1}$ 47. a. $-2x^2/y$; b. $-2x^2y^{-1}$

Exercise 6-3

1. 2   3. 7   5. 2   7. -4 9. 0   11. y^2 13. x^3 15. $4x^3$
17. $3y$ 19. $2x^2y$ 21. 2   23. $2x^2$

Exercise 6-4

1. $3^{1/2}$ 3. $6^{1/3}$ 5. $p^{1/2}$ 7. $x^{1/2}$ 9. $(3x)^{1/3}$ 11. $(6m)^{1/2}$ 13. $m^{3/2}$
15. $5^{3/2}$ 17. $(ax)^{2/3}$ 19. $(x)^{4/3}$ 21. $\sqrt{6}$ 23. $\sqrt[3]{9}$ 25. $\sqrt[4]{x}$
27. $\sqrt{p}$ 29. $\sqrt{3x}$ 31. $\sqrt[4]{5y}$ 33. $\sqrt{6^3}$ or $(\sqrt{6})^3$ 35. $\sqrt[4]{m^3}$
or $(\sqrt[4]{m})^3$ 37. $\sqrt[3]{(4x)^2}$ or $(\sqrt[3]{4x})^2$ 39. $\sqrt[3]{(a^2x)^2}$ or $(\sqrt[3]{a^2x})^2$ 41. 64
43. 16   45. $x^{2/3}$ 47. $m^{13/20}$ 49. $(2a)^1$ or $2a$ 51. $x^{1/8}$ 53. $(ab)^{3/5}$
55. $x^{1/3}y^{1/3}$ 57. $6^{1/2}/x^{1/2}$ 59. $x^{1/3}$ 61. $a^{1/6}b^{1/2}$

Exercise 6-5

1. 4   3. 125   5. 32   7. $5\sqrt{3}$ 9. $2\sqrt{10}$ 11. $3\sqrt[3]{2}$ 13. $3\sqrt[3]{5}$
15. $x^3\sqrt{x}$ 17. $m\sqrt[3]{m^2}$ 19. $7x\sqrt{xy}$ 21. $3xy\sqrt[3]{2y^2}$ 23. $\sqrt{30}$
25. $4\sqrt{3}$ 27. $2\sqrt[3]{9}$ 29. $3x^2y\sqrt{2}$ 31. $36x^2y\sqrt{2y}$ 33. $4xy^2$

Exercise 6-6

1. 4   3. 3   5. $2\sqrt{7}$ 7. $2\sqrt[3]{2}$ 9. $2x$ 11. $q\sqrt{5}$ 13. $\sqrt[3]{9x^2}$
15. $2\sqrt{2}/3$ 17. $\sqrt{10}/2$ 19. $\sqrt{30}/5$ 21. $\sqrt{35}/7$ 23. $\sqrt{6}/3$
25. $2\sqrt{15}/5$ 27. $\sqrt{30x}/6x$ 29. $x\sqrt{5w}/2w$

Exercise 6-7

1. $8\sqrt{2}$ 3. $\sqrt{6}$ 5. $10\sqrt{3} + 6\sqrt{5}$ 7. $6\sqrt{13} + 2\sqrt[3]{13}$ 9. $-\sqrt{3}$
11. $10\sqrt{3}$ 13. $52\sqrt{2}$ 15. $12\sqrt{3}$ 17. $3 + 2\sqrt{2}$ 19. $1 + \sqrt{3}$

21. $\dfrac{3 - \sqrt{14}}{3}$ 23. $\dfrac{2 - \sqrt{2}}{3}$ 25. -1 27. $\dfrac{-1 + \sqrt{3}}{2}$ 29. $-\sqrt{2}/2$

31. $\dfrac{-5 - \sqrt{15}}{5}$

Exercise 6 - 8

1. $3\sqrt{2} - 2$ 3. $4\sqrt{6} + 4\sqrt{3}$ 5. $2\sqrt{6} - 3$ 7. $18\sqrt{2}$ 9. $5 + 3\sqrt{3}$
11. $-4\sqrt{3}$ 13. $3 + \sqrt{3} + \sqrt{6} + \sqrt{2}$ 15. $10 + 5\sqrt{6}$ 17. 1
19. $1 - \sqrt{3}$ 21. $12 - 3\sqrt{2}$ 23. $7 - 4\sqrt{3}$ 25. $\dfrac{13 + 3\sqrt{15}}{2}$
27. $2 - \sqrt{3}$ 29. $11 - 2\sqrt{2}$ 31. $2 + \sqrt{3}$

CHAPTER 6 TEST

1. x^9 2. y^3 3. $16x^4$ 4. $27a^6b^3$ 5. $16a^2/b^4$ 6. 1 7. $6/x^3$
8. $m^2/9n^2$ 9. $10y/x$ 10. $-2x^5y^3$ 11. $y^{17/12}$ 12. 96 13. $x^{7/5}$
14. $8\sqrt{2}$ 15. $2pq^2\sqrt[3]{2p}$ 16. $4x^2\sqrt{x}$ 17. 3 18. $x\sqrt{6}$
19. $\sqrt{6xy}/3y$ 20. $4\sqrt{5}$ 21. $13\sqrt{6}$ 22. $4\sqrt{3}$ 23. 4
24. $-3 + 2\sqrt{2}$ 25. $6 - 3\sqrt{3} - \sqrt{5}$

Exercise 7 - 1

1. $x = 0, x = 1$ 3. $x = 0, x = -3$ 5. $x = -1, x = -3$ 7. $x = 2, x = 3$
9. $x = -3, x = 2$ 11. $x = 3, x = -3$ 13. $x = 4, x = -4$ 15. $z = 0,$
$z = 4$ 17. $x = 0, x = 7$ 19. $x = -1, x = -2$ 21. $y = 10, y = 7$
23. $x = 7, x = -3$ 25. $x = -5, x = 3$ 27. $x = \tfrac{3}{2}, x = -4$ 29. $x = -3$
31. $x = 3, x = \tfrac{2}{3}$ 33. $x = -1, x = 3$ 35. $x = -7, x = 3$ 37. $x = 1$
39. $x = 2, x = 1$ 41. $x = -3$ 43. $x = -2, x = -1$ 45. $x = 2$
47. $x = 2, x = -6$ 49. $x = 2, x = -2$

Exercise 7 - 2

1. $x = 3, x = -3$ 3. $x = 2\sqrt{3}, x = -2\sqrt{3}$ 5. $x = \sqrt{6}, x = -\sqrt{6}$
7. $y = 2, y = -2$ 9. $x = 2, x = -2$ 11. $x = \dfrac{2\sqrt{3}}{3}, x = \dfrac{-2\sqrt{3}}{3}$
13. $y = \dfrac{2\sqrt{10}}{5}, y = \dfrac{-2\sqrt{10}}{5}$ 15. $x = \dfrac{\sqrt{10}}{2}, x = \dfrac{-\sqrt{10}}{2}$ 17. $x = \sqrt{2},$
$x = -\sqrt{2}$ 19. $x = \dfrac{+3\sqrt{2}}{2}, x = \dfrac{-3\sqrt{2}}{2}$ 21. $x = 3 + \sqrt{5}, x = 3 - \sqrt{5}$

23. $y = 6 + \sqrt{33}$, $y = 6 - \sqrt{33}$ 25. $x = \dfrac{5 + \sqrt{17}}{2}$, $x = \dfrac{5 - \sqrt{17}}{2}$

27. $x = \dfrac{4 + \sqrt{14}}{2}$, $x = \dfrac{4 - \sqrt{14}}{2}$ 29. $y = 1$, $y = \frac{1}{2}$

Exercise 7 - 3

1. $x^2 + 5x - 3 = 0$ 3. $x^2 - 6x + 2 = 0$ 5. $3x^2 - 4x + 0 = 0$
7. $y^2 - 3y - 2 = 0$ or $-y^2 + 3y + 2 = 0$ 9. $x^2 + 2x - 1 = 0$ 11. $x = 3$, $x = 2$ 13. $y = -2$, $y = -5$ 15. $x = -5$; $x = 3$ 17. $y = -2$, $y = \frac{1}{3}$
19. $x = -\frac{1}{3}$, $x = \frac{1}{2}$ 21. $x = \dfrac{9 + \sqrt{93}}{2}$, $x = \dfrac{9 - \sqrt{93}}{2}$ 23. $x = \frac{1}{3}$, $x = 1$
25. $x = -\frac{1}{2}$, $x = 3$ 27. $x = \dfrac{-2\sqrt{3}}{3}$, $x = \dfrac{2\sqrt{3}}{3}$ 29. $x = 0$. $x = \frac{3}{2}$
31. $w = \dfrac{\sqrt{71}}{2}$, $w = \dfrac{-\sqrt{71}}{2}$ 33. $x = \dfrac{-\sqrt{17}}{2}$, $x = \dfrac{\sqrt{17}}{2}$ 35. $x = -\frac{1}{2}$, $x = 3$
37. $x = \dfrac{-2\sqrt{3}}{3}$, $x = \dfrac{2\sqrt{3}}{3}$ 39. $x = 0$, $x = \frac{3}{2}$ 41. $x = \dfrac{5 + \sqrt{17}}{4}$, $x = \dfrac{5 - \sqrt{17}}{4}$ 43. $x = \frac{1}{2}$, $x = 1$ 45. $x = 2 + \sqrt{7}$, $x = 2 - \sqrt{7}$

Exercise 7 - 4

1. imaginary   3. not imaginary   5. imaginary   7. $x = 4i$, $x = -4i$
9. $x = 2i\sqrt{3}$, $x = -2i\sqrt{3}$ 11. $y = 8i$, $y = -8i$ 13. $x = 2i\sqrt{6}$, $x = -2i\sqrt{6}$
15. no   17. yes

Exercise 7 - 5

1. $x = \dfrac{3 + i\sqrt{7}}{2}$, $x = \dfrac{3 - i\sqrt{7}}{2}$ 3. $x = 1 + i$, $x = 1 - i$ 5. $y = -2 + i$, $y = -2 - i$ 7. $z = 1 + i\sqrt{2}$, $z = 1 - i\sqrt{2}$ 9. $x = \dfrac{2 + i\sqrt{2}}{2}$, $x = \dfrac{2 - i\sqrt{2}}{2}$
11. $x = -1$, $x = -\frac{2}{3}$ 13. $y = \dfrac{-2 + i}{5}$, $y = \dfrac{-2 - i}{5}$ 15. $x = \dfrac{5 + i\sqrt{23}}{6}$, $x = \dfrac{5 - i\sqrt{23}}{6}$

Exercise 7-6

1. 0 or 9   3. 2 or 7   5. 6 and 4   7. 5 and 3 or -3 and -5
9. 5 and 6   11. -4 and -3 13. 11 ft by 16 ft   15. 24 in.   17. 4 ft
19. 12 in.

Exercise 7-7

1. $x = \frac{1}{4}$ 3. $z = 2, z = -2$ 5. $x = 2, x = -2$ 7. $x = 1$ 9. $x = 1$, $x = 8$ 11. $x = 1$ 13. $x = 2, x = 6$ 15. $y = 6$ 17. $z = 1, z = -\frac{5}{2}$
19. $x = 4, x = -\frac{6}{5}$ 21. $y = 3, y = \frac{1}{5}$ 23. $x = 4$ ($x = -3$ is not a solution)
25. $x = 3$ ($x = -5$ is not a solution)   27. $x = 0$ ($x = 3$ is not a solution)
29. $\frac{1}{2}$ or $-\frac{1}{2}$ 31. $\frac{2}{3}$ or $\frac{3}{2}$ 33. 4 or -4 35. 240 miles per hour, 300 miles per hour

CHAPTER 7 TEST

1. $x = 6, x = -6$ 2. $x = \sqrt{5}, x = -\sqrt{5}$ 3. $x = 6, x = -1$ 4. $y = 4$, $y = 12$ 5. $x = 5, x = -2$ 6. $x = \frac{3}{2}, x = -4$ 7. $y = 3, y = 4$
8. $x = \dfrac{5 + \sqrt{13}}{2}, x = \dfrac{5 - \sqrt{13}}{2}$ 9. $y = 2 + \sqrt{7}, y = 2 - \sqrt{7}$
10. $z = 2 + \sqrt{2}, z = 2 - \sqrt{2}$ 11. $x = \frac{1}{4}, x = 1$ 12. $x = 2i, x = -2i$
13. $x = -5i, x = 5i$ 14. $x = \dfrac{5 + i\sqrt{7}}{4}, x = \dfrac{5 - i\sqrt{7}}{4}$ 15. $y = \dfrac{-3 + i\sqrt{6}}{3}$, $y = \dfrac{-3 - i\sqrt{6}}{3}$ 16. rational, real, complex   17. rational, real, complex
18. natural, rational, real, complex   19. real, complex   20. complex
21. -1 or 6   22. 12 in. by 15 in.   23. $x = -4, x = 6$ 24. $y = -4$, $y = 4$ 25. $\frac{1}{6}$ or 6   26. 50 miles per hour, 60 miles per hour

Exercise 8-1

1.

3.

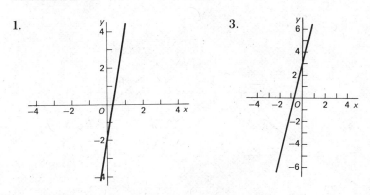

378 Answers to Odd-Numbered Exercises and Chapter Tests

5.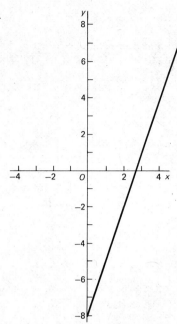

7.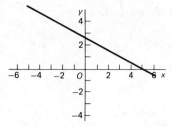

9.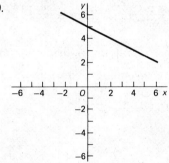

11.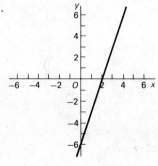

13.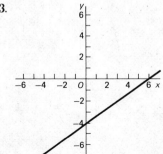

15.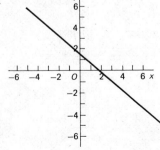

17.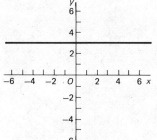

19.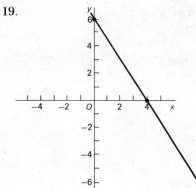

21.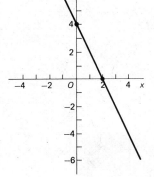

23.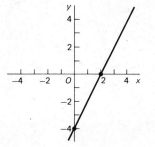

25.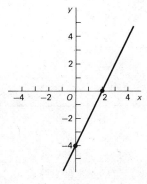

27.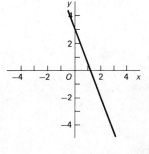

29.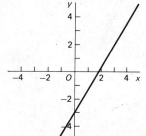

31.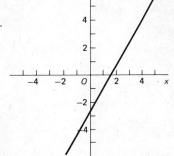

Exercise 8 - 2

1. 2   3. 3   5. −5   7. 2   9. $\frac{1}{13}$ 11. 0   13. 4, −6   15. 3, 2
17. $\frac{1}{2}, -\frac{2}{3}$ 19. −8, 2   21. 2, −3   23. $-\frac{2}{3}, 2$ 25. $-\frac{1}{2}, \frac{1}{6}$ 27. 2, $-\frac{3}{2}$
29. undefined, none

Exercise 8 - 3

1.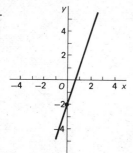

3.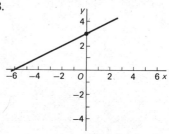

5.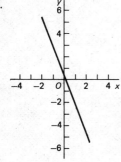

7.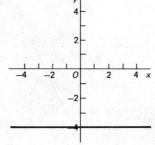

9.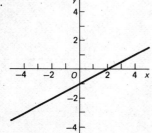

11.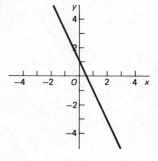

13. 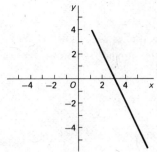   15.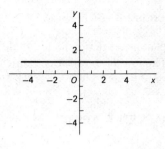

17. 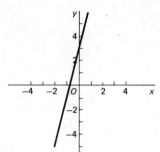   19.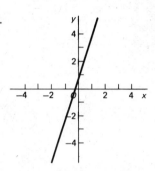

21.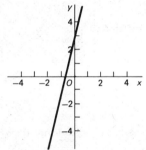

Exercise 8 - 4

1. $x = 1, y = 1$ 3. $x = 2, y = 2$ 5. inconsistent   7. $x = 2, y = 5$
9. $x = \frac{14}{11}, y = \frac{6}{11}$ 11. $x = \frac{10}{3}, y = 2$ 13. $x = \frac{5}{2}, y = -\frac{3}{4}$ 15. $m = \frac{8}{15}$,
$n = \frac{1}{5}$ 17. $x = -3, y = -9$ 19. $x = \frac{5}{4}, y = -\frac{5}{4}$ 21. $w = 2, z = 0$
23. $x = -\frac{52}{7}, y = -\frac{128}{7}$ 25. dependent equations   27. $x = 24, y = 8$

Exercise 8 - 5

1. 17 and 34   3. 21 and 42   5. 10 and 20   7. 8 and 12   9. 18
11. 7 and 14   13. 7 ft by 12 ft   15. 6 dimes   17. $2500 at 6%,
$3500 at 8%   19. 2 and 3

Exercise 8-6

1. The rate of the plane is 125 miles per hour, the wind is 25 miles per hour.
3. The rate of the boat is 5 miles per hour, the current is 1 mile per hour.
5. 40 miles per hour   7. 2.4 miles per hour

CHAPTER 8 TEST

1.
x	y
−4	28/3
−3	8
−2	20/3
−1	16/3
0	4
1	8/3
2	4/3
3	0

2.
x	y
−3	−5
−2	−2
−1	1
0	4
1	7

3.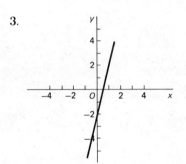

4.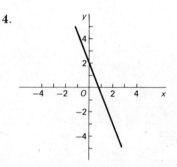

5. x-intercept: 6, y-intercept: −4   6. −3   7. slope: 2, y-intercept: −3

8.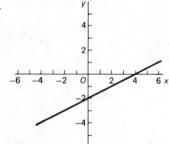

9. $x = 3$, $y = -1$ 10. inconsistent equations (no solutions)   11. 2 and 5
12. 14 and 21   13. John is running at 10 miles per hour and Jack at 15 miles per hour.

Exercise 9 - 1

1.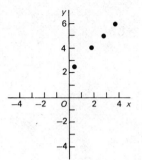

3.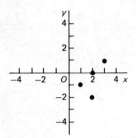

5.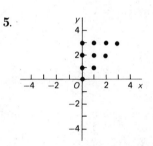

7.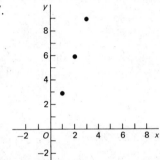

9.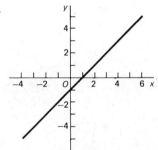

11.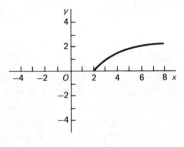

13.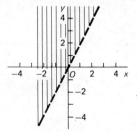

15.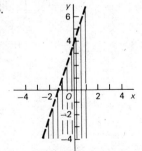

17.
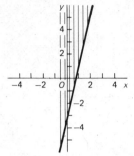

19. domain: all real numbers; range: all real numbers ≥ 4
21. domain: all real numbers ≥ 1; range: all real numbers ≥ 0
23. domain: all real numbers; range: all real numbers ≥ 1

Exercise 9 - 2

1. function   3. function   5. not a function   7. function
9. not a function   11. yes   13. no   15. yes   17. no   19. yes
21. yes   23. no   25. set of all real numbers
27. set of all real numbers except 0   29. set of all real numbers except 4

Exercise 9 - 3

1. a. -1; b. 5; c. -10; d. $3a - 1$ 3. a. -3; b. 1; c. 13; d. $4x^2 - 3$
5. a. -2; b. -5; c. 0   7. a. $x^2 + x - 2$; b. $x^2 + 2xh + h^2 - 3x - 3h$;
c. $2xh + h^2 - 3h$; d. $2x + h - 3$ 9. $6x + 2 + 3h$

Exercise 9 - 4

1. x-intercepts: $-3, 3$; y-intercept: -9 3. x-intercept: -1; no y-intercept
5. x-intercepts: $-2, 2$; y-intercepts: $-2, 2$ 7. x-intercepts: $-1, 1$;
y-intercept: -3 9. x-intercepts: $2, -2$; y-intercepts: $1, -1$
11. symmetric with respect to the y-axis
13. not symmetric with respect to any of x-axis, y-axis, or origin
15. symmetry about x-axis, y-axis, and origin
17. symmetric with respect to the y-axis.
19. symmetric with respect to the x-axis, y-axis, and origin
21. 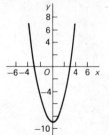   23.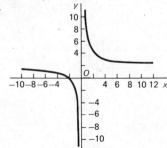

25.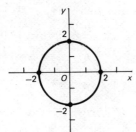

27.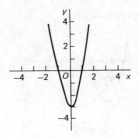

29.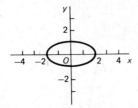

Exercise 9 - 5

1. identity function

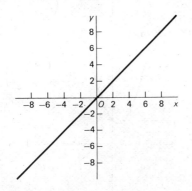

3. linear function

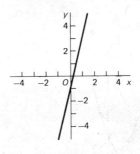

5. linear function

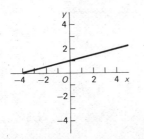

7. absolute value function

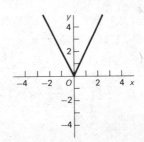

9. quadratic function

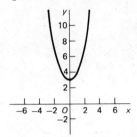

11. linear function

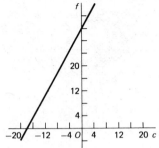

13. $x = 2, x = -2$

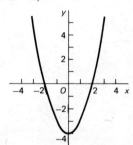

15. $x = 3, x = -3$

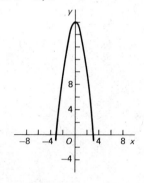

17. $x = -2$

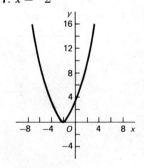

19. no zeros

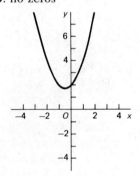

Exercise 9-6

1. a. $15x$; b. $15x$ 3. a. x; b. x 5. yes

7. $g[f(x)] = g[2x - 1] = \dfrac{[2x - 1] + 1}{2} = x;$

$f[g(x)] = f\left[\dfrac{x + 1}{2}\right] = 2\left[\dfrac{x + 1}{2}\right] - 1 = x;$ thus, $f = g^{-1}$ 9. yes

11. $y = \dfrac{x + 1}{3}$

13.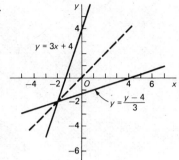

15.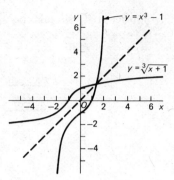

CHAPTER 9 TEST

1.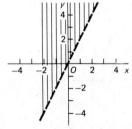

2.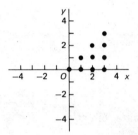

3.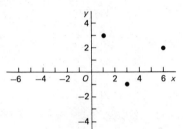

4. domain: all real numbers ≥ 3; range: all real numbers ≥ 0 5. yes
6. no 7. all real numbers except -5 8. 38 9. 1
10. x-intercepts: $-\sqrt{3}$, $\sqrt{3}$; y-intercept: 3; symmetric with respect to the y-axis
11. x-intercepts: 4, -4; y-intercepts: 4, -4; symmetric with respect to x-axis, y-axis, and origin

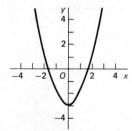

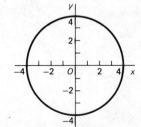

12. x-intercept: 4; y-intercepts: 1, -1; symmetric with respect to the x-axis

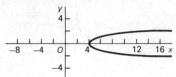

13. linear function

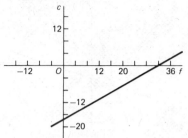

14. identify function

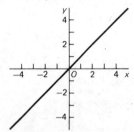

15. absolute value function

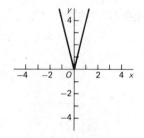

16. quadratic function

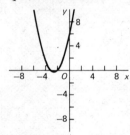

17. $x = 3$, $x = -3$

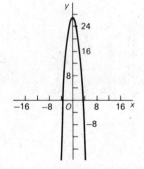

18. yes   **19. a.** $5x^2 + 1$; **b.** $(5x + 1)^2$ **20.** yes   **21.** $y = 2x + 5$

22.

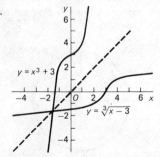

Exercise 10 - 1

1.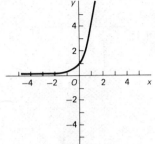

3.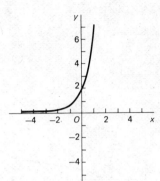

5.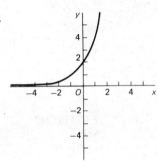

7.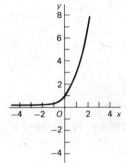

9.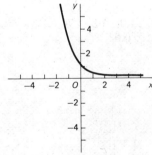

11.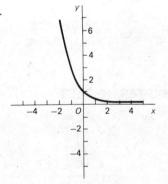

13. 5459

Exercise 10 - 2

1. $2^4 = 16$ 3. $4^{1/2} = 2$ 5. $\log_2 32 = 5$ 7. $\log_4 (\tfrac{1}{4}) = -1$ 9. 3

11. $\frac{1}{2}$

13.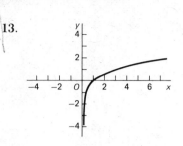

15.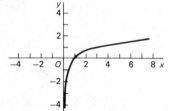

17.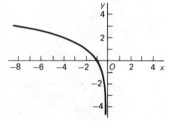

Exercise 10 - 3

1. 4 3. x 5. 1 7. 1 9. 1 11. x 13. 2 15. 1.0792
17. yes 19. 0.5118 21. 2.4082 23. 1.2727 25. 2

Exercise 10 - 4

1. 2.5105 3. 7.5105 − 10 5. 66,000 7. 14.0 9. 4.70 11. 50.0
13. 0.00887 15. 0.000961 17. 0.191 19. 0.0175

CHAPTER 10 TEST

1.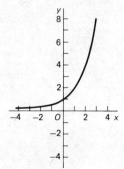

2.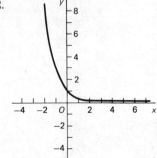

3.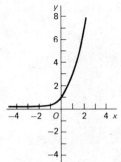

4. $3^4 = 81$ 5. $\log_4 64 = 3$ 6. $\frac{1}{2}$ 7. -1
8. 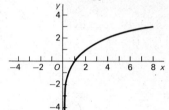   9.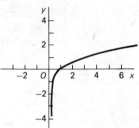

10. 9   11. 1   12. 1.8109   13. $7.8645 - 10$ 14. 0.474   15. 0.251
16. 0.436   17. 685

Exercise 11 - 1

1. yes   3. no   5. no   7. no   9. 2   11. -10 13. $3, -2, 6$
15. yes   17. yes   19. yes

Exercise 11 - 2

1. $3x + 11$, remainder 30   3. $4x + 1$, remainder 7   5. $5x - 13$,
remainder 32   7. $x^2 + 5x + 2$, remainder 4   9. $2x^2 + x - 3$, remainder 9
11. $4x^2 + 4x + 1$, remainder 2   13. $x^2 + \frac{1}{3}x + \frac{37}{9}$, remainder $\frac{19}{27}$ 15. yes
17. yes   19. yes   21. yes   23. no   25. yes

Exercise 11 - 3

1. $\{2, 3\}$ (3 occurs twice)   3. $\{2, 1, -3\}$ 5. $\{2, 3, -1\}$
7. $\{1, 2, -2, -3\}$ 9. $\{2, 2 + 2\sqrt{2}, 2 - 2\sqrt{2}\}$
11. $\left\{-2, \dfrac{3 + i\sqrt{7}}{2}, \dfrac{3 - i\sqrt{7}}{2}\right\}$

Exercise 11-4

1. $\{1, 2\}$ 3. $\{2, -2, 3\}$ 5. $\{-1, 3, -2\}$ 7. $\{3, -2, 4\}$
9. $\left\{2, \dfrac{3+\sqrt{5}}{2}, \dfrac{3-\sqrt{5}}{2}\right\}$ 11. $\{-2, 2+\sqrt{2}, 2-\sqrt{2}\}$ 13. $\{2, 3, \tfrac{1}{2}\}$
15. $\{-2, 4, \tfrac{1}{3}\}$ 17. $\{2, -2, \tfrac{1}{2}\}$ 19. $\left\{2, \dfrac{3+i\sqrt{23}}{8}, \dfrac{3-i\sqrt{23}}{8}\right\}$
21. $\{-3, 2, 1\}$ 23. $\{2, 4, 3, -2\}$ 25. $\left\{3, -1, \dfrac{3+3i\sqrt{3}}{2}, \dfrac{3-3i\sqrt{3}}{2}\right\}$

CHAPTER 11 TEST

1. yes   2. no   3. -7 4. $1, -3, -6$ 5. yes   6. yes
7. $x^2 + 8x + 29$, remainder 117 8. $4x^2 - 8x + 19$, remainder -32
9. not a factor   10. is a solution   11. $\{1, -1, -6\}$
12. $\{1, 2, -3+\sqrt{5}, -3-\sqrt{5}\}$ 13. $\left\{-3, \dfrac{-1+i\sqrt{31}}{4}, \dfrac{-1-i\sqrt{31}}{4}\right\}$
14. $\{-1, 2, 1+\sqrt{3}, 1-\sqrt{3}\}$ 15. $\dfrac{-3+\sqrt{5}}{2}$

Exercise 12-1

1.
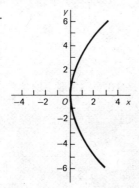

3.
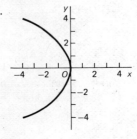

5.
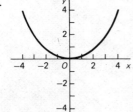

7.
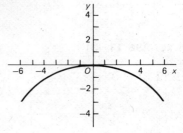

9.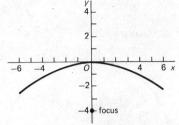

11.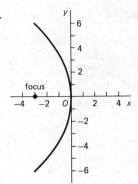

13.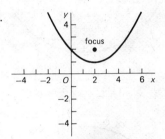

15.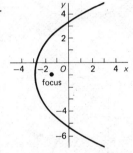

17. $(x-3)^2 = 4(y-1)$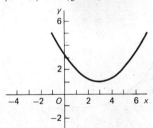

19. $(y+2)^2 = 4(x+5)$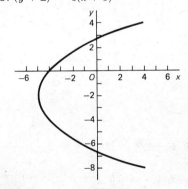

21. $(y-3)^2 = -4(x-1)$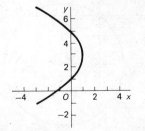

23. $(x+\tfrac{3}{2})^2 = -8(y+\tfrac{1}{2})$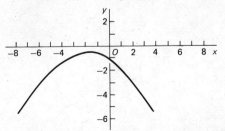

Exercise 12-2

1. $C(0, 0), r = 6$

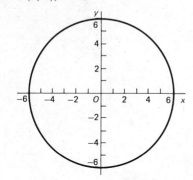

3. $C(0, 0), r = 7$

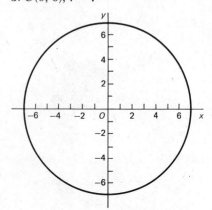

5. $C(1, 3), r = 9$

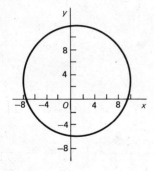

7. $C(-2, -1), r = \sqrt{46}$

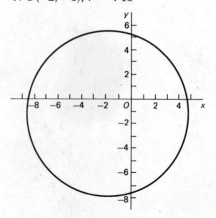

9. $C(1, 1), r = 2$

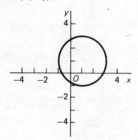

11. $C(3, -4), r = 5$

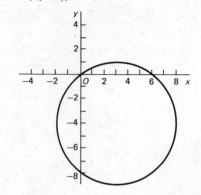

13. $C(\frac{1}{2}, -\frac{3}{2})$, $r = \dfrac{\sqrt{74}}{2}$

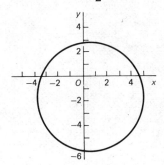

15. $C(\frac{1}{2}, -\frac{3}{2})$, $r = 3$

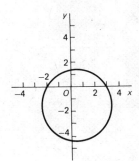

17. yes   19. no

Exercise 12 - 3

1.
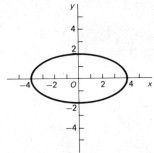

3.
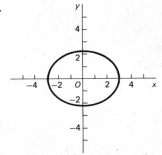

5.
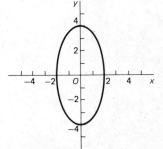

7.
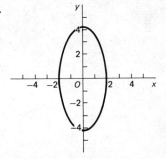

396 Answers to Odd-Numbered Exercises and Chapter Tests

9.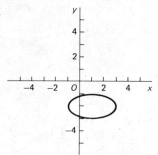

11.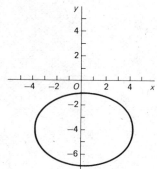

13.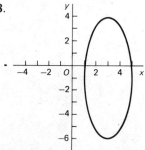

15.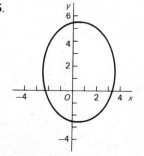

17. $\dfrac{(x-1)^2}{4} + \dfrac{(y+2)^2}{1} = 1$

19. $\dfrac{(x-1)^2}{9} + \dfrac{(y-3)^2}{16} = 1$

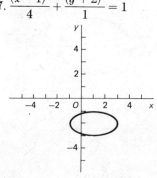

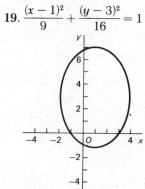

21. $\dfrac{(x-1)^2}{4} + \dfrac{(y-1)^2}{3} = 1$

23. $\dfrac{x^2}{4} + \dfrac{(y+2)^2}{8} = 1$

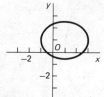

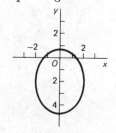

25. ellipse   27. parabola   29. ellipse

Exercise 12-4

1. hyperbola   3. hyperbola   5. parabola   7. hyperbola
9. hyperbola   11. circle

13.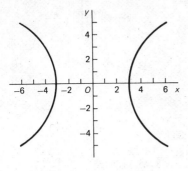

15.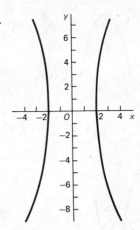

17.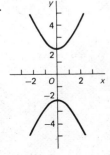

19.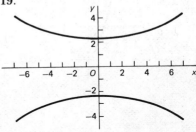

21.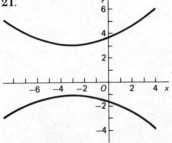

23.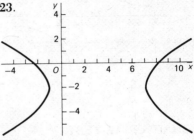

25.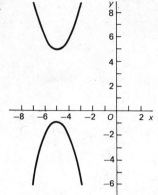

27. $\dfrac{(x-1)^2}{4} - \dfrac{(y-1)^2}{1} = 1$
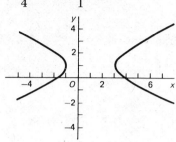

29. $\dfrac{(x+3)^2}{1} - \dfrac{(y-4)^2}{9} = 1$
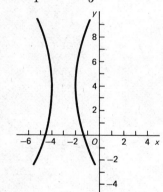

31.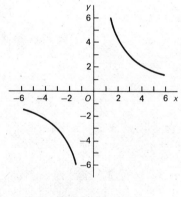

33.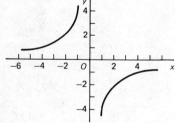

CHAPTER 12 TEST

1. ellipse   2. hyperbola   3. parabola   4. circle

Chapter 12 test **399**

5.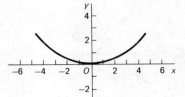

6.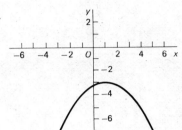

7.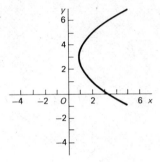

8.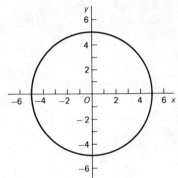

9.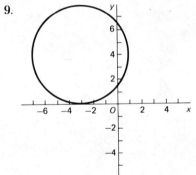

10.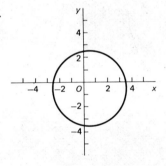

11.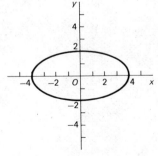

12.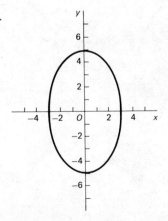

13.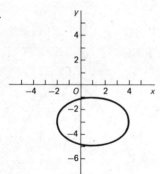

14.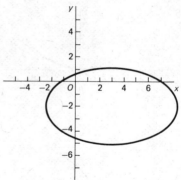

15.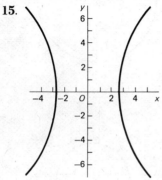

16.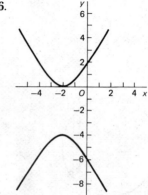

17.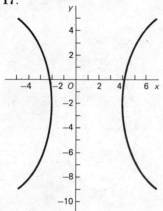

18.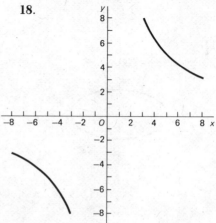

19.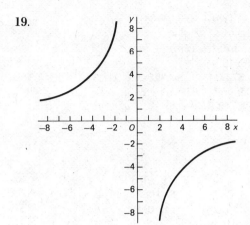

Exercise 13 - 1

1. 3, 6, 9, 12, 15, 18, 21, 24, 27, 30   3. $\frac{1}{3}, \frac{2}{3}, 1, \frac{4}{3}, \frac{5}{3}, 2, \frac{7}{3}, \frac{8}{3}$
5. $-\frac{1}{4}, \frac{1}{4}, -\frac{1}{4}, \frac{1}{4}, -\frac{1}{4}, \frac{1}{4}$ 7. $-\frac{1}{2}, \frac{1}{4}, -\frac{1}{6}, \frac{1}{8}, -\frac{1}{10}, \frac{1}{12}$ 9. $-1, -\frac{1}{4}, -\frac{1}{15}, 0$; 10th term $\frac{1}{20}$ 11. 11, 14, 17   13. $\frac{15}{2}, 9, \frac{21}{2}$ 15. 25   17. $\frac{5}{2}$
19. 15   21. 10   23. 24, 48, 96   25. 24, 16, $\frac{32}{3}$ 27. 320
29. $\frac{27}{2}$ 31. 96   33. $\frac{3}{2}$ 35. 40.5   37. $4096

Exercise 13 - 2

1. $\frac{205}{144}$ 3. $\frac{6687}{840}$ 5. $\frac{25}{4}$ 7. $-\frac{47}{80}$ 9. 100   11. $67\frac{1}{2}$ 13. 30
15. 6   17. 2184   19. $\frac{21}{8}$ 21. 121   23. $\frac{422}{27}$

Exercise 13 - 3

1. yes   3. no   5. no   7. 9   9. $\frac{24}{5}$ 11. 9   13. $\frac{16}{3}$ 15. $\frac{5}{9}$
17. $\frac{67}{99}$

Exercise 13 - 4

1. $x^{18} + 18x^{17}y + 153x^{16}y^2$
3.
```
                    1
                  1   1
                1   2   1
              1   3   3   1
            1   4   6   4   1
          1   5  10  10   5   1
        1   6  15  20  15   6   1
      1   7  21  35  35  21   7   1
    1   8  28  56  70  56  28   8   1
```

5. $x^3 + 3x^2y + 3xy^2 + y^3$ 7. $x^4 - 4x^3y + 6x^2y^2 - 4xy^3 + y^4$
9. $16a^4 + 32a^3b + 24a^2b^2 + 8ab^3 + b^4$
11. $243p^5 - 405p^4q + 270p^3q^2 - 90p^2q^3 + 15pq^4 - q^5$
13. $16x^4 + 96x^3 + 216x^2 + 216x + 243$
15. $729v^6 - 7290v^5 + 30{,}375v^4 - 67{,}500v^3 + 84{,}375v^2 + 56{,}250v + 15{,}625$
17. $32x^5 - 240x^4y + 720x^3y^2 - 1080x^2y^3 + 810xy^4 - 243y^5$
19. $x - 3x^{2/3} + 3x^{1/3} - 1$ 21. $x^{-8} - x^{-6}y + x^{-4}y^2 - x^{-2}y^3 + y^4$

CHAPTER 13 TEST

1. $3, \frac{3}{2}, 1, \frac{3}{4}, \frac{3}{5}, \frac{1}{2}$ 2. $-\frac{1}{2}, \frac{1}{3}, -\frac{1}{4}, \frac{1}{5}, -\frac{1}{6}, \frac{1}{7}$ 3. $8\frac{1}{2}, 10, 11\frac{1}{2}$ 4. -10
5. $20\frac{1}{2}$ or $\frac{41}{2}$ 6. $256, 1024, 4096$ 7. $\frac{1}{2}$ 8. 108 9. $\frac{64}{3}$ or $21\frac{1}{3}$ ft
10. $11\frac{8}{15}$ 11. $-\frac{7}{48}$ 12. 84 13. $46\frac{1}{2}$ 14. $19\frac{3}{8}$ 15. 33
16. $\frac{16}{3}$ or $5\frac{1}{3}$ 17. $\frac{8}{3}$ 18. $\frac{4}{11}$ 19. no
20. $p^4 + 4p^3q + 6p^2q^2 + 4pq^3 + q^4$
21. $32m^5 - 80m^4n + 80m^3n^2 - 40m^2n^3 + 10mn^4 - n^5$

Index

Abscissa, 201
Absolute value, 27, 28
  function, 261, 262
Addition
  associate property of, 21
  commutative property of, 21
  of fractions, 101
  of like terms, 44
  of polynomials, 46
  of radicals, 165, 166
  of signed numbers, 29–31
Addition property of equations, 116
Additive identity, 21
Additive inverse, 22
Algebraic expressions, 42
Algebraic fractions. *See* Fractions
Algebraic symbols, 1
Approximations
  decimal, 13–15, 17–19
Associative property
  of addition, 21
  of multiplication, 21
Asymptotes, 271, 328
Axes
  x, 201
  y, 201

Bar, 25
Binary operations
  order of, 25–27
Binomial formula, 353–356
Binomials, 42–44
  conjugate, 79
  products of, 67, 68
  square of, 76, 77

Characteristic, 283
Circles, 314–317
  general form of equation, 316, 317
  standard form of equation, 315

Coefficients, 42
  numerical, 42
Combining like terms, 45
Combining polynomials, 46, 47
Common denominator
  lowest, 98–100
Common monomial factor, 65–67
Commutative property
  of addition, 21
  of multiplication, 21
Completing the square, 177–180, 311, 316, 323, 330
Complex fractions, 106, 107
Complex numbers, 185–187
Components of a solution, 201
Composite functions, 263, 264
Conic sections, 314
Conjugates
  binomial, 79
Constant term, 43
Coordinates of a point, 201
Coordinate system
  rectangular, 201
Cube of a binomial, 81
Cubes
  table of, 361

Decimal approximations of irrational numbers, 19, 163, 164
Degree
  of a polynomial, 43
  of a term, 43
Denominator
  lowest common, 98–100
Dependent equations, 218
Dependent variable, 242
Descending powers, 69
Differences
  of fractions, 101, 102
  of like terms, 44

404 Index

Differences (*continued*)
  of polynomials, 46, 47
  of radicals, 166
  of signed numbers, 31, 32
  of two cubes, 82
  of two squares, 79, 80
Directrix
  of a parabola, 309
Disjoint sets, 4
Distributive property, 22, 65, 111
Division
  of fractions, 95–96
  of irrational expressions, 161–164
  of monomials, 50–52
  of polynomials, 60–62
  of radicals, 161–164
  of signed numbers, 36–38
  of two powers of the same base, 143, 144
  by zero, 22

e, 18, 273, 274
Ellipses, 318–325
  foci of, 319
  general form of equation, 323
  major axis of, 319
  minor axis of, 319
  standard form of, 319–321
  vertices, 319
Empty set, 2
Equal sets, 3
Equations, 10, 110
  addition property of, 116
  conditional, 110, 111
  dependent, 218
  first-degree, 115
  fractional, 194–197
  graphing, 202
  inconsistent, 218
  linear. *See* Linear equations
  multiplication property of, 116
  in one variable, 110
  quadratic. *See* Quadratic equations
  substitution property of, 115
Equivalent fractions, 99, 100
Equivalent sets, 9
Exponential functions, 271–275
  graphs of, 271–275
  table of values, 364
Exponents, 48–51, 143–150
  fractional, 154–156
  negative, 147–150
  rules of, 143–145, 147
  zero, 147, 148
Expression
  algebraic, 42

Factor theorem, 292
Factoring
  completely, 83–85
  difference of two cubes, 82
  difference of two squares, 80
  by grouping, 85–87
  monomials, 65–67
  perfect binomial cubes, 81, 82
  perfect squares, 76–78
  sum of two cubes, 82
  trinomials, 69–79
First-degree equations. *See* Linear equations
Foci
  of ellipses, 319
  of parabolas, 309
Formulas, 123
Fractional equations, 194–197
Fractions
  addition of, 101
  combining, 102, 103
  complex, 106, 107
  equivalent, 99, 100
  equal, 14
  lowest common denominator of, 98, 99
  lowest terms of, 89
  products of, 93–95
  quotients of, 95, 96
  reducing, 89
  simplification of, 89–92
  subtraction of, 101, 102
Fractional equations, 194–197
Functional notation, 246, 247
Functions, 241–245
  absolute value, 261, 262
  composite, 263, 264
  exponential, 271–275
  graphs of, 243–245, 249–254
  identity, 256
  inverse, 264–268
  linear, 256, 257
  logarithmic, 276–279
  quadratic, 257–261

Graphs
  of exponential functions, 271–275
  of functions, 243–245, 249–254
  using intercepts, 205, 206, 250, 251
  of linear equations, 202–206, 212–216
  of logarithmic functions, 277–279
  of ordered pairs, 235
  of quadratic functions, 258–261
  of rational numbers, 13
  using slopes, 212–216
  using symmetry, 251–254

Index

Half-plane
  closed, 240
  open, 239
Higher-degree polynomials, 290
Horizontal axis, 201
Hyperbola, 326–334
  conjugate axis, 321
  general form, 327
  standard form, 328–330, 332
  transverse axis, 328
  vertices, 328

Identities, 110, 111
Identity
  additive, 21
  multiplicative, 21
Identity function, 256
Imaginary numbers, 185–187
Inconsistent equations, 218
Independent variable, 242
Inequalities, 11
  chain, 11
  involving absolute values, 137–140
  linear, 131–136
Infinite sets, 2
Integers, 10
  graphing, 10
Intercepts, 205, 206, 250, 251
Intersection of sets, 4, 5
Inverse
  additive, 22
  multiplicative, 22
Inverse function, 264–268
Irrational numbers, 17–19
  decimal approximation of, 19
  proof for $\sqrt{2}$, 359

Like terms, 44
  addition of, 44
  combining, 45
  subtraction of, 44
Limit of a sequence, 339
Linear equations
  in one variable, 115–121
    solutions of, 115–121
  in two unknowns, 200–225
    graphs of, 202–206, 212–216
    solution by addition and subtraction, 221–225
    solution by graph, 217–219
    solution by substitution, 220, 221
    solution of, 200–206, 217–225
    word problems, 124–130
Linear functions, 256, 257
Linear inequalities, 131–136, 239
  graphs of, 239, 240
Literal equations, 121–124

Literal factors, 43
Literal numbers, 1
Logarithmic functions, 276–279
  graphs of, 277–279
Logarithms, 279–288
  computations with, 282–288
  proofs of properties, 360
  properties, 280–281
  tables of mantissas, 362, 363
Lowest common denominator, 98–100

Mixture problems, 128–130
Monomials, 42–44
  addition of, 44
  division of, 50–52
  multiplication of, 48, 49
  subtraction of, 44
Motion problems, 230–232
Multinomial, 42
Multiplication
  associative property of, 21
  of binomials, 67, 68
  commutative property of, 21
  of irrational expressions, 156–160
  of monomials, 48, 49
  of polynomials, 59, 60
  of polynomials by monomials, 58
  property of equations, 116
  of signed numbers, 34–36
  of two powers of the same base, 143
Multiplicative identity, 21
Multiplicative inverse, 22

Natural numbers, 2, 9, 10
  graphing, 10
Negative exponents, 147–150
Number line, 10
  integers on, 10
  natural numbers on, 13
  real, 18
Numbers
  absolute value of, 27, 28
  complex, 185–187
  imaginary, 186–187
  irrational, 17–19
  natural, 1, 9, 10
  rational, 13–15
  real, 18, 20, 23
  signed. See Signed numbers
Numerical coefficient, 42

One-to-one correspondence, 9
Order of operations, 25–27
Ordered pairs, 200, 235
Ordinate, 201
Origin, 201

406 Index

Parabolas, 258–261, 308–313
   axis, 259, 309
   focus, 309
   general form of equation, 311
   standard form of equation, 310
   vertex, 309
Parentheses, 24, 54–56
Pascal's triangle, 355
Plotting points, 201
Polynomials, 42
   differences of, 46, 47
   products of, 59, 60
   quotients of, 60–62
   sums of, 46
Power of a power, 145
Powers. See Exponents
Principal roots, 151, 152
Products
   of binomials, 67, 68
   of fractions, 93–95
   of monomials, 48, 49
   of monomials and polynomials, 58
   of polynomials, 59, 60
   to a power, 144
   of powers to the same base, 143
   of signed numbers, 34–36
Pythagorean theorem, 17

Quadrants, 201
Quadratic equations, 174–189
   with complex solutions, 188, 189
   general form, 174, 181
   general form in two variables, 308
   graphs of, 308–333
   solution by completing the square, 177–180
   solution by factoring, 174–176
   solution by formula, 181–184
   word problems, 190–192
Quotients
   of fractions, 60–62
   of monomials, 50–52
   of polynomials, 60–62
   to a power, 144
   of powers to the same base, 143, 144
   of signed numbers, 36–38

Radicals, 154–171
   addition of, 165, 166
   combining, 165–167
   division of, 161–164
   multiplication of, 156–160
   subtraction of, 166
Rationalizing denominators, 162–164, 170, 171

Rational numbers, 13–15
   as decimals, 14, 15, 351, 352
   graphing, 13
Rational solutions to polynomial equations, 301–306
Real numbers, 18, 20–23
   properties of, 20–23
Reciprocal, 22, 149
Rectangular coordinate system, 201
Relations, 235–240
   graphs of, 235–240, 249–254
Remainder theorem, 292
Roots, 151–153
   of equations. See Solutions
   principal, 151, 152

Second-degree equations. See Quadratic equations
Sequences, 336–340
   arithmetic, 337
   converging, 340
   diverging, 340
   geometric, 338, 339
   nth term of, 337, 339
   sum of, 344–346
Series, 347–352
   sum of, 347–351
Sets, 1–6
   disjoint, 4
   empty, 2
   equal, 3
   equivalent, 9
   intersection of, 4, 5
   null, 2
   of ordered pairs, 235
   subsets, 3
   union of, 5, 6
   universal, 4
Signed numbers, 29–38
   addition of, 29–31
   products of, 34–36
   quotients of, 36–38
   subtraction of, 31, 32
Similar terms. See Like terms
Simplifying algebraic expressions, 89–92
Simplifying radical expressions, 156–171
Slope of a line, 207–211
Solutions, 110
   of linear equations in one unknown, 115–121
   of literal equations, 121–124
   partial list of, 200
   of polynomial equations, 298–306
   of quadratic equations, 174–184, 186, 188, 189

Solution of systems of linear equations
  by addition and subtraction, 221–225
  by graphing, 217–219
  by substitution, 220, 221
Square of a binomial, 76, 77
Square roots
  table of, 361
Squares
  table of, 361
Standard form of a quadratic equation, 174
Subsets, 3
Substitution property of equations, 115
Subtraction
  of fractions, 101, 102
  of like terms, 44
  of polynomials, 46, 47
  of radicals, 166
  of signed numbers, 31, 32
Summation notation, 342–344
Sum of sequences, 344–346
Sums of polynomials, 46
Symbolic statements, 111–113
Symbols of grouping, 24–26, 54–56
Symmetry, 251–254
Synthetic division, 294–297
Systems of equations, 217

Terms
  addition of, 44
  coefficients of, 42

  combining like, 45
  like, 44
  subtraction of, 44
Trichotomy law, 11
Trinomial, 42–44
  factoring, 69–79
  perfect squares, 76–78

Union of sets, 5, 6
Universal set, 4

Variable, 43
  dependent, 242
  independent, 242

Word problems, 124–127
  involving fractional equations, 196, 197
  involving linear equations, 124–130
  involving quadratic equations, 190–192
  involving two equations in two unknowns, 226–229

Zero exponents, 147, 148
Zeros of polynomials, 290–293

76 77 78 79 9 8 7 6 5 4 3 2 1